S0-BIF-598

PLINY

NATURAL HISTORY

VIII

LIBRI XXVIII—XXXII

418

PLINY

NATURAL HISTORY

WITH AN ENGLISH TRANSLATION
IN TEN VOLUMES

VOLUME VIII

LIBRI XXVIII–XXXII

BY

W. H. S. JONES, Litt.D., F.B.A.,

HONORARY FELLOW, ST. CATHARINE'S COLLEGE, CAMBRIDGE

CAMBRIDGE, MASSACHUSETTS
HARVARD UNIVERSITY PRESS
LONDON
WILLIAM HEINEMANN LTD
MCMLXXV

American ISBN 0–674–99460–4
British ISBN 0 434 99418 9

First published 1963
Reprinted 1975

Printed in Great Britain

CONTENTS

INTRODUCTION

For the contents of this volume there must be noted the following additions to the authorities already mentioned:

Codex Bambergensis, the oldest manuscript, 10th century, with several correcting hands, styled B.

Codex Toletanus, 13th century, of the same family as V, R, d, styled T.

Green, Peter, *Prolegomena to the study of Magic and superstition in the Natural History of Pliny the Elder*, 1952, a typed doctoral thesis in the Cambridge University Library.

Wolters, X. F. M. G., *Notes on Antique Folklore* based on Pliny's Natural History XXVIII, 22–29, Amsterdam 1935.

Professor E. H. Warmington translated Book XXXII, sections 142–154; and compiled the Index of Fishes. He expresses his grateful thanks to Professor A. C. Andrews of the University of Miami for invaluable help in the identification of aquatic creatures in Pliny; and to members of the staff of the British Museum (Natural History), especially A. Wheeler, I. Galbraith, Miss J. E. King, Dr. Isabella Gordon, Miss A. M. Clark, and W. J. Rees, for bringing the scientific nomenclature up to date.

Second impression (1975): The editor is most grateful to Mr. P. Naiditch of University College London for his vigilant detection of misprints and for numerous suggestions for editorial improvement.

PLINY :
NATURAL HISTORY
BOOK XXVIII

PLINII NATURALIS HISTORIAE

LIBER XXVIII

I. Dicta erat natura omnium rerum inter caelum ac terram nascentium restabantque quae ex ipsa tellure fodiuntur, si non herbarum ac fruticum tractata remedia auferrent traversos ex ipsis animalibus quae sanantur reperta maiore medicina. quid ergo? dixerimus herbas et florum imagines ac pleraque inventu rara ac difficilia, iidem tacebimus quid in ipso homine prosit homini ceteraque genera remediorum inter nos viventia, cum praesertim nisi carenti doloribus morbisque vita ipsa poena fiat? 2 minime vero, omnemque insumemus operam, licet fastidii periculum urgeat, quando ita decretum est, minorem gratiae quam utilitatium vitae respectum habere. quin immo externa quoque et barbaros etiam ritus indagabimus. fides tantum auctores appellet, quamquam et ipsi consensu prope iudicii ista eligere laboravimus potiusque curae rerum quam 3 copiae institimus. illud admonuisse perquam neces-

---

a Or, " to more potent remedies." So Littré.

PLINY: NATURAL HISTORY

BOOK XXVIII

I. I should have finished describing the character *Remedies* of all things growing between heaven and earth, *from animals.* leaving only whatever is dug out of the ground itself, if dealing with remedies derived from plants and shrubs did not make me digress to the wider sphere of medicines *a* obtained from the very living creatures that themselves are healed. Well then, shall I, who have described plants and forms of flowers, including many rare things that are difficult to find, say nothing about the benefits to man that are to be found in man himself, nothing about the other kinds of remedies that live among us, especially as life itself becomes a punishment for those who are not free from pains and diseases? Surely I must, and I shall devote all my care to the task, although I realize the •isk of causing disgust, since it is my fixed determination to have less regard for popularity than for benefiting human life. Furthermore, my investigations will include foreign things and even outlandish customs; belief here can appeal only to authority, although I myself also, when choosing my detail, have striven to find views almost universally believed, and I have stressed careful research rather than abundance of material. One

3

sarium est, dictas iam a nobis naturas animalium et quae cuiusque essent inventa—neque enim minus profuere medicinas reperiendo quam prosunt praebendo—nunc quae in ipsis auxilientur indicari, neque illic in totum omissa, itaque haec esse quidem alia, illis tamen conexa.

4 II. Incipiemus autem ab homine ipsum sibi exquirente,[1] inmensa statim difficultate obvia. sanguinem quoque gladiatorum bibunt ut viventibus poculis comitiales [morbi],[2] quod spectare facientes in eadem harena feras quoque horror est. at, Hercule, illi ex homine ipso sorbere efficacissimum putant calidum spirantemque et vivam[3] ipsam animam ex osculo vulnerum, cum plagis omnino ne[4] ferarum quidem admoveri ora mos sit humanus.[5] alii medullas crurum quaerunt et cerebrum infantium.

5 nec pauci apud Graecos singulorum viscerum membrorumque etiam sapores dixere omnia persecuti ad resigmina unguium, quasi vero sanitas videri possit feram ex homine fieri morboque dignum in ipsa medicina, egregia, Hercules, frustratione, si non prosit. aspici humana exta nefas habetur, quid

---

[1] exquirente *Urlichs* : exquirentes RdE : exquirentis V.

[2] morbi *in uncis Mayhoff.   Sed cf.* § 7 *et* § 35.

[3] vivam *Detlefsen*: unam *codd.*: una *Warmington.*

[4] omnino ne *Mayhoff* : omne V[2]Er : ne *Gelenius, Detlefsen.*

[5] mos sit humanus *Mayhoff* : fas sit. humanas *Detlefsen.* mos Tf : mus V[1]R : mus fas V[2] : fas Er : humanus *omnes codd.*

---

[a] See VIII. §§ 97 foll. and XXV. §§ 89 foll.

[b] This seems to refer to the difficulty discussed in §§ 10 foll. Perhaps the rest of the chapter is an afterthought of

thing it is very necessary to point out: I have already described [a] the natures of living creatures and the discoveries we owe to each (for they did no less good by discovering medicines than they do by supplying them), I am now showing what help is to be found in the creatures themselves. I did not entirely leave out this then; so although the new matter is different, it is yet intimately connected with the old.

II. But I shall begin with man seeking aid for himself out of himself, and at the outset there will meet us a most baffling puzzle.[b] The blood too of gladiators is drunk by epileptics as though it were a draught of life, though we shudder with horror when in the same arena we look at even the beasts doing the same thing. But, by Heaven!, the patients think it most effectual to suck from a man himself warm, living blood, and putting their lips to the wound [c] to drain the very life, although it is not the custom of men to apply their mouths at all to the wounds even of wild beasts. Others seek to secure the leg-marrow and the brain of infants. Not a few among the Greeks have even spoken of the flavour of each organ and limb, going into all details, not excluding nail parings; just as though it could be thought health for a man to become a beast, and to deserve disease as punishment in the very process of healing.[d] And, by Heaven!, well deserved is the disappointment if these remedies prove of no avail. To look at human entrails is considered sin; what

Remedies from man.

Pliny; Mayhoff, while reading *quoque* in his text, suggests *quippe* in his textual notes.

[c] Perhaps, "by kissing the wounds," or, as Littré, "from the gaping wounds."

[d] Or: "for the very remedies he adopts."

5

6 mandi? quis ista invenit, Osthane? tecum enim
res erit, eversor iuris humani monstrorumque artifex
qui primus ea condidisti, credo, ne vita tui oblivis-
ceretur. quis invenit singula membra humana
mandere? qua coniectura inductus? quam potest
medicina ista originem . habuisse? quis veneficia
innocentiora fecit quam remedia? esto, barbari
externique ritus invenerant, etiamne Graeci suas
7 fecere has artes? extant commentationes Demo-
criti ad aliud noxii hominis ex capite ossa plus
prodesse, ad alia amici et hospitis. iam vero
vi interempti dente gingivas in dolore scariphari
Apollonius efficacissimum scripsit, Meletos oculorum
suffusiones felle hominis sanari. Artemon calvaria
interfecti neque cremati propinavit aquam e fonte
noctu comitialibus morbis. ex eadem suspendio
interempti catapotia fecit contra canis rabiosi
8 morsus Antaeus. atque etiam quadrupedes homine[1]
sanavere, contra inflationes boum perforatis cornibus
inserentes ossa humana, ubi homo occisus esset aut
crematus siliginem quae pernoctasset suum morbis
dando. procul a nobis nostrisque litteris absint
ista. nos auxilia dicemus, non piacula, sicubi lactis
puerperarum usus mederi poterit, sicubi saliva
9 tactusve corporis ceteraque similia. vitam quidem

---

[1] homine *Pintianus, Mayhoff*: homines *codd., Detlefsen.*

---

[a] A Persian Magus of the early fifth century B.C. to whom
were attributed many works on oriental magic.
[b] Possibly, " guess-work."
[c] Diogenes Laertius attributes to this philosopher works on
medicine and regimen, and probably many spurious works
also were foisted on him.
[d] Probably a physician who lived in the first century B.C.
[e] An unknown.

must it be to eat them? Who was the first, Osthanes,[a] to think of such devices as yours? For it is you who must bear the blame, you destroyer of human rights and worker of horrors; you were their first founder, in order, I suppose, to perpetuate your memory. Who first thought of chewing one by one human limbs? What soothsaying[b] guided him? What origin could your medical practices have had? Who made magic potions more innocent than their remedies? Granted that foreigners and barbarians had discovered the rites, did the Greeks also make these arts their own? There is extant a treatise of Democritus[c] stating that one complaint is more benefited by bones from the head of a criminal, and other complaints by those of a friend or guest. Moreover, Apollonius[d] put in writing that to scrape sore gums with the tooth of a man killed by violence is most efficacious, and Meletos[e] that the gall of a human being cures cataract. Artemon[f] treated epilepsy with draughts of water drawn from a spring by night and drunk out of the skull of a man killed but not cremated. From the skull of a man hanged Antaeus[g] made pills to cure the bites of a mad dog. Even quadrupeds too have been cured by remedies taken from a man; to cure flatulence in oxen their horns have been pierced and human bones inserted; for sick pigs wheat has been given which had remained for a whole night where a man had been killed or cremated. Far from me and my writings be such horrors. I shall speak not of sins but of aids, such as when will prove an effective remedy the milk of lying-in women, or human saliva, or contact with a human body, and the like. I do

---

[f] An unknown. [g] An unknown.

non adeo expetendam censemus ut quoquo modo
trahenda sit. quisquis es talis,[1] aeque moriere,
etiam cum [2] obscaenus vixeris aut nefandus.
quapropter hoc primum quisque in remediis animi
sui habeat, ex omnibus bonis quae homini tribuit
natura nullum melius esse tempestiva morte, idque
in ea optimum quod illam sibi quisque praestare
poterit.

10   III. Ex homine remediorum primum maximae
quaestionis et semper incertae est, polleantne [3]
aliquid verba et incantamenta carminum. quod
si verum est, homini acceptum fieri oportere con-
veniat, sed viritim sapientissimi cuiusque respuit
fides, in universum vero omnibus horis credit vita
nec sentit. quippe victimas caedi sine precatione
11 non videtur referre aut deos rite consuli. praeterea
alia sunt verba inpetritis, alia depulsoriis, alia
commendationis, videmusque certis precationibus
obsecrasse [4] summos magistratus et, ne quod ver-
borum praetereatur aut praeposterum dicatur, de
scripto praeire aliquem rursusque alium custodem
dari qui adtendat, alium vero praeponi qui favere
linguis iubeat, tibicinem canere, ne quid aliud ex-
audiatur, utraque memoria insigni, quotiens ipsae

---

[1] *Comma ante* talis *trans.* Mayhoff.
[2] etiam cum *multi codd., vulg., Detlefsen* : etiam quam
VT : tamquam *Mayhoff.*
[3] polleantne VRdTf *Mayhoff* : valeantne Er *vulg., Detlefsen.*
[4] obsecrasse] obsecrare *coni. Mayhoff.*

---

[a] With Mayhoff's reading : " Whoever you are, as such
you will die, just as if your life will have been one of foulness
or sin."

not indeed hold that life ought to be so prized that by any and every means it should be prolonged. You holding this view, whoever you are, will none the less die, even though you may have lived longer through foulness or sin.[a] Wherefore let every man consider that first among the remedies for his soul is this: that of all the blessings given to man by Nature none is greater than a timely death, and herein the brightest feature is that each man can have the power to bestow it on himself.

III. Of the remedies derived from man, the first *Have words* raises a most important question, and one never *power?* settled: have words and formulated incantations any effect? If they have, it would be right and proper to give the credit to mankind. As individuals, however, all our wisest men reject belief in them, although as a body the public at all times believes in them unconsciously. In fact the sacrifice of victims without a prayer is supposed to be of no effect; without it too the gods are not thought to be properly consulted. Moreover, there is one form of words for getting favourable omens, another for averting evil, and yet another for a commendation. We see also that our chief magistrates have adopted fixed formulas for their prayers; that to prevent a word's being omitted or out of place a reader dictates beforehand the prayer from a script; that another attendant is appointed as a guard to keep watch, and yet another is put in charge to maintain a strict silence; that a piper plays so that nothing but the prayer is heard. Remarkable instances of both kinds of interference are on record: cases when the noise of actual ill omens has ruined the prayer, or when a mistake has been made in the prayer itself; then sud-

dirae obstrepentes nocuerint quotiensve precatio
erraverit, sic repente extis adimi capita vel corda
12 aut geminari victima stante.   durat inmenso exemplo
Deciorum patris filiique quo se devovere carmen,
extat Tucciae Vestalis incesti deprecatio qua usa
aquam in cribro tulit anno urbis DXVIIII. boario
vero in foro Graecum Graecamque defossos aut
aliarum gentium cum quibus tum res esset etiam
nostra aetas vidit. cuius sacri precationem qua
solet praeire XVvirum collegii magister si quis
legat, profecto vim carminum fateatur, ea omnia
adprobantibus DCCCXXX annorum eventibus.
13 Vestales nostras hodie credimus nondum egressa
urbe mancipia fugitiva retinere in loco precatione,
cum, si semel recipiatur ea ratio et deos preces
aliquas exaudire aut ullis moveri verbis, confitendum
sit de tota coniectatione. prisci quidem nostri
perpetuo talia prodidere, difficillimumque ex his
etiam fulmina elici, ut suo loco docuimus.
14   IV. L. Piso primo annalium auctor est Tullum
Hostilium regem ex Numae libris eodem quo illum
sacrificio Iovem caelo devocare conatum, quoniam
parum rite quaedam fecisset, fulmine ictum, multi
vero magnarum rerum fata et ostenta verbis per-

---

a See Livy VIII. 9 and X. 28.
b See Valerius Maximus VIII. 1.
c 145 B.C.
d Plutarch *Roman Questions* 283.
e Or: " all magical charms must be accepted."
f See Book II. § 140.
g Consul in 133 B.C. and an opponent of the Gracchi.

denly the head of the liver, or the heart, has disappeared from the entrails, or these have been
doubled, while the victim was standing. There has
come down to us a striking example of ritual in that
with which the Decii,[a] father and son, devoted themselves; extant too is the plea of innocence uttered by
the Vestal Tuccia[b] when, accused of unchastity, she
carried water in a sieve, in the year of the City six
hundred and nine.[c] Our own generation indeed even
saw buried alive in the Cattle Market a Greek man
and a Greek woman, and victims from other peoples
with whom at the time we were at war.[d] The prayer
used at this ceremony is wont to be dictated by the
Master of the College of the Quindecimviri, and if one
reads it one is forced to admit that there is power in
ritual formulas, the events of eight hundred and thirty
years showing this for all of them. It is believed
today that our Vestal virgins by a spell root to the
spot runaway slaves, provided they have not left the
City bounds, and yet, if this view is once admitted,
that the gods hear certain prayers, or are moved
by any form of words, the whole question must be
answered in the affirmative.[e] Our ancestors, indeed,
reported such wonders again and again, and that,
most impossible of all, even lightning can be brought
by charms from the sky, as I have mentioned[f] on
the proper occasion.

IV. Lucius Piso[g] in the first Book of his *Annals*
tells us that King Tullus Hostilius used the same
sacrificial ritual as Numa, which he found in Numa's
books, in an attempt to draw Jupiter down from the
sky, and was struck by lightning because he made
certain mistakes in the ceremony; many indeed
assure us that by words the destinies and omens of

15 mutari. cum in Tarpeio fodientes delubro funda-
menta caput humanum invenissent, missis ob id ad
se legatis Etruriae celeberrimus vates Olenus Calenus
praeclarum id fortunatumque cernens interrogatione
in suam gentem transferre temptavit. scipione prius
determinata templi imagine in solo ante se: Hoc
ergo dicitis, Romani? hic templum Iovis optimi
maximi futurum est, hic caput invenimus? constan-
tissima annalium adfirmatione transiturum fuisse
fatum in Etruriam, ni praemoniti a filio vatis legati
Romani respondissent: Non plane hic sed Romae
16 inventum caput dicimus. iterum id accidisse tradunt, cum in fastigium eiusdem delubri praeparatae
quadrigae fictiles in fornace crevissent, et iterum
17 simili modo retentum augurium. haec satis sint
exemplis ut appareat ostentorum vires et in nostra
potestate esse ac prout quaeque accepta sint ita
valere. in augurum certe disciplina constat neque
diras neque ulla auspicia pertinere ad eos qui
quamcumque [1] rem ingredientes observare se ea
negaverint, quo munere divinae indulgentiae maius
nullum est. quid? non et legum ipsarum in duo-
18 decim tabulis verba sunt: qui fruges excantassit, et
alibi: qui malum carmen incantassit? Verrius Flaccus
auctores ponit quibus credatur [2] in obpugnationibus

---

[1] qui quamcumque *coni. Mayhoff*: quicumque *Detlefsen*:
qui quamque *Mayhoff in textu*, RdE *vulg.*: quicquam quae V.
[2] credatur *Warmington*: credat *codd.*

---

[a] Perhaps " obviously."
[b] See *Remains of Old Latin* (Loeb) vol. III, pp. 474, 475 and
478, 479.
[c] A distinguished writer of the latter part of the first
century B.C. He wrote on history and antiquities, dying in
the reign of Tiberius.

mighty events are changed. During the digging
of foundations for a shrine on the Tarpeian Hill
there was discovered a human head. For an interpretation envoys were sent to Olenus of Cales, the
most distinguished seer of Etruria. Perceiving that
the sign portended glory and success, Olenus tried
by questioning to divert the blessing to his own
people. He first traced with his staff the outline of a
temple on the ground in front of him, and then
asked: " Is this then, Romans, what you say?
' Here will be the temple of Jupiter, All-good and
Almighty; here we found the head?'" The *Annals*
most firmly insists that the destiny of Rome would
have passed to Etruria, had not the Roman envoys,
forewarned by the seer's son, replied: " Not exactly [a]
here, but it was in Rome that we say the head was
found." It is said that the same thing happened
again when a clay four-horse chariot, designed for the
roof of the same shrine, grew larger in the furnace,
and once more in a similar way was the happy
augury retained. Let these instances suffice to show
that the power of omens is really in our own control, and that their influence is conditional upon the
way we receive each. At any rate, in the teaching of
the augurs it is a fundamental principle that neither
evil omens nor any auspices affect those who at the
outset of any undertaking declare that they take
no notice of them; no greater instance of the divine
mercy could be found than this boon. Again, in
the actual laws of the Twelve Tables we find also
these words: [b] " Whoever shall have bewitched the
crops," and in another place: " whoever shall have
cast an evil spell." Verrius Flaccus [c] cites trustworthy
authorities to show that it was the custom, at the

ante omnia solitum a Romanis sacerdotibus evocari
deum cuius in tutela id oppidum esset promittique
illi eundem aut ampliorem apud Romanos cultum.
et durat in pontificum disciplina id sacrum, constatque ideo occultatum in cuius dei tutela Roma esset,
19 ne qui hostium simili modo agerent. defigi quidem
diris deprecationibus nemo non metuit. hoc pertinet
ovorum quae exorbuerit quisque calices coclearum-
que protinus frangi aut isdem coclearibus perforari.
hinc Theocriti apud Graecos, Catulli apud nos
proximeque Vergilii incantamentorum amatoria imi-
tatio. multi figlinarum opera rumpi credunt tali
modo, non pauci etiam serpentes; ipsas recanere
et hunc unum illis esse intellectum contrahique
Marsorum cantu etiam in nocturna quiete. etiam [1]
parietes incendiorum deprecationibus conscribuntur.
20 neque est facile dictu externa verba atque ineffabilia
abrogent fidem validius an Latina inopinata et [2]
quae inridicula videri cogit animus semper aliquid
inmensum exspectans ac dignum deo movendo,
21 immo vero quod numini imperet. dixit Homerus
profluvium sanguinis vulnerato femine Ulixen inhi-
buisse carmine, Theophrastus ischiadicos sanari,
Cato prodidit luxatis membris carmen auxiliare,

---

[1] etiam *multi codd. Detlefsen* : iam d, *Mayhoff.*
[2] et *post* Latina *trans. Mayhoff.*

---

[a] See *Idyll* II.
[b] See *Eclogues* VIII.   The Catullus passages are not extant.
[c] Referring to the so-called *Ephesia grammata* and gibberish
of many incantations.
[d] See *Odyssey* XIX. 457, where it is not Odysseus, but
Autolycus and his sons that effect the cure.
[e] See Athenaeus XIV. 18.

very beginning of a siege, for the Roman priests
to call forth the divinity under whose protection
the besieged town was, and to promise him the
same or even more splendid worship among the
Roman people. Down to the present day this
ritual has remained part of the doctrine of the
Pontiffs, and it is certain that the reason why the
tutelary deity of Rome has been kept a secret
is to prevent any enemy from acting in a similar
way. There is indeed nobody who does not fear
to be spell-bound by imprecations. A similar feeling makes everybody break the shells of eggs or
snails immediately after eating them, or else
pierce them with the spoon that they have used.
And so Theocritus [a] among the Greeks, Catullus
and quite recently Virgil [b] among ourselves, have
represented love charms in their poems. Many
believe that by charms pottery can be crushed,
and not a few even serpents; that these themselves
can break the spell, this being the only kind of
intelligence they possess; and by the charms of the
Marsi they are gathered together even when asleep
at night. On walls too are written prayers to avert
fires. It is not easy to say whether our faith is more
violently shaken by the foreign, unpronounceable
words,[c] or by the unexpected Latin ones, which
our mind forces us to consider absurd, being always
on the look-out for something big, something ade-
quate to move a god, or rather to impose its will on
his divinity. Homer said that by a magic formula
Ulysses [d] stayed the haemorrhage from his wounded
thigh; Theophrastus [e] that there is a formula to
cure sciatica; Cato [f] handed down one to set dis-

[f] See Cato CLX.

M. Varro podagris. Caesarem dictatorem post unum ancipitem vehiculi casum ferunt semper ut primum consedisset, id quod plerosque nunc facere scimus, carmine ter repetito securitatem itinerum aucupari solitum.

22 V. Libet hanc partem singulorum quoque conscientia coarguere. cur enim primum anni incipientes [1] diem laetis precationibus invicem faustum ominamur? cur publicis lustris etiam nomina victimas ducentium prospera eligimus? cur effascinationibus adoratione peculiari occurrimus, alii Graecam Nemesin invocantes, cuius ob id Romae simulacrum in Capitolio est, quamvis Latinum 23 nomen non sit? cur ad mentionem defunctorum testamur memoriam eorum a nobis non sollicitari? cur inpares numeros ad omnia vehementiores credimus, idque in febribus dierum observatione intellegitur? cur ad primitias pomorum haec vetera esse dicimus, alia nova optamus? cur sternuentes salutamus, quod etiam Tiberium Caesarem, tristissimum, ut constat, hominum in vehiculo exegisse tradunt, et aliqui nomine quoque consalutare re-24 ligiosius putant? quin et absentes tinnitu aurium praesentire sermones de se receptum est. Attalus adfirmat, scorpione viso si quis dicat duo, cohiberi nec vibrare ictus, et quoniam scorpio admonuit, in

---

[1] incipientes V(?)E *Detlefsen* : incipientis *Mayhoff.*

---

[a] See Varro *R.R.* I. ii. 27.

[b] Or (Wolters), " their rest is not being disturbed."

[c] Or, " the more scrupulous think that they must."

[d] Probably not Attalus III, King of Pergamus, who died in 133 B.C. Perhaps an unknown physician. See Wolters, p. 52.

[e] " *Africa* was personified, in the time of Hadrian, as a woman, represented in divers ways on bronze coins, with a scorpion in her hand or on her head " (Wolters, p. 56).

located limbs, Marcus Varro [a] one for gout. The dictator Caesar, after one serious accident to his carriage, is said always, as soon as he was seated, to have been in the habit of repeating three times a formula of prayer for a safe journey, a thing we know that most people do today.

V. I should like to reinforce this part of my argument by adding an appeal to the personal feeling of the individual. Why on the first day of the year do we wish one another cheerfully a happy and prosperous New Year? Why do we also, on days of general purification, choose persons with lucky names to lead the victims? Why do we meet the evil eye by a special attitude of prayer, some invoking the Greek Nemesis, for which purpose there is at Rome an image of the goddess on the Capitol, although she has no Latin name? Why on mentioning the dead do we protest that their memory is not being attacked by us? [b] Why do we believe that in all matters the odd numbers are more powerful, as is implied by the attention paid to critical days in fevers? Why at the harvest of the first-fruits do we say: " These are old," and pray for new ones to take their place? Why do we say " Good health " to those who sneeze? This custom according to report even Tiberius Caesar, admittedly the most gloomy of men, insisted on even in a carriage, and some think it more effective [c] to add to the salutation the name of the sneezer. Moreover, according to an accepted belief absent people can divine by the ringing in their ears that they are the object of talk. Attalus [d] assures us that if on seeing a scorpion one says " Two," it is checked and does not strike. The mention of scorpions [e]

Why are we super-stitious?

17

Africa nemo destinat aliquid nisi praefatus Africam, in ceteris vero gentibus deos ante obtestatus ut velint. nam si mensa adsit,[1] anulum ponere translatitium videmus, quoniam etiam mutas[2] religiones pollere 25 manifestum est. alius saliva post aurem digito relata sollicitudinem animi propitiat. pollices, cum faveamus, premere etiam proverbio iubemur. in adorando dextram ad osculum referimus totumque corpus circumagimus, quod in laevum fecisse Galliae religiosius credunt. fulgetras poppysmis adorare 26 consensus gentium est. incendia inter epulas nominata aquis sub mensam profusis abominamur. recedente aliquo ab epulis simul verri solum aut bibente conviva mensam vel repositorium tolli inauspicatissimum iudicatur. Ser. Sulpicii principis viri commentatio est quamobrem mensa linquenda[3] non sit, nondum enim plures quam convivae numerabantur. nam sternumento revocari ferculum mensamve, si non postea gustetur aliquid, inter diras

---

[1] mensa adsit VRd, *Mayhoff*: mens adflicta sit *Detlefsen*.

[2] mutas *Sillig*: multas *codd.*: quin etiam mutas . . . est; nam si mensa adsit *Wolters*.

[3] linquenda *codd.*: admovenda *Wolters, qui* nondum . . . numerabantur *in uncis ponit*.

---

[a] Mayhoff would emend this dubious Plinian *nam* to *iam*, which is an improvement, but to transpose the clauses of this sentence (with Wolters) makes it possible to give *nam* its usual meaning: "Moreover, it is clear that actions even without words have powers, for it is a universal custom, we see, etc."

reminds me that in Africa nobody decides on anything without first saying " Africa," whereas among
all other peoples a man prays first for the approval
of the gods. But[a] when a table is ready it is a
universal custom, we see, to take off one's ring, since
it is clear that scrupulous actions, even without words,
have their powers. Some people, to calm mental
anxiety, carry saliva with the finger to behind the ear.
There is even a proverb that bids us turn down[b] our
thumbs to show approval. In worshipping we
raise our right hand to our lips and turn round our
whole body, the Gauls considering it more effective[c]
to make the turn to the left. All peoples agree in
worshipping lightning by clucking with the tongue. *Superstitions*
If during a banquet fires have been mentioned we *at table.*
avert the omen by pouring water under the table.
It is supposed to be a most unlucky sign for the floor
to be swept while a diner is leaving the banquet, or
for a table or dumb-waiter to be removed while a
guest is drinking. Servius Sulpicius,[d] a noble Roman,
has left an essay on why we should not leave the
table ;[e] for in his day it was not the custom to have
more tables than there were guests ; for if a course
or a table is recalled by a sneeze and nothing of it
tasted afterwards, it is considered an evil portent, as

[b] See Mayor on Juvenal III. 36. Wolters translates
premere " to enclose."
[c] So Wolters, making *religiosius* objective. Perhaps,
however, it is subjective, meaning " more devout."
[d] A contemporary of Cicero, who took part in the troublous
politics of the period.
[e] A difficult sentence. Wolters reads *admovenda* for
linquenda and brackets *nondum . . . numerabantur* as a gloss.
He also brackets *aut omnino non esse*. Much of the difficulty
of this passage comes from the ambiguity of the word *mensa*.
See the additional note A, page 563.

27 habetur, aut omnino non[1] esse. haec instituere illi
qui omnibus negotiis horisque interesse credebant
deos, et ideo placatos etiam vitiis nostris reliquerunt.
quin et repente conticescere convivium adnotatum
est[2] non nisi in pari praesentium numero, isque
famae labor est ad quemcumque eorum pertinens.
cibus etiam e manu prolapsus reddebatur[3] utique
per mensas, vetabantque munditiarum causa deflare,
et sunt condita auguria, quid loquenti cogitantive
id acciderit, inter execratissima, si pontifici accidat
dicis causa epulanti. in mensa utique id reponi
28 adolerique ad Larem piatio est. medicamenta
priusquam adhibeantur in mensa forte deposita
negant prodesse. ungues resecari nundinis Romanis
tacenti atque a digito indice multorum persuasione[4]
religiosum est, capillum vero contra defluvia ac
dolores capitis XVII luna atque XXVIIII. pagana
lege in plerisque Italiae praediis cavetur ne mulieres
per itinera ambulantes torqueant fusos aut omnino
detectos ferant, quoniam adversetur id omnium spei,
29 praecipue frugum. M. Servilius Nonianus princeps

---

[1] non *Gelenius*: nam E : inane *fere omnes codd.*, *Mayhoff,
qui lacunam post* habetur *indicat*: *del.* aut . . . esse *Wolters.*
[2] est *codd.* : set *Mayhoff.*
[3] *Ante* reddebatur *addit* non *Wolters.*
[4] multorum persuasione *Mayhoff* : mulierum peculiare
Detlefsen : multorum pecuniae *codd. Fortasse* opinione
(*Haupt*).

---

[a] This could mean : "either considered an evil portent or
none at all." (Warmington.)
[b] Littré says " malgré nos vices."
[c] So Bostock and Riley, and also Wolters, but Littré has :
" de l'un quelconque d'entre eux."
[d] The emendation of Wolters : " used not to be put back,"
is more in accordance with customs elsewhere.

is to eat nothing at all.[a] These customs were established by those of old, who believed that gods are present on all occasions and at all times, and therefore left them to us reconciled even in our faults.[b] Moreover, it has been remarked that a sudden silence falls on a banquet only when the number of those present is even, and that it portends danger to the reputation of each [c] of them. Food also that fell from the hand used to be put back [d] at least during courses, and it was forbidden to blow off, for tidiness, any dirt; [e] auguries have been recorded from the words or thoughts of the diner who dropped food, a very dreadful omen being if the Pontiff should do so at a formal dinner. In any case putting it back on the table and burning it before the Lar counts as expiation.[f] Medicines set down by chance on a table before being used are said to lose their efficacy. To cut the nails on the market days at Rome in *Various other super-* silence, beginning with the forefinger, is a custom *stitions.* many people feel binding on them; while to cut the hair on the seventeenth day of the month and on the twenty-ninth prevents its falling out as well as headaches. A country rule observed on most Italian farms forbids women to twirl their spindles while walking along the road, or even to carry them uncovered, on the ground that such action blights the hopes of everything, especially the hope of a good harvest. Marcus Servilius Nonianus,[g] a leading

---

[e] Wolters thinks that *deflare* here means, " to remove." Perhaps: " blow off any crumbs to tidy up." So Warmington.

[f] Wolters translates " as sin." He says that *piatio* here is the same as *piaculum*, holding that dropped food was left where it was.

[g] Consul A.D. 35, died 59, and known personally to Pliny, who mentions him several times.

civitatis non pridem in metu lippitudinis, priusquam ipse eam nominaret aliusve ei praediceret, duabus litteris Graecis PA chartam inscriptam circumligatam lino subnectebat collo, Mucianus ter consul eadem observatione viventem muscam in linteolo albo, his remediis carere ipsos lippitudine praedicantes. carmina quidem extant contra grandines contraque morborum genera contraque ambusta, quaedam etiam experta, sed prodendo obstat ingens verecundia in tanta animorum varietate. quapropter de his ut cuique libitum fuerit opinetur.

30 VI. Hominum monstrificas naturas et veneficos aspectus diximus in portentis gentium et multas animalium proprietates, quae repeti supervacuum est. quorundam hominum tota corpora prosunt, ut ex his familiis quae sunt terrori serpentibus tactu ipso levant percussos suctuve madido,[1] quorum e genere sunt Psylli Marsique et qui Ophiogenes vocantur in insula Cypro, ex qua familia legatus Evagon nomine a consulibus Romae in dolium serpentium coniectus experimenti causa circum-
31 mulcentibus linguis miraculum praebuit. signum eius familiae est, si modo adhuc durat, vernis temporibus odoris virus. atque eorum sudor quoque

---

[1] madido E *Detlefsen*: modo *Mayhoff*: tumodo R: tumido *multi codd.*

---

[a] These letters have no hidden meaning; "they probably belong to the abracadabra of magic" (Wolters). Perhaps they were intended to be the last two letters of it.
[b] C. Licinius Mucianus was consul for the third time in A.D. 72. In 68-69 he was governor of Syria with a command of four legions. See Tacitus *Histories* I. 10.
[c] See Book VII. §§ 13 foll.

citizen of Rome, who was not so long ago afraid of ophthalmia, used to tie round his neck, before he mentioned the disease himself or any one else spoke to him about it, a sheet of paper fastened with thread, on which were written the two Greek letters rho and alpha; [a] Mucianus,[b] three times consul, following the same observance, used a living fly in a white linen bag. Both avowed that by these remedies they themselves were kept free from ophthalmia. We certainly still have formulas to charm away hail, various diseases, and burns, some actually tested by experience, but I am very shy of quoting them, because of the widely different feelings they arouse. Wherefore everyone must form his own opinion about them as he pleases.

VI. Persons possessed of powers of witchcraft and of the evil eye, along with many peculiar characteristics of animals, I have spoken of [c] when dealing with marvels of the nations; it is superfluous to go over the ground again. Of certain men the whole bodies are beneficent, for example the members of those families that frighten serpents. These by a mere touch or by wet suction [d] relieve bitten victims. In this class are the Psylli, the Marsi, and the Ophiogenes, as they are called, in the island of Cyprus. An envoy from this family, by name Evagon, was at Rome thrown by the consuls as a test into a cask of serpents, which to the general amazement licked him all over. A feature of this family, if it still survives, is the foul smell of its members in spring. Their sweat also, not only

People with magic powers.

[d] There is much to be said for Mayhoff's *modo,* " only." But *madido* suggests that much fluid was drawn from the wound. Salmasius in fact conjectured *umido.*

23

medebatur, non modo saliva. nam in insula Nili
Tentyri nascentes tanto sunt crocodilis terrori ut
vocem quoque eorum fugiant. horum omnium
generum insita[1] repugnantia interventum quoque
mederi constat, sicuti adgravari vulnera introitu
eorum qui umquam fuerint serpentium canisve dente
32 laesi. iidem gallinarum incubitus, pecorum fetus
abortu vitiant. tantum remanet virus ex accepto
semel malo ut venefici fiant venena passi. remedio
est ablui prius manus eorum aquaque illa eos quibus
medearis inspergi. rursus a scorpione aliquando
percussi numquam postea a crabronibus, vespis
33 apibusve feriuntur. minus miretur hoc qui sciat
vestem a tineis non attingi quae fuerit in funere,
serpentes aegre praeterquam laeva manu extrahi.
e Pythagorae inventis non temere fallere,[2] in-
positivorum nominum inparem vocalium numerum
clauditates oculive orbitatem ac similes casus dextris
adsignare partibus, parem laevis. ferunt difficiles
partus statim solvi, cum quis tectum in quo sit gravida
transmiserit lapide vel missili ex his qui tria animalia
singulis ictibus interfecerint, hominem, aprum,
34 ursum. probabilius id facit hasta velitaris evulsa
corpori hominis, si terram non attigerit. eosdem
enim inlata effectus habet. sic et sagittas corpori
eductas, si terram non attigerint, subiectas cubantibus

---

[1] insita *Mayhoff*: in sua *codd.*
[2] fallere] *Mayhoff* fallare *coni., ut* arbitrere *XI* § 82.

---

[a] *I.e.* to disease, poison etc.
[b] The *Thesaurus* gives *impositus* and *inditus* as equivalents
of *impositivus*. A *nomen impositivum* would be any name

their saliva, had curative powers. But the natives of Tentyris, an island on the Nile, are such a terror to the crocodiles that these run away at the mere sound of their voice. All these peoples, so strong their natural antipathy,[a] can, as is well known, effect a cure by their very arrival, just as wounds grow worse on the entry of those who have ever been bitten by the tooth of snake or dog. The latter also addle the eggs of a sitting hen, and make cattle miscarry; so much venom remains from the injury once received that the poisoned are turned into poisoners. The remedy is for their hands to be first washed in water, which is then used to sprinkle on the patients. On the other hand, those who have once been stung by a scorpion are never afterwards attacked by hornets, wasps or bees. He may be less surprised at this who knows that moths do not touch a garment that has been worn at a funeral, and that snakes are with difficulty pulled out of their holes except with the left hand. One of the dis- *Various* coveries of Pythagoras will not readily deceive you: *kinds of* that an uneven number of vowels in given [b] names *magic power.* portends lameness, blindness, or similar disability, on the right side, an even number of vowels the same disabilities on the left. It is said that difficult labour ends in delivery at once, if over the house where is the lying-in woman there be thrown a stone or missile that has killed with one stroke each three living creatures —a human being, a boar, and a bear. A successful result is more likely if a light-cavalry spear is used, pulled out from a human body without the ground being touched. The result indeed is the same if the

other than those the individual could not avoid (*e.g.* the family name).

25

amatorium esse Orpheus et Archelaus scribunt,
quin et comitiales morbos sanari cibo e carne ferae
occisae eodem ferro quo homo interfectus sit.
quorundam partes medicae sunt, sicuti diximus de
Pyrrhi regis pollice, et Elide solebat ostendi Pelopis
scapula,[1] quam eburneam adfirmabant. naevos in
facie tondere religiosum habent etiam nunc multi.

35 VII. Omnium vero in primis ieiunam salivam
contra serpentes praesidio esse docuimus, sed et
alios efficaces eius usus recognoscat vita. despuimus
comitiales morbos, hoc est contagia regerimus.
simili modo et fascinationes repercutimus dextraeque
36 clauditatis occursum. veniam quoque a deis spei
alicuius audacioris petimus in sinum spuendo, et iam [2]
eadem ratione terna despuere precatione [3] in omni
medicina mos est, atque ita effectus adiuvare,
incipientes furunculos ter praesignare ieiuna saliva.
mirum dicimus, sed experimento facile: si quem
paeniteat ictus eminus comminusve inlati et statim

---

[1] scapula quam *Gronovius, Detlefsen, qui lacunam indicat*:
os ulnamque eam *Mayhoff*: *pro* scapula *varia* (ostilnam,
ostiliam, ostiliani) *codd.*

[2] et iam *Detlefsen, Mayhoff, qui* etenim *vel* multis etiam
coni.: etiam Er: *om. plerique codd.*

[3] precatione *Urlichs, Mayhoff*: deprecatione *Detlefsen,
vulg.*: praedicatione *codd.*: *an* pro precatione?

---

[a] Many spurious works of a medical nature were attributed
to the Orpheus of mythology.

[b] Archelaus was possibly the Greek poet living in Egypt,
some of whose epigrams are in the *Anthology*.

[c] See Book VII. § 20.

[d] Pausanias (V. 13, 4) says that the bone was the ὠμοπλάτη
(shoulder blade), and that it had disappeared (ἠφάνιστο)
by his time. Mayhoff's conjecture would mean " elbow."

[e] Mayhoff brackets the last sentence, which seems out of
place.

spear is carried indoors. So too, as Orpheus [a] and Archelaus [b] write, arrows drawn out of a body and not allowed to touch the ground act as a love-charm upon those under whom when in bed they have been placed. Moreover, add these authorities, epilepsy is cured by food taken from the flesh of a wild beast killed by the same iron weapon that has killed a human being. Some men have healing powers confined to parts of their body. We have mentioned the thumb of King Pyrrhus,[c] and at Elis there used to be shown a shoulder blade [d] of Pelops, which was stated to be of ivory. Many men even today have scruples about cutting hair from moles on the face.[e]

VII. I have however pointed out that the best of all safeguards against serpents is the saliva of a fasting human being, but our daily experience may teach us[f] yet other values of its use. We spit on epileptics in a fit, that is, we throw back infection.[g] In a similar way we ward off witchcraft and the bad luck that follows meeting a person lame in the right leg. We also ask forgiveness of the gods for a too presumptuous hope by spitting into our bosom; the same reason again accounts for the custom, in using any remedy, of spitting on the ground three times by way of ritual,[h] thus increasing its efficacy, and of marking early incipient boils three times with fasting saliva. It is surprising, but easily tested, that if one is sorry for a blow, whether inflicted by hand or by a missile, and at once

Remedial uses of human saliva.

---

[f] Or, " should examine."

[g] From *hoc* to *regerimus* may be a gloss.

[h] A curious ablative. Perhaps *pro precatione* or *cum precatione*. Spitting three times is a regular part of preparing or giving medicine or treatment.

expuat in mediam manum qua percussit, levatur
ilico in percusso culpa.[1] hoc saepe delumbata
quadrupede adprobatur statim a tali remedio correcto
37 animalis ingressu. quidam vero adgravant ictus
ante conatum simili modo saliva in manum ingesta.
credamus ergo et lichenas leprasque ieiunae inlitu
adsiduo arceri, item lippitudines matutina cottidie
velut inunctione, carcinomata †malo terrae subacto,†[2]
cervicis dolores saliva ieiuni dextra manu ad dextrum
poplitem relata, laeva ad sinistrum, si quod animal
38 aurem intraverit et inspuatur, exire. inter amuleta
est editae quemque urinae inspuere, similiter in
calciamentum dextri pedis priusquam induatur,
item cum quis transeat locum in quo aliquod periculum
adierit. Marcion Zmyrnaeus, qui de simplicibus
effectibus scripsit, rumpi scolopendras marinas sputo
tradit, item rubetas aliasque ranas, Ofilius serpentes,
si quis in hiatum earum expuat, Salpe torporem
sedari quocumque membro stupente, si quis in sinum
expuat aut si superiores palpebrae saliva tangantur.[3]
39 nos si haec et illa[4] credamus rite fieri, extranei
interventu aut, si dormiens spectetur infans, a

[1] culpa *codd.* : poena *vulg.* : *Mayhoff* plaga *coni.*
[2] malo terrae subacto] *Mayhoff* terra ea subacta *coni.*
sed putat locum nondum sanatum esse.
[3] superiores palpebrae saliva tangantur *ego* : superiores
palpebras saliva tangat. cur *Mayhoff* : superior palpebra
multi codd. : tangantur (Vr), tangatur, tangant *codd.*
[4] Nos si haec et illa *Hermolaus Barbarus* : eo magis
Detlefsen : non et *Mayhoff* : nos aut eos *codd.*

[a] See critical note and *Index of Plants* in vol. VII. There
is perhaps a lacuna, or *subacto* may be corrupt.

spits into the palm of the hand that gave the wound,
the resentment of the victim is immediately softened.
Corroborative evidence is often seen in draught
animals; when the animal has been flogged to lameness, after the remedy of spitting has been tried,
it at once resumes its pace. Some persons indeed
add force to their blows in a similar way by spitting
into the hand before making their effort. Let us
therefore believe that lichens too and leprous sores
are kept in check by continual application of fasting saliva, as is also ophthalmia by using saliva
every morning as eye ointment, carcinomata by
kneading earth apple [a] with saliva, and pains in the
neck by applying fasting saliva with the right
hand to the right knee and with the left hand to the
left knee; let us also believe that any insect that
has entered the ear, if spat upon, comes out. It
acts as a charm for a man to spit on the urine he has
voided; similarly to spit into the right shoe before
putting it on, also when passing a place where one
has run into some danger. Marcion of Smyrna,[b]
who wrote on the virtues of simples, tells us that the
sea scolopendra bursts if spat upon, as do also
bramble and other toads. Ofilius [c] says that serpents too burst if one spits into their open mouths,
and Salpe [d] that sensation in any numbed limb is
restored by spitting into the bosom, or if the upper
eyelids are touched with saliva. If we hold these
beliefs, we should also believe that the right course,
on the arrival of a stranger, or if a sleeping baby
is looked at, is for the nurse to spit three times at

[b] An unknown.
[c] Perhaps an error for Opilius, which is read by the MS d.
[d] A woman of Lemnos who wrote on the diseases of women.

nutrice terna adspui ? [1] quamquam [2] religione tutatur et Fascinus, imperatorum quoque, non solum infantium custos, qui deus inter sacra Romana a Vestalibus colitur et currus triumphantium sub his pendens defendit medicus invidiae, iubetque eosdem respicere [3] similis medicina linguae, ut sit exorata a tergo Fortuna gloriae carnifex.

40   VIII. Morsus hominis inter asperrimos quosque numeratur. medentur sordes ex auribus ac, ne quis miretur, etiam scorpionum ictibus serpentiumque statim inpositae, melius ex percussi auribus. produnt ita et reduvias sanari, serpentium vero ictum contusi dentis humani farina.

41   IX. Capillus puero qui primum decisus est podagrae inpetus dicitur levare circumligatus, et in totum inpubium inpositus. virorum quoque capillus canis morsibus medetur ex aceto et capitum volneribus ex oleo aut vino; si credimus, a revulso cruci quartanis, conbustus utique capillus carcinomati. pueri qui primus ceciderit dens, ut terram non attingat, inclusus in armillam et adsidue in bracchio 42 habitus muliebrium locorum dolores prohibet. pollex in pede praeligatus proximo digito tumores inguinum

---

   [1] adspui *codd. et edd.* : despui *C. F. W. Müller.*
   [2] quamquam E *Detlefsen* : in os ? quamquam *Mayhoff* : quamquam illos VRdT.
   [3] respicere *Gronovius* : recipere *codd.*

---

   [a] With the reading *despui*, "on the ground"; with Mayhoff's reading, "in the baby's face," or "mouth."
   [b] Fascinus was the spirit or daemon of the phallus, an emblem of which was hung round the necks of infants to keep away evil influences. An image was also attached to the car of a triumphant general, in which, too, was a slave, who bade him look back, saying: *respice post te, hominem te memento.* See Juvenal X. 41.

her charge.[a]   And yet the baby is further under the divine protection of Fascinus,[b] guardian not only of babies but of generals, a deity whose worship, part of the Roman religion, is entrusted to the Vestals; hanging under the chariots of generals at their triumphs he defends them as a physician from jealousy, and the similar physic of the tongue bids them look back, so that at the back Fortune, destroyer of fame, may be won over.[c]

VIII. The bite of a human being is considered *Human bites.* to be a most serious one.   It is treated with ear wax, and (let no one be surprised) this, if applied locally at once, is also good for the stings of scorpions and for the bites of serpents, being more efficacious if taken from the ears of the sufferer.   Hangnails too are said to be cured in this way; the bite of serpents by a human tooth ground to powder.

IX. The hair cut off first from a child's head, if *Use of hair etc.* tied round the affected part,[d] is said to relieve attacks of gout, as does the application of the hair of all, generally speaking, who have not arrived at puberty. The hair of adult men also, applied with vinegar, is good for dog bites, with oil or wine for wounds on the head.   If we believe it, the hair of a man torn from the cross is good for quartan ague; burnt hair is certainly good for carcinoma.   The first tooth of a child to fall out, provided that it does not touch the ground, if set in a bracelet and worn constantly on a woman's arm, keeps pain away from her private parts.   If the big toe is tied to the one next to it,

---

[c] Or, "kept away from behind."

[d] Mayhoff puts a semicolon at *circumligatus* and a comma only at *inpositus.*

sedat, in manu dextera duo medii lino leviter colligati destillationes atque lippitudines arcent. quin et eiectus lapillus calculoso alligatus supra pubem levare ceteros dicitur ac iocineris etiam dolores et celeritatem partus facere. adicit Granius efficaciorem ad hoc esse ferro exemptum. partus accelerat hic mas ex quo quaeque conceperit, si cinctu suo soluto feminam cinxerit, dein solverit adiecta precatione se vinxisse, eundem et soluturum, atque abierit.

43   X. Sanguine ipsius hominis ex quacumque parte emisso efficacissime anginam inlini tradunt Orpheus et Archelaus, item ora comitiali morbo conlapsorum, exurgere enim protinus. quidam, si pollices pedum pungantur eaeque guttae si ferantur [1] in faciem, aut si virgo dextro pollice attingat, hac coniectura

44 censentes virgines carnes edendas. Aeschines Atheniensis excrementorum cinere anginis mede-batur et tonsillis uvisque et carcinomatis. hoc medicamentum vocabat botryon. multa genera morborum primo coitu solvuntur primoque femin-arum mense aut, si id non contingit, longinqua fiunt maximeque comitiales. quin et a serpente, a scorpione percussos coitu levari produnt, verum feminas venere ea laedi. oculorum vitia fieri

---

[1] si ferantur *Urlichs, Detlefsen* : referantur *Mayhoff* : se ferantur V : sēferantur R.

---

[a] An unknown.   [b] See List of Diseases.

swellings in the groin are relieved; if the two middle fingers of the right hand are lightly tied together with a linen thread, catarrhs and ophthalmia are kept away. Again, a stone voided by a sufferer from bladder trouble, if attached above the pubes, is said to relieve other similar patients as well as pains in the liver, and also to hasten child-birth. Granius [a] adds that the stone is more effective for the last purpose if it has been cut out by an iron knife. If the man by whom a woman has conceived unties his girdle and puts it round her waist, and then unties it with the ritual formula: " I bound, and I too will unloose," then taking his departure, child-birth is made more rapid.

X. The blood let from any part of the patient himself makes, we are told by Orpheus and Archelaus, a very efficacious application for quinsy; [b] efficacious too if applied to the mouth of those who have fainted in an epileptic fit, for they rise up immediately. Some say the big toes should be pricked and the drops of blood applied to the face, or that a virgin should touch it [c] with her right thumb; hence their conclusion that epileptics should eat virgin meat. Aeschines the Athenian [d] used the ash of excrements for quinsy, sore tonsils, sore uvula, and carcinomata. This medicament he called botryon. Many kinds of illness are cleared up by the first sexual intercourse, or by the first menstruation; if they do not, they become chronic, especially epilepsy. Moreover, it is held that snake bites and scorpion stings are relieved by intercourse, but that the act does harm to the woman. They say that neither ophthalmia nor other eye troubles afflict those who, when they wash

Use of blood etc.

---

[c] Or, " the patient."   [d] An unknown.

negant nec lippire eos qui, cum pedes lavant, aqua inde ter oculos tangant.

45 XI. Inmatura morte raptorum manu strumas, parotidas, guttura tactu sanari adfirmant, quidam vero cuiuscumque defuncti, dumtaxat sui sexus, laeva manu aversa. et ligno fulgure icto reiectis post terga manibus demorderi aliquid et ad dentem qui doleat admoveri remedio esse produnt. sunt qui praecipiant dentem suffiri dente hominis sui sexus, et eum qui caninus vocetur insepulto exemp-

46 tum adalligari. terram e calvaria psilotrum esse palpebrarum tradunt, herba vero, si qua ibi genita sit, commanducata dentes cadere, ulcera non serpere osse hominis circumscripta. alii e tribus puteis pari mensura aquas miscent et prolibant novo fictili, relicum dant in tertianis accessu febrium bibendum. iidem in quartanis fragmentum clavi a cruce involutum lana collo subnectunt, aut spartum e cruce, liberatoque condunt caverna quam sol non attingat.

47 XII. Magorum haec commenta sunt, ut [1] cotem qua ferramenta saepe exacuta sint subiectam ignari cervicalibus de [2] veneficio deficientis evocare indicium, ut ipse dicat quid sibi datum sit et ubi et quo tempore, auctorem tamen non nominare. fulmine utique

---

[1] sunt, ut] sunt qui V : *Mayhoff* sicuti *coni.*
[2] de] e *coni. Mayhoff, vel delendum putat.*

---

[a] Or, "after a cure has been effected."
[b] Possibly "sorcery," "magic potion." Cf. Book XXV. § 10.

their feet, touch the eyes three times with the water they have used.

XI. We are assured that the hand of a person *Magical* carried off by premature death cures by a touch *cures.* scrofulous sores, diseased parotid glands, and throat affections; some however say that the back of any dead person's left hand will do this if the patient is of the same sex. A piece bitten off from wood struck by lightning by a person with hands thrown behind his back, if it is applied to an aching tooth, is a remedy we are told for the pain. Some prescribe fumigation of the tooth with a human tooth from one of the same sex, and to use as an amulet a dog-tooth taken from an unburied corpse. Earth taken out of a skull acts, it is said, as a depilatory for the eye-lashes, while any plant that has grown in the skull makes, when chewed, the teeth fall out, and ulcers marked round with a human bone do not spread. Some mix in equal quantities water from three wells, pour a libation from new earthenware, and give the rest to be drunk, at the rise of temperature, by sufferers from tertian ague. These also wrap up in wool and tie round the neck of quartan patients a piece of a nail taken from a cross, or else a cord taken from a crucifixion, and after the patient's neck has been freed[a] they hide it in a hole where the sunlight cannot reach.

XII. Here are some lies of the Magi, who say that *Marvellous* a whetstone on which iron tools have been often *remedies of* sharpened, if placed without his knowledge under *the Magi.* the pillows of a man sinking from the effects of poisoning,[b] actually makes him give evidence about what has been given him, where and when, but not the name of the criminal. It is certainly a fact

percussum circumactum in vulnus hominem loqui
48 protinus constat. inguinibus medentur aliqui liceum
telae detractum alligantes novenis septenisve nodis,
ad singulos nominantes viduam aliquam atque ita
inguini adalligantes. liceo et clavum aliudve quod
quis calcaverit alligatum ipsos iubent gerere, ne sit
dolori vulnus. verrucas abolent a vicensima luna in
limitibus supini ipsam intuentes ultra caput manibus
porrectis et quicquid adprehendere eo fricantes.
49 clavum corporis, cum cadit stella si quis destringat
vel[1] cito sanari aiunt, cardinibus ostiorum aceto
adfusis lutum fronti inlitum capitis dolorem sedare,
item laqueum suspendiosi circumdatum temporibus.
si quid e pisce haeserit faucibus, cadere demissis in
aquam frigidam pedibus, si vero ex aliis ossibus,
inpositis capiti ex eodem vase ossiculis, si panis
haereat, ex eodem in utramque aurem addito pane.
50 XIII. Quin et sordes hominis in magnis fecere
remediis quaestuosorum gymnasia[2] Graecorum,
quippe ea strigmenta molliunt, calfaciunt, discutiunt,
conplent, sudore et oleo medicinam facientibus.
volvis inflammatis contractisque admoventur. sic
et menses cient, sedis inflammationes et condylomata
leniunt, item nervorum dolores, luxata, articulorum
51 nodos. efficaciora ad eadem strigmenta a balneis, et

[1] vel *codd.*: vellere *Detlefsen* § 61 *coll.*
[2] quaestuosorum gymnasia *vulg.*, *Detlefsen*: quaestus
gymnici *Mayhoff*: quaestivo gimnit VR: quaestorum
gymnasia d.

---

[a] Or, " recovers his power of speech."
[b] Celsus (V. 11) says that *sordes ex gymnasio* is a discutient.

that the victim of lightning, if turned upon the wounded side, at once begins to speak.[a] Some treat affections of the groin by tying with nine or seven knots a thread taken from a web, at each knot naming some widow, and so attach it to the groin as an amulet. To prevent a wound's being painful they prescribe wearing as an amulet, tied on the person with a thread, the nail or other object that he has trodden on. Warts are removed by those who, after the twentieth day of the month, lie face upwards on a path, gaze at the moon with hands stretched over their head, and rub the wart with whatever they have grasped. If a corn or callus is cut when a star is falling, they say that it is very quickly cured, and that applying to the forehead the mud obtained by pouring vinegar over a front door's hinges relieves headaches, as does also the rope used by a suicide if tied round the temples. Should a fish bone stick in the throat, they say that it comes out if the feet are plunged into cold water; if however it is another kind of bone, bits of bone from the same pot should be applied to the head; if it is a piece of bread that sticks, pieces from the same loaf must be placed in either ear.

XIII. Moreover, important remedies have been *Human off-* made by the profit-seeking Greeks even with human *scourings.* offscouring from the gymnasia; for the scrapings from the bodies soften, warm, disperse,[b] and make flesh, sweat and oil forming an ointment. This is used as a pessary for inflammation and contraction of the uterus. So used it is also an emmenagogue; it soothes inflammations of the anus and condylomata, likewise pains of the sinews, dislocations, and knotty joints. More efficacious for the same purposes

37

ideo miscentur suppuratoriis medicamentis. nam illa quae sunt ex ceromate permixta caeno articulos tantum molliunt, calfaciunt, discutiunt efficacius,
52 sed ad cetera minus valent. excedit fidem inpudens cura qua sordes virilitatis contra scorpionum ictus singularis remedii celeberrimi auctores clamant, rursus in feminis quas [1] infantium alvo editas in utero ipso contra sterilitatem subdi censent, meconium vocant. immo etiam ipsos gymnasiorum rasere parietes, et illae quoque sordes excalfactoriam vim habere dicuntur, panos discutiunt, ulceribus senum puerorumque et desquamatis ambustisve inlinuntur.

53 XIV. Eo minus omitti convenit ab animo hominis pendentes medicinas. abstinere cibo omni aut potu, alias vino tantum aut carne, alias balneis, cum quid eorum postulet valetudo, in praesentissimis remediis habetur. his adnumeratur exercitatio, intentio vocis, ungui, fricari cum ratione. vehemens enim fricatio spissat, lenis mollit, multa adimit corpus, auget modica. in primis vero prodest ambulatio, gestatio et ea pluribus modis, equitatio stomacho et coxis
54 utilissima, phthisi navigatio, longis morbis locorum mutatio, item somno sibi mederi aut lectulo aut rara vomitione. supini cubitus oculis conducunt, at proni tussibus, in latera adversum destillationes.

[1] quas *codd.*: aquas *coni. Warmington.*

are scrapings from the bath, and so these are in-gredients of ointments for suppurations. But those that have wax salve in them, and are mixed with mud, are more efficacious only for softening joints, for warming and for dispersing, but for all other purposes they are less powerful. Shameless beyond belief is the treatment prescribed by very famous authorities, who proclaim that male semen is an excellent antidote to scorpion stings, holding on the other hand that a pessary for women made from the faeces of babies voided in the uterus itself is a cure for barrenness; they call it meconium. Moreover, they have scraped the very walls of the gymnasia, and these offscourings are said to have great warming properties; they dis-perse superficial abscesses, and are applied as oint-ment to the sores of old people and children, as well as to excoriations and burns.

XIV. It would be all the less seemly to pass over the remedies that are in the control of a man's will. *Remedies depending on the will.* To fast from all food and drink, sometimes only from wine or meat, sometimes from baths, when health demands such abstinence, is held to be one of the most sovereign remedies. Among the others are physical exercise, voice exercises, anointing, and massage if carried out with skilled care; for violent massage hardens, gentle softens, too much reduces flesh and a moderate amount makes it. Especially beneficial however are walking, carriage rides of various kinds, horse riding, which is very good for the stomach and hips, a sea voyage for consumption, change of locality for chronic diseases, and self-treatment by sleep, lying down, and occasional emetics. Lying on the back is good for the eyes, on the face for coughs, and on either side for catarrhs. Aristotle

Aristoteles et Fabianus plurimum somniari circa ver et autumnum tradunt, magisque supino cubitu, at prono nihil, Theophrastus celerius concoqui dextri
55 lateris incubitu, difficilius a supinis. sol quoque remediorum maximum ab ipso sibi praestari potest, sicuti linteorum strigiliumque vehementia. perfundere caput calida ante balnearum vaporationem et postea frigida, saluberrimum intellegitur, item praesumere et cibis et interponere frigidam eiusdemque potu [1] somnos antecedere et, si libeat, interrumpere. notandum nullum animal aliud calidos
56 potus sequi ideoque non esse naturales. mero ante somnos colluere ora propter halitus, frigida matutinis inpari numero ad cavendos dentium dolores, item posca oculos contra lippitudines, certa experimenta sunt, sicut totius corporis valetudinem iuvari [2] varietate victus inobservata. Hippocrates tradit non prandentium celerius senescere exta. verum id remediis cecinit, non epulis, quippe multo utilissima est temperantia in cibis. L. Lucullus hanc de se praefecturam servo dederat, ultimoque probro manus in cibis triumphali seni deiciebatur vel in Capitolio epulanti, pudenda re servo suo facilius parere quam sibi.

---

[1] potu *codd.* : potus *Detlefsen.*
[2] valetudinem iuvari *Dal.*, *Sillig*, *Detlefsen* : valetudini in *Mayhoff*; valetudini *aut.* valetudinē in *codd.*

and Fabianus tell us that dreaming is most common around spring and autumn, and especially when we lie on the back; when we lie on the face there are no dreams at all. Theophrastus says that quicker digestion results from lying on the right side, more difficult digestion from lying on the back. Sunshine too, best of remedies, we can administer to ourselves, as we can the vigorous use of towels and scrapers. To bathe the head with hot water before the hot steam of the bath, and with cold water after it, is understood to be very healthful; so it is to drink cold water before a meal and at intervals during it, and to take a draught of the same before going to sleep, breaking your sleep, if you like, in order to drink. It should be observed that no animal except man likes hot drinks, which is evidence that they are unnatural. Experience plainly shows that it is good before sleeping to rinse the mouth with neat wine as a safeguard against offensive breath, and with cold water an uneven number of times in the morning to keep off toothache; that to bathe the eyes in vinegar and water prevents ophthalmia, and that general health is promoted by an unstudied variety of regimen. Hippocrates teaches that the habit of not taking lunch makes the internal organs age more rapidly; in this aphorism, however, he is thinking of remedies, not encouraging gluttony, for by far the greatest aid to health is moderation in food. L. Lucullus gave charge over himself to a slave to enforce control, and he, an old man who had celebrated a triumph, suffered the very deep disgrace of having his hand kept away from the viands even when feasting in the Capitol, with the added shame of obeying his own slave more readily than himself.

57   XV. Sternumenta pinna gravedinem emendant,
et si quis mulae nares, ut tradunt, osculo attingat,
sternutamenta et singultum. ad hoc Varro suadet
palmam alterna[1] manu scalpere, plerique anulum e
sinistra in longissimum dextrae digitum transferre,
in aquam ferventem manus mergere. Theophrastus
senes laboriosius sternuere dicit.

58   XVI. Venerem damnavit Democritus ut in qua
homo alius exiliret ex homine, et, Hercules, raritas
eius utilior. athletae tamen torpentes restituuntur
venere, vox revocatur, cum e candida declinat in
fuscam. medetur et lumborum dolori, oculorum
hebetationi, mente captis ac melancholicis.

59   XVII. Adsidere gravidis, vel cum remedium alicui
adhibeatur, digitis pectinatim inter se inplexis
veneficium est, idque conpertum tradunt Alcmena
Herculem pariente, peius, si circa unum ambove
genua, item poplites alternis genibus inponi. ideo
haec in consiliis ducum potestatiumve fieri vetuere
maiores velut omnem actum inpedientia, vetuere
60 vero et sacris votisve simili modo interesse, capita
autem aperiri aspectu magistratuum non venerationis
causa iussere, sed, ut Varro auctor est, valetudinis,

----

[1] alterna R, *Gelenius, Mayhoff*: in altera *multi codd.,*
Detlefsen: alterutra *coni. Warmington.*

----

a Or, " discomfort."

XV. Sneezing caused by a feather relieves a *Sneezing.* cold in the head, and sneezing and hiccough are relieved by touching with the lips, it is said, the nostrils of a mule. For sneezing Varro advises us to scratch the palm of each hand with the other; most people advise us to transfer the ring from the left hand to the longest finger of the right, and to dip the hands into very hot water. Theophrastus says that old people sneeze with greater difficulty *a* than others.

XVI. Sexual intercourse was disapproved of by *Sexual* Democritus, as being merely the act whereby one *intercourse.* human being springs from another. Heaven knows, the less indulgence in this respect the better. Athletes, however, when sluggish regain by it their activity, and the voice, when it has lost its clearness and become husky, is restored. It cures pain in the loins, dulness of vision, unsoundness of mind and melancholia.

XVII. To sit in the presence of pregnant women, *Various* or when medicine is being given to patients, with the *unlucky* fingers interlaced comb-wise, is to be guilty of sorcery, *and lucky* a discovery made, it is said, when Alcmena was *acts.* giving birth to Hercules. The sorcery is worse if the hands are clasped round one knee or both, and also to cross the knees first in one way and then in the other. For this reason our ancestors forbade such postures at councils of war or of officials, on the ground that they were an obstacle to the transaction of all business. They also forbade them, indeed, to those attending sacred rites and prayers; but to uncover the head at the sight of magistrates they ordered, not as a mark of respect, but (our authority is Varro) for the sake of health, for the habit of baring the head

quoniam firmiora consuetudine ea fierent. cum
quid oculo inciderit, alterum conprimi prodest, cum
aqua dextrae auriculae, sinistro pede exultari capite
in dextrum umerum devexo, invicem e diversa aure.
si tussim concitet saliva, in fronte ab alio adflari, si
iaceat uva, verticem [1] morsu alterius suspendi, in
cervicium dolore poplites fricare [2] aut cervicem in
61 poplitum, pedes in humo deponi, si nervi in his
cruribusve tendantur in lectulo, aut si in laeva parte
id accidat, sinistrae plantae pollicem dextra manu
adprehendi, item e diverso, extremitates corporis
velleribus perstringi contra horrores sanguinemve
narium inmodicum,[3] . . . lino vel papyro principia
genitalium, femur medium ad cohibenda urinae
profluvia, in stomachi solutione pedes pressare [4] aut
62 manus in ferventem aquam demitti. iam et sermoni
parci multis de causis salutare est. triennio Maece-
natem Melissum [5] accepimus silentium sibi impera-
visse a convolsione reddito sanguine. nam eversos
scandentesque ac iacentes si quid ingruat contraque
ictus spiritum cohibere singularis praesidii est, quod
63 inventum esse animalis docuimus. clavum ferreum
defigere in quo loco primum caput fixerit corruens

[1] verticem VdT, *Mayhoff*: a vertice R (?) E *vulg.*,
Detlefsen.
[2] fricari *velit Sillig.*
[3] *Post* inmodicum *lacunam indicat Mayhoff.*
[4] pressari *velit Sillig.*
[5] Melissi iussu *coni. Mayhoff.*

---

[a] With the reading *a vertice,* " to hold him up suspended
by the top of his head with another's teeth," a difficult feat,
one would think.

[b] Mayhoff's lacuna, filled up by *item circumligari,* would
mean : " to tie round with thread or papyrus."

gives it greater strength. When something has fallen into the eye, it does good to press down the other; when water gets into the right ear, to jump with the left leg, leaning the head towards the right shoulder; if into the left ear, to jump in the contrary way; if saliva provokes a cough, for another person to blow on the forehead; if the uvula is relaxed, for another to hold up the top of the head [a] with his teeth; if there is pain in the neck, to rub the back of the knees, and to rub the neck for pain in the back of the knees; to plant the feet on the ground for cramp in feet or legs when in bed; or if the cramp is on the left side to seize with the right hand the big toe of the left foot and *vice versa*; to rub the extremities with pieces of fleece to stop shivers or violent nose-bleeding; . . .[b] with linen or papyrus the tip of the genitals and the middle of the thigh to check incontinence of urine; for weakness of the stomach to press together the feet or dip the hands into very hot water. Moreover, to refrain from talking is healthful for many reasons. Maecenas Melissus,[c] we are told, imposed a three-year silence on himself because of spitting of blood after convulsions. But if any danger threatens those thrown down, climbing, or prostrate, and as a guard against blows, to hold the breath is an excellent protection, a discovery which, I have stated,[d] we owe to an animal. To drive an iron nail into the place first [e]

---

[c] The conjecture of Mayhoff would mean: "Maecenas, on the recommendation of Melissus," i.e., of his medical attendant.

[d] See Book VIII. § 138.

[e] Or, possibly: "into the place struck by the front of his head."

morbo comitiali absolutorium eius mali dicitur.
contra renum aut lumborum, vesicae cruciatus in
balnearum soliis pronos urinam reddere mitigatorium
habetur. vulnera nodo Herculis praeligare mirum
64 quantum ocior medicina est, atque etiam cottidiani
cinctus tali nodo vim quandam habere utilem dicuntur,
quippe cum Herculaneum prodiderit numerum
quoque quaternarium Demetrius condito volumine, et
quare quaterni cyathi sextariive non essent potandi.
contra lippitudines retro aures fricare prodest et
lacrimosis oculis frontem. augurium ex homine ipso
est non timendi mortem in aegritudine quamdiu
oculorum pupillae imaginem reddant.

65 XVIII. Magna et urinae non ratio solum sed
etiam religio apud auctores invenitur digestae in
genera, spadonum quoque ad fecunditatis veneficia.
verum ex his quae referre fas sit inpubium puerorum
contra salivas aspidum quas ptyadas vocant, quoniam
venena in oculos hominum expuant, contra oculorum
albugines, obscuritates, cicatrices, argema, palpebras
et cum ervi farina contra adustiones, contra aurium
pura vermiculosque, si decoquatur ad dimidias partes
cum porro capitato novo fictili. vaporatio quoque ea

---

^a A difficult knot with no ends to be seen.
^b Possibly a physician who lived about 200 B.C. Nothing
else is known of him.
^c It is difficult to bring out the contrast between *ratio*
and *religio* without suggesting notions of which Pliny, and
perhaps the Romans generally, were ignorant. Possibly
the former refers to a property supposed to be understood

struck by the head of an epileptic in his fall is said
to be deliverance from that malady. For severe
pain in the kidneys, loins or bladder, it is supposed
to be soothing if the patient voids his urine while
lying on his face in the tub of the bath. To tie up
wounds with the Hercules knot [a] makes the healing
wonderfully more rapid, and even to tie daily the
girdle with this knot is said to have a certain useful-
ness, for Demetrius [b] wrote a treatise in which he
states that the number four is one of the prerogatives
of Hercules, giving reasons why four cyathi or sextarii
at a time should not be drunk. For ophthalmia it is
good to rub behind the ears, and for watery eyes the
forehead. From the patient himself it is a reliable
omen that, as long as the pupils of his eyes reflect
an image, a fatal end to an illness is not to be feared. *Medical uses of urine.*

XVIII. Our authorities attribute to urine also great
power, not only natural but supernatural; [c] they
divide it into kinds, using even that of eunuchs to
counteract the sorcery that prevents fertility. But
of the properties it would be proper to speak of I
may mention the following :—the urine of children
not yet arrived at puberty is used to counteract the
spittle of the ptyas, an asp so called because it spits
venom into men's eyes; for albugo,[d] dimness, scars,
argema,[d] and affections of the eyelids; with flour
of vetch for burns; and for pus or worms in the ear
if boiled down to one half with a headed leek in
new earthenware. Its steam too is an emmena-

(i.e. normal), and the latter to one mysterious and not understood (abnormal). Of course there are other meanings of
religio, which may be objective or subjective.

    [d] For *albugo* and *argema* see List of Diseases. The *ptyas*
(from πτύω) = the spitting asp.

66 menses feminarum ciet. Salpe fovet illa[1] oculos
firmitatis causa, inlinit sole usta cum ovi albo,
efficacius struthocameli, binis horis. hac et atra-
menti liturae abluuntur. virilis podagris medetur
argumento fullonum, quos ideo temptari eo morbo
negant. veteri miscetur cinis ostreorum adversus
eruptiones in corpore infantium et omnia ulcera
67 manantia. ea exesis, ambustis, sedis vitiis, rhagadiis et scorpionum ictibus inlinitur. obstetricum
nobilitas non alio suco efficacius curari pronuntiavit
corporum pruritus, nitro addito ulcera capitum,
porrigines, nomas, praecipue genitalium. sua cuique
autem, quod fas sit dixisse, maxime prodest, confestim
perfuso canis morsu, echinorumque spinis inhaeren-
tibus[2] in spongea lanisve inposita aut adversus rabidi
canis morsus cinere ex ea subacto, contraque serpentium ictus. nam contra scolopendras mirum proditur vertice tacto urinae suae gutta liberari protinus
laesos.
68 XIX. Auguria valetudinis ex ea traduntur, si
mane candida, dein rufa sit, illo modo concoquere,
hoc concoxisse significatur. mala signa rubrae,
pessima nigrae, mala bullantis. crassa,[3] in qua quod
subsidit si album est, significat circa articulos aut
viscera dolorem inminere, eadem viridis morbum

---

[1] *Post* illa *add.* cum E: cum luteo *C. Brakman (Mnemosyne*
1930).
[2] inhaerentibus] *Post hoc verbum et codd.*: *del. vult Mayhoff*:
ego delevi.
[3] crassa *Mayhoff*: crassae *aut* et crassae *codd.*

---

[a] Mayhoff thinks that there is a lacuna, e.g. " and honey."
[b] Fullers used it in their work.
[c] With the reading *crassae* " thick " will be an epithet
applied to the bubbling urine.

gogue. Salpe would foment the eyes with urine [a] to strengthen them, and would apply it for two hours at a time to sun-burn, adding the white of an egg, by preference that of an ostrich. Urine also takes out ink blots. Men's urine relieves gout, as is shown by the testimony of fullers,[b] who for that reason never, they say, suffer from this malady. Old urine is added to the ash of burnt oyster-shells to treat rashes on the bodies of babies, and for all running ulcers. Pitted sores, burns, affections of the anus, chaps, and scorpion stings, are treated by applications of urine. The most celebrated midwives have declared that no other lotion is better treatment for irritation of the skin, and with soda added for sores on the head, dandruff, and spreading ulcers, especially on the genitals. Each person's own urine, if it be proper for me to say so, does him the most good, if a dog-bite is immediately bathed in it, if it is applied on a sponge or wool to the quills of an urchin that are sticking in the flesh, or if ash kneaded with it is used to treat the bite of a mad dog, or a serpent's bite. Moreover, for scolopendra bite a wonderful remedy is said to be for the wounded person to touch the top of his head with a drop of his own urine, when his wound is at once healed.

XIX. Urine gives us symptoms of general health: if in the morning it is clear, becoming tawny later, the former means that coction is still going on, the latter that it is complete. A bad symptom is red urine, a bad one also when it bubbles, and the worst of all when it is very dark. Thick [c] urine, in which what sinks to the bottom is white, means that there is pain coming on about the joints or in the region of the bowels; if it is green, that the bowels

49

viscerum, pallida bilis, rubens sanguinis. mala et
in qua veluti furfures atque nubeculae apparent.
69 diluta quoque alba vitiosa est, mortifera vero crassa
gravi odore et in pueris tenuis ac diluta. Magi
vetant eius causa contra solem lunamque nudari
aut umbram cuiusquam ab ipso respergi. Hesiodus
iuxta obstantia reddi suadet, ne deum aliquem nudatio
offendat. Osthanes contra mala medicamenta omnia
auxiliari promisit matutinis suam cuique instillatam
in pedem.
70 XX. Quae ex mulierum corporibus traduntur ad
portentorum miracula accedunt, ut sileamus divisos
membratim in scelera abortus, mensum piacula
quaeque alia non obstetrices modo verum etiam
ipsae meretrices prodidere, capilli si crementur,
odore serpentes fugari, eodem nidore vulvae morbo
71 strangulatas respirare, cinere eo quidem, si in testa
sint cremati vel cum spuma argenti, scabritias
oculorum ac prurigines emendari, item verrucas et
infantium ulcera, cum melle capitis quoque vulnera
et omnium ulcerum sinus, addito melle ac ture
panos, podagras, cum adipe suillo sacrum ignem,
sanguinem sisti, inlito item[1] formicationes corporum.
72 XXI. De lactis usu convenit dulcissimum esse
mollissimumque et in longa febre coeliacisque
utilissimum, maxime eius quae iam infantem re-

---

[1] item *Mayhoff*: et in *codd.*: et *vulg.*

---

[a] *Works and Days* ll. 727 foll.
[b] A Magus who accompanied Xerxes on his expedition
against Greece. See Book XXX. § 8, and the long article in
Pauly, *s.v.* Ostanes.
[c] See XXVIII § 85 *tactis omnino menstruo postibus inritas
fieri Magorum artes.* It is however possible that the other
meaning of *piaculum* ("crime") is intended here. *Cf.*
many remarks in Chapter XXIII.

are diseased. Pale urine means diseased bile, red urine diseased blood. Bad urine also is that in which is to be seen as it were bran, and cloudiness. Watery, pale, urine also is unhealthy, but thick, foul-smelling urine indicates death, as does thin, watery urine from children. The Magi say that when making urine one must not expose one's person to the face of the sun or moon, or let drops fall on anyone's shadow. Hesiod [a] advises us to urinate facing an object that screens, lest our nakedness should offend some deity. Osthanes [b] assured people that protection against all sorcerers' potions is secured by letting one's own morning urine drip upon the foot.

XX. Some reported products of women's bodies *Remedies* should be added to the class of marvels, to say nothing *from* of tearing to pieces for sinful practices the limbs of *women.* still-born babies, the undoing of spells by the menstrual fluid,[c] and the other accounts given not only by midwives but actually by harlots. For example: that the smell of burnt woman's hair keeps away serpents, and the fumes of it make women breathe naturally who are choking with hysteria; this same ash indeed, from hair burnt in a jar, or used with litharge, cures roughness and itch of the eyes, as well as warts and sores on babies; that with honey it cures also wounds on the head and the cavities made by any kind of ulcer, with honey and frankincense, superficial abscesses and gout; that with lard it cures erysipelas and checks haemorrhage, and that when applied it cures also irritating rashes on the body.

XXI. As to the use of woman's milk, it is agreed that it is the sweetest and most delicate of all, very useful in long fevers and coeliac disease, especially the milk of a woman who has already weaned her

51

moverit. et in malacia stomachi, in febribus rosioni-
busque efficacissimum experiuntur, item mammarum
collectionibus cum ture, oculo ab ictu cruore suffuso
et in dolore aut epiphora, si inmulgeatur, plurimum
prodest, magisque cum melle et narcissi suco aut
turis polline, superque in omni usu efficacius eius
quae marem enixa sit multoque efficacissimum eius
quae geminos pepererit mares et si vino ipsa cibisque
73 acrioribus abstineat. mixto praeterea ovorum candido liquore madidaque lana frontibus inpositum [1]
fluctiones oculorum suspendit. nam [2] si rana saliva
sua oculum asperserit, praecipuum est remedium,
et contra morsum eiusdem bibitur instillaturque.
eum qui simul matris filiaeque lacte inunctus sit
liberari omni oculorum metu in totam vitam adfirmant.
aurium quoque vitiis medetur admixto modice oleo
aut, si ab ictu doleant,[3] anserino adipe tepefactum.
si sit odor gravior, ut plerumque fit longis vitiis,
74 diluto melle lana includitur. et contra regium
morbum in oculis relictum instillatur cum elaterio.
peculiariter valet potum contra venena quae data
sint e marino lepore, bupraesti,[4] aut [5] ut Aristoteles
tradit, dorycnio,[6] et contra insaniam quae facta sit
hyoscyami potu. podagris quoque iubent inlini
cum cicuta, alii cum oesypo et adipe anserino,

---

[1] inpositum *codd.*: inposita *coni. Mayhoff.*
[2] nam *codd.*: etiam *coni. Mayhoff.*
[3] *Ante* anserino *an* cum *addendum?*
[4] bupraesti] *varia codd.*: bupraestim *Detlefsen.*
[5] aut] mutatim *multi codd.*: del. *Detlefsen*: aut etiam
Mayhoff.
[6] dorycnio *Mayhoff*: dorycnium *Detlefsen*: *varia codd.*

---

[a] See *Index of Plants* in vol. VII.
[b] Perhaps some species of cantharides.

baby. For nausea of the stomach, in fevers, and for gnawing pains, it is found most efficacious, also with frankincense for gatherings on the breasts. It is very beneficial to an eye that is bloodshot from a blow, in pain, or suffering from a flux, if it is milked straight into it, more beneficial still if honey is added and juice of narcissus *a* or powdered incense. For all purposes, moreover, a woman's milk is more efficacious if she has given birth to a boy, and much the most efficacious is hers, who has borne twin boys and herself abstains from wine and the more acrid foods. Mixed moreover with liquid white of eggs, and applied to the forehead on wool soaked in it, it checks fluxes of the eyes. But if a toad has squirted its fluid into the eye it is a splendid remedy; for the bite also of the toad it is drunk and poured in drops into the wound. It is asserted that one who has been rubbed with the milk of mother and daughter together never needs to fear eye trouble for the rest of his life. Affections of the ears also are successfully treated by the milk mixed with a little oil, or, if there is any pain from a blow, warmed with goose grease. If there is an offensive smell from the ears, as usually happens in illnesses of long standing, wool is put into them soaked in milk in which honey has been dissolved. When jaundice has left traces remaining in the eyes, the milk together with elaterium is dropped into them. A draught of woman's milk is especially efficacious against the poison of the sea-hare, of the buprestis, *b* or, as Aristotle tells us, of dorycnium, and for the madness caused by drinking henbane. Combined with hemlock it is also prescribed as a liniment for gout; others make it up with the suint of wool and goose

53

qualiter et vulvarum doloribus inponitur. alvum
etiam sistit potum, ut Rabirius scribit, et menses
75 ciet. eius vero quae feminam enixa sit ad vitia
tantum in facie sananda praevalet. pulmonum
quoque incommoda lacte mulieris sanantur, cui si
admisceatur inpubis pueri urina et mel Atticum,
omnia coclearium singulorum mensura, † marmora † [1]
quoque aurium eici invenio. eius quae marem peperit lacte gustato canes rabiosos negant fieri.
76 XXII. Mulieris quoque salivam ieiunam potentem
diiudicant cruentatis oculis et contra epiphoras, si
ferventes anguli oculorum subinde madefiant, efficacius, si cibo vinoque se pridie ea abstinuerit.
invenio et fascia mulieris alligato capite dolores
minui.
77 XXIII. Post haec nullus est modus. iam primum
abigi grandines turbinesque contra fulgura ipsa mense
nudato. sic averti violentiam caeli, in navigando
quidem tempestates etiam sine menstruis. ex ipsis
vero mensibus, monstrificis alias, ut suo loco indicavimus, dira et infanda vaticinantur, e quibus dixisse
non pudeat, si in defectus lunae solisve congruat
vis [2] illa, inremediabilem fieri, non segnius et in silente
luna, coitusque tum maribus exitiales esse atque

---

[1] marmora *codd.*, *vulg.* : pura *Detlefsen coll.* § 65 : vermes
Mayhoff, qui etiam harenas renium, *pro* marmora aurium : *pro*
marmora *coni.* murmura *Warmington*.

[2] vis *vulg.*, *Mayhoff* : pestis *Detlefsen* : is VR : *om.* dx.

---

[a] None of the emendations of the corrupt *marmora* seems
likely. Perhaps Mayhoff's suggestion of *harenas renium*
(" gravel expelled from the bladder ") is the best. I translate
Mayhoff's *vermes*.
[b] See Book VII. § 64.

grease, in the form that is also used as an application for pains of the uterus. A draught also acts astringently upon the bowels, as Rabirius writes, and is an emmenagogue. The milk of a woman however who has borne a girl is excellent, but only for curing spots on the face. Lung affections also are cured by woman's milk, and if Attic honey is mixed with it and the urine of a child before puberty, a single spoonful of each, I find that worms *a* too are driven from the ears. The mother of a boy gives a milk a taste of which, they say, prevents dogs from going mad.

XXII. The saliva too of a fasting woman is judged to be powerful medicine for bloodshot eyes and fluxes, if the inflamed corners are occasionally moistened with it, the efficacy being greater if she has fasted from food and wine the day before. I find that a woman's breast-band tied round the head relieves headache.

XXIII. Over and above all this there is no limit to woman's power. First of all, they say that hail-storms and whirlwinds are driven away if menstrual fluid is exposed to the very flashes of lightning; that stormy weather too is thus kept away, and that at sea exposure, even without menstruation, prevents storms. Wild indeed are the stories told of the mysterious and awful power of the menstruous discharge itself, the manifold magic of which I have spoken of in the proper place.*b* Of these tales I may without shame mention the following: if this female power should issue when the moon or sun is in eclipse, it will cause irremediable harm; no less harm if there is no moon; at such seasons sexual intercourse brings disease and death upon the

55

78 pestiferos, purpuram quoque eo tempore ab his
pollui, tanto vim esse maiorem, quocumque autem
alio menstruo si nudatae segetem ambiant, urucas
et vermiculos scarabaeosque ac noxia alia decidere.
Metrodorus Scepsius in Cappadocia inventum prodit
ob multitudinem cantharidum; ire ergo per media
arva retectis super clunes vestibus. alibi servatur
ut nudis pedibus eant capillo cinctuque dissoluto.
cavendum ne id oriente sole faciant, sementim enim
arescere, item novella[1] tactu in perpetuum laedi,
rutam et hederam res medicatissimas ilico mori.

79 multa diximus de hac violentia, sed praeter illa certum
est apes tactis alvariis fugere, lina, cum coquantur,
nigrescere, aciem in cultris tonsorum hebetari, aes
contactu grave virus odoris accipere et aeruginem,
magis si descrescente luna id accidat, equas, si sint
gravidae, tactas abortum pati, quin et aspectu omnino,
quamvis procul visas, si purgatio illa post virginitatem

80 prima sit aut in virgine aetatis sponte. nam et[2]
bitumen in Iudaea nascens sola hac vi superari
filo vestis contactae docuimus. nec igni quidem

---

[1] novella *multi codd.*, *Mayhoff*: novella prata *Detlefsen*:
novella ta V[1].

[2] nam et *Detlefsen*: manet (*cum priore sententia*) *Mayhoff*:
nam ut V: nam V[2] r: ut d T. *Coni. etiam* eveniat *Mayhoff*.

---

[a] It should be noticed how often the word *vis* occurs in this
chapter. It is curiously like the "mana" or "orenda" of
modern students of folklore. See the article *Kultus* in Pauly.

[b] It is hard to see how the readings of the MSS. have arisen,
whatever reading or emendation we adopt. Mayhoff's
manet would be more attractive were not *prima sit* the natural
continuation of the clause introduced by *aut*. Is it possible

man; purple too is tarnished then by the woman's touch. So much greater then is the power [a] of a menstruous woman. But at any other time of menstruation, if women go round the cornfield naked, caterpillars, worms, beetles and other vermin fall to the ground. Metrodorus of Scepsos states that the discovery was made in Cappadocia owing to the plague there of Spanish fly, so that women walk, he says, through the middle of the fields with their clothes pulled up above the buttocks. In other places the custom is kept up for them to walk barefoot, with hair dishevelled and with girdle loose. Care must be taken that they do not do so at sunrise, for the crop dries up, they say, the young vines are irremedially harmed by the touch, and rue and ivy, plants of the highest medicinal power, die at once. I have said much about this virulent discharge, but besides it is certain that when their hives are touched by women in this state bees fly away, at their touch linen they are boiling turns black, the edge of razors is blunted, brass contracts copper rust and a foul smell, especially if the moon is waning at the time, mares in foal if touched miscarry, nay the mere sight at however great a distance is enough, if the menstruation is the first after maidenhood, or that of a virgin who on account of age is menstruating naturally for the first time. But the bitumen [b] also that is found in Judaea can be mastered only by the power of *Menstrual* this fluid, as I have already stated,[c] a thread from an *fluid.* infected dress is sufficient. Not even fire, the all-con-

that the last two syllables of *bitumen*, spelt backwards (*nem ut*), are responsible?

[c] See Book VII. § 65, a portion of Pliny's work from which many of the statements made here are repeated.

vincitur quo cuncta, cinisque etiam ille, si quis aspargat lavandis vestibus, purpuras mutat, florem coloribus adimit. ne ipsis quidem feminis malo suo inter se inmunibus abortus facit inlitu, aut si omnino praegnas

81 supergradiatur. quae Lais et Elephantis inter se contraria prodidere de abortivis,[1] carbone e radice brassicae vel myrti vel tamaricis in eo sanguine extincto, itemque asinas tot annis non concipere quot grana hordei contacta ederint, quaeque alia nuncupavere monstrifica aut inter ipsas pugnantia, cum haec fecunditatem fieri isdem modis quibus sterilitatem illa praenuntiaret, melius est non credere.

82 Bithus Durrachinus hebetata aspectu specula recipere nitorem tradit isdem aversa rursus contuentibus, omnemque vim talem resolvi, si mullum piscem secum habeant, multi vero inesse etiam remedia tanto malo, podagris inlini, strumas et parotidas et panos, sacros ignes, furunculos, epiphoras tractatu mulierum earum leniri, Lais et Salpe canum rabiosorum morsus et tertianas quartanasque febres menstruo in lana arietis nigri argenteo bracchiali

83 incluso, Diotimus Thebanus vel omnino vestis ita infectae portiuncula ac vel licio[2] bracchiali inserto.[3]

---

[1] abortivis *codd.*, *Detlefsen* : abortivo *post vet. Dal.*, *Mayhoff.*

[2] licio *Caesarius*, *Mayhoff* : pellicio d Tr, *vulg.*, *Detlefsen* : pelicio V R.

[3] inserto T *Mayhoff* : inserte, inserta, insertae *codd.* : insertae *vulg.*, *Detlefsen.*

---

[a] An unknown.

[b] Authoress of poems admired by Tiberius. Perhaps the lady that Galen says wrote on the subject of cosmetics.

[c] An unknown.

[d] See note on § 38.

[e] An unknown.

quering, overcomes it; even when reduced to ash, if sprinkled on clothes in the wash, it changes purples and robs colours of their brightness. Nor are women themselves immune to the effect of this plague of their sex; a miscarriage is caused by a smear, or even if a woman with child steps over it. Lais [a] and Elephantis [b] do not agree in their statements about abortives, the burning root of cabbage, myrtle, or tamarisk extinguished by the menstrual blood, about asses' not conceiving for as many years as they have eaten grains of barley contaminated with it, or in their other portentous or contradictory pronouncements, one saying that fertility, the other that barrenness is caused by the same measures. It is better not to believe them. Bithus [c] of Dyrrhachium says that a mirror which has been tarnished by the glance of a menstruous woman recovers its brightness if it is turned round for her to look at the back, and that all this sinister power is counteracted if she carries on her person the fi•h called red mullet. Many however say that even this great plague is remedial; that it makes a liniment for gout, and that by her touch a woman in this state relieves scrofula, parotid tumours, superficial abscesses, erysipelas, boils and eye-fluxes. Lais and Salpe [d] hold that the bite of a mad dog, tertians, and quartans are cured by the flux on wool from a black ram enclosed in a silver bracelet; Diotimus [e] of Thebes says that even a bit, nay a mere thread,[f] of a garment contaminated in this way and enclosed in the bracelet,

---

[f] With the reading *pellicio*: " even a bit of a contaminated garment inserted in a leather strap round the arm." There is something attractive about this reading, for which almost as much could be said as for *licio*.

Sotira obstetrix tertianis quartanisque efficacissimum
dixit plantas aegri subterlini, multoque efficacius ab
ipsa muliere et ignorantis, sic et comitiales excitari.
Icatidas medicus quartanas finiri coitu, incipientibus
84 dumtaxat menstruis, spopondit. inter omnes vero
convenit, si aqua potusque formidetur a morsu canis,
supposita tantum calici lacinia tali, statim metum
eum discuti, videlicet praevalente sympathia illa
Graecorum, cum rabiem canum eius sanguinis
gustatu incipere dixerimus. cinere eo iumentorum
omnium [1] ulcera sanari certum est addita caminorum
farina et cera, maculas autem e veste eas non nisi
85 eiusdem urina ablui, cinerem per se rosaceo mixtum
feminarum praecipue capitis dolores sedare inlitum
fronti, asperrimamque vim profluvii eius esse per se
annis virginitate resoluta. id quoque convenit, quo
nihil equidem libentius crediderim, tactis omnino
menstruo postibus inritas fieri Magorum artes,
86 generis vanissimi, ut aestimare licet. ponam enim
vel modestissimum e promissis eorum, ex homine
siquidem resigmina unguium e pedibus manibusque
cera permixta, ita ut dicatur tertianae, quartanae
vel cotidianae febri remedium quaeri, ante solis
ortum alienae ianuae adfigi iubent ad remedia in
his morbis, quanta vanitate, si falsum est, quanta
vero noxia, si transferunt morbos! innocentiores ex

[1] omnium *codd.* : omnia *Mayhoff, fortasse recte.*

---

[a] An unknown.
[b] For *sympathia* see XXIV. § 1.
[c] For transference see XXX. § 64 and E. Stemplinger
Antique und moderne Volkmedizin, p. 66.

is sufficient. The midwife Sotira has said that it is a
very efficacious remedy for tertians and quartans to
smear with the flux the soles of the patient's feet,
much more so if the operation is performed by the
woman herself without the patient's knowledge,
adding that this remedy also revives an epileptic
who has fainted. Icatidas [a] the physician assures
us that quartans are ended by sexual intercourse,
provided that the woman is beginning to menstruate.
All are agreed that, if water or drink is dreaded after
a dog-bite, if only a contaminated cloth be placed
beneath the cup, that fear disappears at once, since
of course that sympathy, as Greeks call it, has an all-
powerful effect, for I have said that dogs begin to go
mad on tasting that blood. It is a fact that, added
to soot and wax, the ash of the flux when burnt heals
the sores of all draught-animals, but menstrual
stains on a dress can be taken out only by the urine
of the same woman, that the ash, mixed with nothing
but rose oil, if applied to the forehead, relieves headache, especially that of women, and that the power
of the flux is most virulent when virginity has been
lost solely through lapse of time. This also is agreed,
and there is nothing I would more willingly believe,
that if door-posts are merely touched by the menstrual discharge, the tricks are rendered vain of the
Magi, a lying crowd, as is easily ascertained. I will
give the most moderate of their promises: take the
parings of a patient's finger nails and toe nails, mix
with wax, say that a cure is sought for tertian,
quartan or quotidian fever, and fasten them before
sunrise on another man's door as a cure for these
diseases. What a fraud if they lie! What wicked-
ness if they pass the disease on! [c] Less guilty are

his omnium digitorum resigmina unguium ad
cavernas formicarum abici iubent eamque quae prima
coeperit trahere correptam subnecti collo, ita discuti
morbum.

87 XXIV. Haec sunt quae retulisse fas sit ac pleraque
ex his non nisi honore dicto, reliqua intestabilia,
infanda, ut festinet oratio ab homine fugere. in
ceteris claritates animalium aut operum sequemur.
elephanti sanguis, praecipue maris, fluctiones omnes
88 quas rheumatismos vocant sistit. ramentis eboris
cum melle Attico, ut aiunt, nubeculae in facie, scobe
paronychia tolluntur. proboscidis tactu capitis dolor
levatur, efficacius si et sternuat. dextra [1] pars
proboscidis cum Lemnia rubrica adalligata inpetus
libidinum stimulat. sanguis et syntecticis prodest,
iocurque comitialibus morbis.

89 XXV. Leonis adipes cum rosaceo cutem in facie
custodiunt a vitiis candoremque. sanant et adusta
nivibus articulorumque tumores. Magorum vanitas
perunctis adipe eo faciliorem gratiam apud populos
regesve promittit, praecipue tamen eo pingui quod
90 sit inter supercilia, ubi esse nullum potest. similia
dentis, maxime a dextera parte, villique e rostro
inferiore promissa sunt. fel aqua addita claritatem
oculis inunctis facit et cum adipe eiusdem comitiales
morbos discutit levi gustu et ut protinus qui sumpsere

---

[1] *Warmington coni.* sternuat a dextra (*aut* ad dextram).
pars *etc.*

---

[a] See the List of Diseases.
[b] Does this mean a small piece taken from a dead animal?
At any rate the sentence is queer, and one suspects corruption,
or else a *lacuna* after *proboscidis*. Warmington's suggestion
is a good one: " sneezes to the right. A bit of the trunk
etc." The triangular tip of the trunk is still regarded by

those of them who tell us to cut all the nails, throw the parings near ant holes, catch the first ant that begins to drag a paring away, tie it round the neck, and in this way the disease is cured.

XXIV. This is all the information it would be right for me to repeat, most of which also needs an apology from me. As the rest of it is detestable and unspeakable, let me hasten to leave the subject of remedies from man. Taking the other animals I shall try to find what is striking either in them or in their effects.

The blood of an elephant, particularly that of the male, checks all the fluxes that are called *rheumatismi*.[a] *Remedies from the elephant.* Ivory shavings with Attic honey are said to remove dark spots on the face, and ivory dust whitlows. By the touch of the trunk headache is relieved, more successfully if the animal also sneezes. The right side of the trunk [b] used as an amulet with the red earth of Lemnos is aphrodisiac. The blood too is good for consumption, and the liver for epilepsy.

XXV. Lion fat with rose oil preserves fairness of complexion and keeps the face free from spots; *Remedies from the lion.* it also cures frost-bite and swollen joints. The lying Magi promise those rubbed with this fat a readier popularity with peoples and with kings, especially when the fat is that between the brows, where no fat can be. Similar promises are made about the possession of a tooth, especially one from the right side, and of the tuft beneath the muzzle. The gall, used with the addition of water as a salve, improves vision, and if lion fat is added a slight taste cures epilepsy, provided that those who have taken it at

the Burmese as aphrodisiac. See *Elephant Bill*, by J. H. Williams.

cursu id digerant. cor in cibo sumptum quartanis medetur, adips cum rosaceo cotidianis febribus. perunctos eo bestiae fugiunt, resistere etiam insidiis videtur.

91 XXVI. Cameli cerebrum arefactum potumque ex aceto comitialibus morbis aiunt mederi, item fel cum melle potum, hoc et anginae, cauda arefacta solvi alvum, fimi cinere crispari capillum. cum oleo et dysintericis prodest inlitus cinis potusque quantum tribus digitis capiatur, et comitialibus morbis. urinam fullonibus utilissimam esse tradunt itemque ulceribus manantibus—barbaros constat servare eam quinquennio et heminis pota [1] ciere alvum—saetas e cauda contortas et sinistro bracchio allgiatas quartanis mederi.

92 XXVII. Hyaenam Magi ex omnibus animalibus in maxima admiratione posuerunt, utpote cui et ipsi magicas artes dederint vimque qua alliciat ad se homines mente alienatos. de permutatione sexus annua vice diximus, ceteraque de monstrifica natura eius; nunc persequemur quaecumque medicinis 93 produntur. praecipue pantheris terrori esse traditur, ut ne conentur quidem resistere, et aliquid e corio eius habentem non adpeti, mirumque dictu, si pelles utriusque contrariae suspendantur, decidere pilos

---

[1] pota d *vulg. Mayhoff*: potae V *Detlefsen*: potam *Sillig*. *Mayhoff* barbaros servare *cum* manantibus *coniungit*. *Coni*. hemina *Warmington*.

---

[a] Mayhoff would put a full stop not after *capillum* but after *oleo*. He refers to Dioscorides *Euporista* I 91 (97): ἀπόπατος καμήλου καεῖσα καὶ σὺν ἐλαίῳ καταπλασθεῖσα. This, however, refers to an ointment for making children's hair beautiful and thick, not to one for making any hair curly. Of course some greasy base is usually necessary for the application of any powder.

once aid its digestion by running. The heart taken as a food cures quartans; the fat with rose oil cures quotidians. Wild beasts run away from those smeared with it, and it is supposed to protect even from treachery.

XXVI. They say that a camel's brain, dried and taken in vinegar, cures epilepsy, as does the gall taken with honey, this being also a remedy for quinsy; that the tail when dried is laxative, and that the ash of the burnt dung makes the hair curl.[a] This ash applied with oil is also good for dysentery, as is a three-finger pinch taken in drink, and also for epilepsy. They say that the urine is very useful to the fullers, and for running ulcers—it is a fact that foreigners keep it for five years, and use hemina-doses as a purgative—and that the tail hairs plaited into an amulet for the left arm cure quartan fevers. *Remedies from the camel.*

XXVII. The Magi have held in the highest admiration the hyaena of all animals, seeing that they have attributed even to an animal magical skill and power,[b] by which it takes away the senses and entices men to itself. I have spoken [c] of its yearly change of sex and its other weird characteristics; now I am going to speak of all that is reported about its medicinal properties. It is said to be a terror to panthers in particular, so that a panther does not even attempt to resist an hyaena; that a person carrying anything made of hyaena leather is not attacked, and, marvellous to relate, if the skins of each are hung up opposite to one another the hairs *Remedies from the hyaena.*

[b] In Chapter XXIII were, it seems, several instances of *vis* in the sense of " mana."

[c] See VIII. § 105. For the change of sex, Ovid *Metamorphoses* XV. 409 foll.

65

pantherae; cum fugiant venantem, declinare ad
dexteram ut praegressi hominis vestigia occupent;
quod si successerit, alienari mentem ac vel ex equo
hominem decidere; at si in laevam detorserit,
deficientis argumentum esse celeremque capturam,
facilius autem capi, si cinctus suos venator flagellum-
que inperitans equo septenis alligaverit nodis.

94 mox, ut est sollers ambagibus vanitas Magorum,
capi iubent geminorum signum transeunte luna
singulosque prope pilos servari; capitis dolori
inligatam cutem prodesse quae fuerit in capite eius;
lippitudini fel inlitum frontibus aut, ne omnino
lippiatur, decoctum cum mellis Attici cyathis tribus
et croci uncia inunctum; sic et caligines discuti et

95 suffusiones; claritatem excitari melius inveterato
medicamento, adservari autem in Cypria pyxide;
eodem sanari argema, scabritias, excrescentia in
oculis, item cicatrices, glaucomata vero iocineris
recentis inassati sanie cum despumato melle inunctis.
dentes eius dentium doloribus tactu prodesse vel
alligatos ordine,[1] umeros umerorum et lacertorum
doloribus; eiusdem dentes, si de sinistra parte rostri,
inligatos pecoris aut capri pelle stomachi cruciatibus,

96 pulmones in cibo sumptos coeliacis, ventriculis [2]

[1] ordine, humeros *vulg.*, *Detlefsen* : numeri ordine *Mayhoff* :
humeri (umeri) ordine *codd. An* numeri ordine, humeros ?
[2] ventriculis *codd.* : vel ventriculi *Mayhoff*.

---

[a] With Mayhoff's reading, "the shoulders" should be
omitted. This reading keeps the order of words in the MSS.,
but the sympathetic (or imitative) magic disappears.

of the panther fall off. When an hyaena is running
away from the hunter, any swerve it makes to the
right has for its object stepping in the man's tracks
as he now goes in front. If it succeeds, the
man is deranged and even falls off his horse.
Should however the hyaena swerve to the left, it
is a sign of failing strength and a speedy capture;
this will be easier however if the hunter tie his
girdle with seven knots, and seven in the whip
with which he controls his horse. The Magi go on
to recommend, so cunning are the evasions of the
fraudulent charlatans, that the hyaena should be
captured when the moon is passing through the
constellation of the Twins, without, if possible,
the loss of a single hair. They add that the skin
of its head if tied on relieves headache; that the
gall if applied to the forehead cures ophthalmia,
preventing it altogether if an ointment is made of gall
boiled down with three cyathi of Attic honey and
one ounce of saffron, and that the same prescription
disperses film and cataract. They say that clear
vision is secured better if the medicament is kept till
old, but it must be in a box of copper; the same is
a cure for argema, scabbiness, excrescences and
scars on the eyes, but opaqueness needs an ointment
made with gravy from fresh roasted liver added to
skimmed honey. They add that hyaena's teeth
relieve toothache by the touch of the corresponding
tooth, or by using it as an amulet, and the shoulders [a]
relieve pains of the shoulders and arm muscles; that
the animal's teeth (but they must be from the left
side of the muzzle), wrapped in sheep skin or goat
skin, are good for severe pains in the stomach, the
lungs taken as food for coeliac disease, and their

cinerem cum oleo inlitum; nervis medullas e dorso cum oleo vetere ac felle; febribus quartanis iocur degustatum ter ante accessiones; podagris spinae cinerem cum lingua et dextro pede vituli marini addito felle taurino, omnia pariter cocta atque inlita hyaenae pelle; in eodem morbo prodesse et fel
97 cum lapide Assio; tremulis, spasticis, exilientibus et quibus cor palpitet aliquid ex corde coctum mandendum ita ut reliquae partis cinis cum cerebro hyaenae inlinatur; pilos etiam auferri hac conpositione inlita aut per se felle, evulsis prius quos renasci non libeat; sic et palpebris inutiles tolli; lumborum doloribus carnes e lumbis edendas inlinendasque cum oleo; sterilitatem mulierum emendari oculo cum glycyrrhiza et aneto sumpto in cibo, promisso intra
98 triduum conceptu. contra nocturnos pavores umbrarumque terrorem unus e magnis dentibus lino alligatus succurrere narratur. suffiri furentes eodem et circumligari ante pectus cum adipe renium aut iocinere aut pelle[1] praecipiunt. mulieri candida a pectore hyaenae caro et pili septem[2] et genitale cervi, si inligentur dorcadis pelle, e[3] collo suspensa
99 continere partus promittuntur; venerem stimulare genitalia ad sexus suos[4] in melle sumpta, etiamsi

[1] pelle *codd.*: felle *coni. Mayhoff, fortasse recte.*
[2] septem *codd.*: septeni *Mayhoff.*
[3] e *add. Mayhoff*: *om. codd.*
[4] ad sexus suos *codd.*: ab sexu suo *coni. Mayhoff.*

---

[a] The power of the number three is superior to the imitative magic of the " four " that we should expect for quartans.
[b] See XXXVI. § 131 for the *sarcophagus lapis* found at Assos in the Troad.
[c] Mayhoff's *felle* for the *pelle* of the MSS. is most attractive. A few words later on *pelle* occurs, and might easily cause the change from *felle* to *pelle.*

ash, applied with oil, for pain in the belly; that sinews are soothed by its spinal marrow with its gall and old oil, quartan fevers relieved by three [a] tastes of the liver before the attacks, gout by the ash of the spine, with the tongue and right foot of a seal added to bull's gall, all being boiled together and applied on hyaena skin. In the same disease the gall of the hyaena (so they say) with the stone of Assos [b] is beneficial; adding that those afflicted with tremors, spasms, jumpiness, and palpitation, should eat a piece of the heart boiled, but the rest must be reduced to ash and hyaena's brain added to make an ointment; that an application of this mixture or of the gall by itself removes hairs, those not wanted to grow again must first be pulled out; by this method unwanted eye-lashes are removed; that for pains in the loins flesh of an hyaena's loins should be eaten and used as an ointment with oil; that barrenness in women is cured by an eye taken in food with liquorice and dill, conception being guaranteed within three days. For night terrors and fear of ghosts one of the large teeth tied on with thread as an amulet is said to be a help. They recommend fumigation with such a tooth for delirium, and to tie one round in front of the patient's chest, adding fat from the kidneys, or a piece of liver, or of skin.[c] A woman is guaranteed never to miscarry if, tied round her neck in gazelle leather, she wears white flesh from a hyaena's breast, seven hyaena's hairs, and the genital organ of a stag. A hyaena's genitals taken in honey stimulate desire for their own sex,[d] even when men hate inter-

---

[d] Mayhoff's *a sexu suo* would mean "from homosexuality."

viri mulierum coitus oderint; quin immo totius
domus concordiam eodem genitali et articulo spinae
cum adhaerente corio adservatis constare. hunc [1]
spinae [2] articulum sive [3] nodum Atlantion vocant;
est autem primus. in comitialium quoque remediis
100 habent eum. adipe accenso serpentes fugari dicunt;
maxilla comminuta in aneso et in cibo sumpta horrores
sedari; eodem suffitu mulierum menses evocari.
tantumque est vanitatis ut, si ad bracchium alligetur
superior e dextra parte rostri dens, iaculantium ictus
deerraturos negent. palato eiusdem arefacto et
cum alumine Aegyptio calefacto ac ter in ore per-
mutato faetores et ulcera oris emendari, eos vero
qui linguam in calciamento sub pede habeant non
101 latrari a canibus; sinistra parte cerebri naribus
inlita morbos perniciosos mitigari sive hominum sive
quadripedum; frontis corium fascinationibus resistere, cervicis carnes, sive mandantur sive arefactae
bibantur, lumborum doloribus; nervis a dorso
armisque suffiendos nervorum dolores, pilos rostri
admotos mulierum labris amatorium esse; iocur in
102 potu datum torminibus et calculis mederi. nam cor
in cibo potuve sumptum omnibus doloribus corporum
auxiliari, lienem lienibus, omentum ulcerum inflammationibus cum oleo, medullas doloribus spinae
et nervorum lassitudini; renium nervos potos in

[1] hunc r : hinc *rel. codd.*

[2] spinae *rel. codd.* : ruinae r.

[3] sive *codd.* : scite *Mayhoff, qui etiam lacunam ante* sive
coni.

---

[a] The text is very uncertain, but Mayhoff's *scite* (" cleverly ")
can hardly be right. The variant *ruinae* shows that the source
of corruption lies very deep.

course with women; nay the peace of the whole household is assured by keeping in the home these genitals and a vertebra with the hide still adhering to them. This vertebra or joint they call the Atlas joint; [a] it is the first. They consider it too to be one of the remedies for epilepsy. They add that burning hyaena fat keeps serpents away; that the jawbone, pounded in anise and taken in food, relieves fits of shivering, and that fumigation with it is an emmenagogue. They lie so grossly as to declare that, if an upper tooth from the right side of the muzzle is tied to the arm of a man, his javelin will never miss its mark. They say too that the palate of a hyaena, dried, and warmed with Egyptian alum, [b] cures foul breath and ulcers in the mouth, if the mixture is renewed three times; that those however who carry a hyaena's tongue in their shoe under the foot never have dogs bark at them; that if a part of the left side of the brain is smeared on patients' nostrils dangerous diseases are relieved, whether of man or quadruped; that the hide of the forehead averts the evil eye, and the flesh of the neck, whether eaten, or dried and taken in drink, is good for lumbago; that sinews from the back and shoulders should be used for fumigating painful sinews; that hairs from the muzzle, applied to a woman's lips, act as a love-charm; that the liver given in drink cures colic and stone in the bladder. But they add that the heart, taken either in food or in drink, gives relief from all pains of the body, the spleen from those of the spleen, the caul with oil from inflamed ulcers, and the marrow from pains of the spine and of tired sinews; that the kidney sinews

[b] For *alumen* see Spencer's *Celsus* vol. II p. xviii.

vino cum ture fecunditatem restituere ademptam
veneficio; vulvam cum mali Punici dulcis cortice
in potu datam prodesse mulierum vulvae; adipe a
lumbis suffiri parientes difficulter et statim parere; e
dorso medullam adalligatam contra vanas species
103 opitulari, spasticis genitale e maribus suffitu, item
lippientibus; ruptis et contra inflammationes,
servatos[1] pedes tactu, laevos dexteris partibus,
dexteros laevis; sinistrum pedem superlatum par-
turienti letalem esse, dextro inlato facile eniti.
membranam quae fel continuerit cardiacis potam in
vino vel in cibo sumptam[2] succurrere; vesicam in
vino potam contra urinae incontinentiam; quae
104 autem in vesica inventa sit urina, additis oleo ac
sesamis et melle haustam prodesse stomachi acrimoniae[3] veteri. costarum primam et octavam
suffitu ruptis salutarem esse; ex spina vero parturientibus ossa; sanguinem cum polenta sumptum
torminibus; eodem tactis postibus ubicumque
Magorum infestari artes, non elici deos nec conloqui,

[1] servatos *codd.*: adversos *Mayhoff, qui etiam* alternos
coni.: fervefactos *coni. Sillig.*

[2] *Post* sumptam *habent* contra (r *excepto*) *codd.*: *post*
contra *lacunam indicat Mayhoff.*

[3] stomachi acrimoniae *Mayhoff*: acrimoniae *Caesarius*:
aegrimoniae *Gelenius, Hermolaus Barbarus*: aegrimonio
codd.

[a] A semicolon at *lippientibus* improves the run of this
sentence.

[b] The *servatos* of the MSS. can hardly be right, but it just
makes sense, and the proposed emendations are not convincing.

taken with frankincense in wine restore fertility
lost through sorcery; that the uterus with the rind
of a sweet pomegranate given in drink is good for the
uterus of women; that the fat from the loins, used
in fumigation, gives even immediate delivery to
women in difficult labour; that the spinal marrow
used as an amulet is a help against hallucinations,
and fumigation with the male organ against spasms,
as well as ophthalmia; [a] that for ruptures and inflammations a help is the touch of an hyaena's feet, which
are kept for the purpose,[b] of the left foot for affections on the right side, and of the right foot for
affections on the left side; that the left foot, drawn
across [c] a woman in labour, causes death, but the right
foot laid on [c] her easy delivery. The Magi say that
the membrane enclosing the gall, taken in wine or
in the food, is of use in cardiac affections; that the
bladder taken in wine relieves incontinence of urine,
and the urine found in the bladder, drunk with oil,
sesame, and honey added, relieves chronic acidity
of the stomach; [d] that the first or [e] eighth rib,
used in fumigation, is curative for ruptures, but the
spinal bones are so for women in labour; the blood
taken with pearl barley is good for colic, and if the
door-posts are everywhere touched with this blood,
the tricks of the Magi are made ineffective, for they
can neither call down the gods nor speak with them,

---

[c] Littré, I think wrongly, translates as though *superlatum*
and *inlatum* meant the same thing.

[d] Mayhoff's emendation, bold as it is, is strongly supported
by *acrimonia stomachi* in XXIII. § 142; otherwise, to keep
the idea of like curing like, one would be tempted to emend
to *urinae acrimoniae veteri*.

[e] This is probably the meaning of the *et* in this clause
because of the singular *salutarem* in the predicate.

sive lucernis sive pelvi, sive aqua sive pila, sive quo
alio genere temptetur; carnes si edantur, contra
rabidi canis morsus efficaces esse, etiamnum iocur
efficacius. carnes vel ossa hominis quae in ventriculo
105 occisae inveniantur suffitu podagricis auxiliari;
si ungues inveniantur in his, mortem alicuius capien-
tium significari; excrementa sive ossa reddita, cum
interematur, contra magicas insidias pollere; fimum
quod in intestinis inventum sit arefactum ad dysintericos valere, potum inlitumque cum adipe anserino toto corpore opitulari laesis malo medicamento;
a cane vero morsis adipem inlitum et corium sub-
stratum; rursus tali sinistri cinere decocto cum
106 sanguine mustelae perunctos omnibus odio venire;
idem fieri oculo decocto. super omnia est quod
extremam fistulam intestini contra ducum ac
potestatium iniquitates commonstrant et ad suc-
cessus petitionum iudiciorumque ac litium eventus,
si omnino [1] aliquis secum habeat; eiusdem caverna
in sinistro lacerto alligata si quis mulierem prospiciat,
amatorium esse tam praesens ut ilico sequatur;
eiusdem loci pilorum cinerem ex oleo inlitum viris
qui sint probrosae mollitiae severos, non modo
pudicos mores induere.

[1] omnino *Mayhoff*: omnino tantum *codd.*

---

[a] For another list of apparatus see XXX. § 14 *aqua et
sphaeris et aere et stellis et lucernis ac pelvibus securibusque.*
Some of the articles are suggestive of modern fortune-telling.

whether they try lamps, bowl, water, globe,[a] or any other means; that to eat the flesh neutralizes the bites of a mad dog, the liver being still more efficacious. They add that the flesh or bones of a man found in the stomach of an hyaena when killed relieve gout by fumigation; that if finger nails are found in them it is a sign of death for one of the hunters; that excrement or bones, voided when the beast is being killed, can prevail against the insidious attacks of sorcerers; that dung found in the intestines is, when dried, excellent for dysentery, and, taken in drink and applied with goose grease, gives relief anywhere in the body to the victims of noxious drugs; that for dog-bites, however, rubbing with the fat as ointment, and lying on the skin, are helpful; that on the other hand those rubbed with the ash of the left pastern bone, boiled down with weasel's blood, incur universal hatred, the same effect being produced by a decoction of the eye. Over and above all these things they assert that the extreme end of the intestine prevails against the injustices of leaders and potentates, bringing success to petitions and a happy issue to trials and lawsuits if it is merely kept on the person; that the anus, worn as an amulet on the left arm, is so powerful a love-charm that, if a man but espies a woman, she at once follows him; that the hairs also of this part, reduced to ashes, mixed with oil, and used as ointment on men guilty of shocking effeminacy, make them assume, not only a modest character, but one of the strictest morality.[b]

[b] This remarkable chapter, throwing light as it does on folk-medicine and ancient superstitions, calls for a longer note than can be printed in the text. See Additional Note B (p. 563).

107 XXVIII. Proxime fabulosus est crocodilus ingenio [1]
quoque, ille cui vita in aqua terraque communis.
duo enim genera eorum. illius e dextra maxilla
dentes adalligati dextro lacerto coitus, si credimus,
stimulant, canini dentes febres statas arcent ture
repleti—sunt enim cavi—ita ne diebus quinque ab
aegro cernatur qui adalligaverit. idem pollere et
ventre exemptos lapillos adversus febrium horrores
108 venientes tradunt. eadem de causa Aegypti perungunt adipe aegros suos. alter illi similis, multum
infra magnitudine, in terra tantum odoratissimisque
floribus vivit. ob id intestina eius diligenter ex-
quiruntur iucundo nidore referta; crocodileam
vocant, oculorum vitiis utilissimam cum porri suco
109 inunctis et contra suffusiones vel caligines. inlita
quoque ex oleo cyprino molestias in facie nascentes
tollit, ex aqua vero morbos omnes quorum natura
serpit in facie, nitoremque reddit. lentigines tollit
ac varos maculasque omnes, et contra comitiales
morbos bibitur ex aceto mulso binis obolis. adposita
menses ciet. optima quae candidissima et friabilis
minimeque ponderosa, cum teratur inter digitos,
110 fermentescens. lavatur ut cerussa. adulterant amylo
aut Cimolia, sed maxime ⟨sturnorum fimo quos⟩ [2]
captos oryza tantum pascunt. felle inunctis oculis
ex melle contra suffusiones nihil utilius praedicant.

---

[1] ingenio (ingento V) *codd.*: ingens *Harduinus. Post*
magnitudine (*l. 11*) ingenio quoque *transferre velit Warmington,*
fortasse recte.

[2] sturnorum fimo quos *Ianus, Detlefsen, ex Dioscoride*
(*II 80*), *sed* qui r *Gelenius, Mayhoff:* sui VRd *vulg.*

---

[a] Hardouin's ingenious conjecture would mean : " and he is a
huge creature, and amphibious."

[b] Jan's addition is due to Dioscorides II 80 : δολίζουσι δὲ αὐτὴν
ψάρας ὀρύζῃ τρέφοντες καὶ τὴν ἄφοδον ὁμοίαν οὖσαν πωλοῦντες.

XXVIII. Almost as legendary is the crocodile, *Crocodiles.*
in its nature ᵃ also—I mean the famous one, which
is amphibious; for there are two kinds of crocodiles.
His teeth from the right jaw, worn as an amulet
on the right arm, are (if we believe it) aphrodisiac,
while the dog teeth, stuffed with frankincense
(for they are hollow), drive away the intermittent
fevers if the sick man can be kept for five days
from seeing the person who fastened them on.
It is said that pebbles taken from his belly have
a similar power to check feverish shivers as they
come on. For the same reason the Egyptians rub
their sick with its fat. The other kind of crocodile
is similar to this, though much smaller in size, living
only on land and eating very sweet-scented flowers.
Its intestines therefore are much in demand, being
filled with fragrant stuff called crocodilea, which
with leek juice makes a very useful salve for affections
of the eyes, and to treat cataract or films. Applied
also with cyprus oil crocodilea removes blotches
appearing on the face, with water indeed all those
diseases the nature of which is to spread over the face,
and it also clears the complexion. It removes
freckles, pimples, and all spots; two-oboli doses
are taken in oxymel for epilepsy, and a pessary
made of it acts as an emmenagogue. The best kind
is very shiny, friable, and extremely light, ferment-
ing when rubbed between the fingers. It is washed
in the same way as white lead. They adulterate it
with starch or Cimolian chalk, but mostly with
the dung of starlings,ᵇ which they catch and feed
on nothing but rice. We are assured that there is no
more useful remedy for cataract than to anoint the
eyes with crocodile's gall and honey. They say

intestinis et reliquo corpore eius suffiri vulva laborantes salutare tradunt, item velleribus circumdari
vapore eiusdem infectis. corii utriusque cinis ex
aceto inlitus his partibus quas secari opus sit aut
nidor cremati sensum omnem scalpelli aufert.
111 sanguis utriusque claritatem visus inunctis . . .[1]
cicatrices oculorum emendat. corpus ipsum excepto
capite pedibusque elixum manditur ischiadicis tus-
simque veterem sanat, praecipue in pueris, item
lumborum dolores. habent et adipem quo tactus
pilus defluit. hic perunctos a crocodilis tuetur,
instillaturque morsibus. cor adnexum in lana ovis
nigrae cui nullus alius colos incursaverit et primo
partu genitae quartanas abigere dicitur.
112   XXIX. Iungemus illis simillima et peregrina aeque
animalia, priusque chamaeleonem peculiari volumine
dignum existimatum Democrito ac per singula
membra desecratum,[2] non sine magna voluptate
nostra cognitis proditisque mendaciis Graecae vanitatis. similis et magnitudine est supra dicto crocodilo, spinae tantum acutiore curvatura et caudae
113 amplitudine distans.[3] nullum animal pavidius existimatur et ideo versicoloris esse mutationis. vis eius
maxima contra accipitrum genus. detrahere enim
supervolantem ad se traditur et voluntarium praebere

---

[1] *Lacunam indicavi*: dat *Detlefsen*: excitat *Mayhoff,*
qui etiam facit *coni.*
  [2] desecratum R d: dissertatum *coni. Mayhoff.*
  [3] distans] " *Locus nondum sanatus* " *Mayhoff.*

---

[a] Does the *et* mean " or " ? The phrase is a queer one,
unless it means that the body used in the fumigation should
contain the intestines, which are essential for a cure.

that fumigation with the intestines and [a] the rest of its body is of benefit to women with uterine trouble, as it is to wrap them up in a fleece impregnated with its steam. Ashes from burning the skin of either kind of crocodile, applied in vinegar to the parts in need of surgery, or even the fumes, cause no pain to be felt from the lancet. The blood of either kind, if the eyes are anointed with it, improves the vision and removes eye scars. The body itself, boiled without the head and feet, is eaten for sciatica and cures chronic cough, especially that of children, as well as lumbago. Crocodiles also have a fat, a touch of which makes hair fall out. Used as embrocation this protects from crocodiles, and is poured by drops into their bites. The heart, tied on in the wool of a black sheep, the first-born of its mother, the wool having no other colour intermixed, is said to drive away quartan fevers. [b]

XXIX. To these animals I will add others very *Chamaeleon.* like them and equally foreign, taking first the chamaeleon, thought by Democritus worthy of a volume to itself, each part of the body receiving separate attention. It afforded me great amusement to read an exposure of Greek lies and fraud. The chamaeleon is also as big as the crocodile just mentioned, [c] differing only in the greater curve of the spine and in the size of its tail. People think it the most timid of animals, and that it is for this reason it continually changes its colour. Over the hawk family it has very great power, for as a hawk flies overhead, it is brought down to the chamaeleon,

---

[b] Quartans were supposed to be caused by black bile. See Hippocrates, *Nature of Man*, ch. XV (Loeb IV, p. 41).
[c] I.e. the land animal of § 108.

lacerandum ceteris animalibus. caput eius et guttur,
si roboreis lignis accendantur, imbrium et tonitruum
concursus facere Democritus narrat, item iocur in
114 tegulis ustum. reliqua ad veneficia pertinentia quae
dicit, quamquam [1] falsa existimantes, omittemus
praeterquam ubi inrisu coarguendum: [2] dextro
oculo, si viventi eruatur, albugines oculorum cum
lacte capríno tolli, lingua adalligata pericula puer-
perii; eundem salutarem esse parturientibus, si sit
domi, si vero inferatur, perniciosissimum. linguam,
si viventi exempta sit, ad iudiciorum eventus pollere,
cor adversus quartanas inligatum lana nigra primae
115 tonsurae. pedem e prioribus dextrum pelle hyaenae
adalligatum sinistro bracchio contra latrocinia terro-
resque nocturnos pollere, item dextram mamillam [3]
contra formidines pavoresque; sinistrum vero pedem
torreri in furno cum herba quae aeque chamaeleon
vocetur, additoque unguento pastillos eos [4] in ligneo
vas conditos praestare, si credimus, ne cernatur ab
116 aliis qui id habeat. armum dextrum ad vincendos
adversarios vel hostes valere, utique si abiectos
eiusdem nervos calcaveris—sinistrum umerum [5] quibus
monstris consecret, qualiter somnia quae velis et
quibus velis mittantur, pudet referre—somnia ea
dextro pede resolvi, sicut sinistro latere lethargos quos

---

[1] quamquam *codd., edd.*: tanquam *vet. Dal.*
[2] coarguendum d(?) *Gelenius*: coarguent eum *Mayhoff*:
coarguentium VR *vulg.*
[3] mamillam *codd. edd.*: maxillam *vet. Dal.*
[4] eos *codd.*: factos *coni. Mayhoff.*
[5] umerum *codd. Detlefsen*: vero *Mayhoff*: mirum *vulg.*

---

[a] And therefore harmless.
[b] Perhaps " chamaeleon; " *eundem* is ambiguous.

they say, and made an unresisting prey for other
animals to tear. Democritus relates that its head
and throat, if burnt on logs of oak, cause storms
of rain and thunder, as does the liver if burnt
on tiles. The rest of what he says is of the
nature of sorcery, and although I think that it is
untrue,[a] I shall omit all, except where something
must be refuted by being laughed at; examples are
as follow. The right eye, plucked from the living
animal and added to goat's milk, removes white
ulcers on the eyes; the tongue, worn as an amulet,
the perils of childbirth. The same eye,[b] if in the
house, is favourable to childbirth; if brought in,
very dangerous. The tongue, taken from the living
animal, controls the results of cases in the courts;
the heart, tied on with black wool of the first shear-
ing, overcomes quartan fevers. The right front
foot, tied as an amulet to the left arm by hyaena
skin, is powerful protection against robbery and
terrors of the night, and the right teat [c] against fears
and panic. The left foot however is roasted in a
furnace with the plant that also is called chamaeleon,
an unguent is added, and the lozenges thus made
are stored away in a wooden vessel and, if we believe
it, make the owner invisible to others. The right
shoulder has power to overcome adversaries and
public enemies, especially if a person throws away
sinews of the same animal and treads on them. But
as to the left shoulder, I am ashamed to repeat the
grotesque magic that Democritus assigns to it; how
any dreams you like may be sent to any person you
like; how these dreams are dispelled by the right
foot, just as the torpor caused by the right foot is

[c] The conjecture *maxillam* will mean " jaw."

fecerit dexter. sic [1] capitis dolores insperso vino
in quo latus alterutrum maceratum sit sanari. si
feminis sinistri vel pedis cinere misceatur lac suillum,
117 podagricos fieri [2] inlitis pedibus. felle glaucomata
et suffusiones corrigi prope creditur tridui inunctione,
serpentes fugari ignibus instillato, mustelas contrahi
in aquam coiecto, corpori vero inlito detrahi pilos.
idem praestare narrat iocur cum ranae rubetae
pulmone inlitum, praeterea iocinere amatoria dissolvi,
melancholicos autem sanari, si ex corio chamaeleonis
sucus herbae Heleniae bibatur, intestina et fartum
eorum, cum animal id nullo cibo vivat, cum [3] simi-
arum urina inlita inimicorum ianuae odium omnium
118 hominum his conciliare; cauda flumina et aquarum
impetus sisti, serpentes soporari; eadem medicata
cedro et murra inligataque gemino ramo palmae
percussam aquam discuti, ut quae intus sint omnia
appareant, utinamque eo ramo contactus esset
Democritus, quoniam ita loquacitates inmodicas
promisit inhiberi. palamque est virum alias sagacem
et vitae utilissimum nimio iuvandi mortales studio
prolapsum.
119 XXX. Ex eadem similitudine est scincus—et
quidam terrestrem crocodilum esse dixerunt—
candidior autem et tenuiore cute. praecipua tamen

---

[1] sic d T *Detlefsen* : set *Mayhoff* : sit V R : *del. vulg.*
[2] fieri *codd. edd.*: liberari *vel* sanari *coni. Mayhoff*: refici
vel sanos fieri *Warmington.*
[3] cum] *Add. Detlefsen* : *post* urina *add.* una *Mayhoff.*

---

[a] Probably some emendation is required meaning " cured."
[b] Littré thinks that Pliny is here giving both the Greek
word (*glaucoma*) and the Latin (*suffusio*) for one disease of the
eye.
[c] A plain instance of *vero* introducing the climax of a list.

dispelled by the left flank. In this way headache
is cured by sprinkling on the head wine in which
either side of a chamaeleon has been soaked. If
sow's milk is mixed with the ash of the left thigh
or foot, gout is caused *a* by rubbing the feet with the
mixture. It is practically a current belief that
anointing the eyes for three days with the gall is a
cure for opaqueness of the eye and cataract,*b* that
serpents run away if the gall is dropped into fire,
that weasels run together when it is thrown into
water, while *c* hairs are removed from the body when
it is rubbed therewith. Democritus relates that the
same result comes from applying the liver with the
lung of the bramble toad; that moreover the liver
makes of no effect love charms and philtres, curing
melancholy also if the juice of the herb helenium
is drunk in a chamaeleon's skin; that the intestines
and their content (although the animal lives without
food) with the urine of apes, if smeared on the door
of an enemy, brings on him the hatred of all men;
that by its tail rivers and rushing waters are stayed
and serpents put to sleep; that the tail also, if
treated with cedar and myrrh and tied on to a twin
palm-branch, divides the water struck with it, so that
all within becomes plain. Would that Democritus
had been touched with such a branch, seeing that he
assures us that by it wild talk is restrained! It is
clear that a man, in other respects of sound judgement
and of great service to humanity, fell very low
through his over-keenness to help mankind.

XXX. A similar animal is the scincos *d*—and *The scincos.*
indeed it has been styled the land crocodile—but it is
paler, and with a thinner skin. The chief difference,

d Not the lizard now called the skink but a larger one.

differentia dinoscitur a crocodilo squamarum serie
a cauda ad caput versa. maximus Indicus, deinde
Arabicus. adferuntur salsi. rostrum eius et pedes
in vino albo poti cupiditates veneris accendunt,
utique cum satyrio et erucae semine singulis drachmis
omnium ac piperis duabus admixtis. ita pastilli
120 singularum drachmarum bibantur. per se laterum
carnes obolis binis cum murra et pipere pari modo
potae efficaciores ad idem creduntur. prodest et
contra sagittarum venena, ut Apelles tradit, ante
posteaque sumptus. in antidota quoque nobilia
additur. Sextius plus quam drachmae pondere
in vini hemina potum perniciem adferre tradit,
praeterea eiusdem [1] decocti ius cum melle sumptum
venerem inhibere.

121   XXXI. Est crocodilo cognatio quaedam amnis
eiusdem geminique victus cum hippopotamio, re-
pertore detrahendi sanguinis, ut diximus, plurimo
autem super Saiticam praefecturam. huius corii
cinis cum aqua inlitus panos sanat, adips frigidas
febres, item fimum suffitu, dentes e parte laeva
dolorem dentium scarifatis gingivis. pellis eius e

---

[1] eiusdem *codd*. : lentium *Gesner e Dioscoride* II 66.

---

[a] I.e. with no other part of the beast added.
[b] A native of Thasos mentioned by Galen.
[c] Sextius Niger, " who wrote in Greek," as Pliny says in
his list of authorities, was a writer on *materia medica*. He is
mentioned by both Dioscorides and Galen. Some scholars
believe that Pliny drew much of his information from this
source, as he never mentions Dioscorides.
[d] The reason for Gesner's emendation *lentium* is that
Dioscorides in his account of the σκίγκος (II 66 Wellmann)

however, between it and the crocodile is in the
arrangement of the scales, which are turned from the
tail towards the head. The Indian is the biggest
scincos, next coming the Arabian. They import
them salted. Its muzzle and feet, taken in white
wine, are aphrodisiac, especially with the addition
of satyrion and rocket seed, a single drachma of all
three and two drachmae of pepper being compounded. One-drachma lozenges of the compound
should be taken in drink. Two oboli of the flesh of
the flanks by itself,[a] taken in drink with myrrh and
pepper in similar proportions, are believed to be
more efficacious for the same purpose. It is also
good for the poison of arrows, as Apelles [b] informs us,
if taken before and after the wound. It is also an
ingredient of the more celebrated antidotes.
Sextius [c] says that more than a drachma by weight,
taken in a hemina of wine, is a fatal dose, and that
moreover the broth of a scincos [d] taken with honey is
antaphrodisiac.

XXXI. There is a kind of relationship between *Hippo-*
the crocodile and the hippopotamus, for they both *potamus.*
live in the same river and both are amphibious. The
hippopotamus, as I have related,[e] was the discoverer
of bleeding, and is most numerous above the pre-
fecture of Sais. His hide, reduced to ash and applied
with water, cures superficial abscesses; the fat and
likewise the dung chilly agues by fumigation, and the
teeth on the left side, if the gums are scraped with
them, aching teeth. The hide from the left side of
his forehead, worn as an amulet on the groin, is an

says: ἀναπαύεσθαι δὲ τὴν ἐπίτασιν τῆς προθυμίας φακοῦ ἀφεψήματι
μετὰ μέλιτος πινομένῳ.
  e Book VIII. § 96.

sinistra parte frontis inguini adalligata venerem inhibet, eiusdem cinis alopecias explet. testiculi drachma ex aqua contra serpentes bibitur. sanguine pictores utuntur.

122 XXXII. Peregrinae sunt et lynces, quae clarissime quadripedum omnium cernunt. ungues earum omnes cum corio exuri efficacissime in Carpatho insula tradunt. hoc cinere poto propudia virorum, eiusdemque aspersu feminarum libidines inhiberi, item pruritus corporum, urina stillicidia vesicae. itaque eam protinus terra pedibus adgesta obruere traditur. eadem autem et iugulorum dolori monstratur in remedio.

123 XXXIII. Hactenus de externis. nunc praevertemur ad nostrum orbem, primumque communia animalium remedia atque eximia dicemus, sicuti e lactis usu. utilissimum cuique maternum. [concipere nutrices exitiosum est, hi sunt enim infantes qui colostrati appellantur, densato lacte in casei speciem. est autem colostra prima a partu spongea densitas lactis.] [1] maxime autem alit quodcumque humanum, mox caprinum, unde fortassis fabulae Iovem ita nutritum dixere. dulcissimum ab hominis camelinum, efficacissimum ex asinis. magnorum animalium

124 et corporum facilius redditur. stomacho adcommodatissimum caprinum, quoniam fronde magis quam

---

[1] *uncos ego posui.*

---

[a] I think that this sentence belongs elsewhere, perhaps after § 72. Another possibility is that Pliny forgot what he said in XI. § 237, where he calls *colostratio* an ailment caused by the young's taking mother's milk too soon. If Pliny wrote *concipere . . . speciem*, the next sentence, *est autem . . . lactis*, might be a scribe's marginal correction, which was

antaphrodisiac; the same reduced to ash restores hair lost through mange. A drachma of a testicle is taken in water for snake bite. The blood is used by painters.

XXXII. The lynx too is a foreign animal, and has keener sight than any other quadruped. On the island of Carpathus all their nails, with the hide, make, it is said, a very efficacious medicine when reduced to ash by burning. They say that these ashes taken in drink by men check shameful conduct, and sprinkled on women lustful desire; that they also cure irritation of the skin and that the urine cures strangury. And so, as is said, the animal at once covers it with earth by scratching with his paws. This urine is also prescribed for pain in the throat. *Lynx.*

XXXIII. Hitherto I have dealt with things foreign, but will now turn to the Roman world, speaking first of remedies common to all animals and excellent in quality, such as milk and its uses. Mother's milk is for èverybody the most beneficial. [It is very bad for women to conceive while nursing; their nurselings are called *colostrati*, the milk being thick like cheese. But colostra is the first milk given after delivery, and is thick and spongy.]ᵃ But any woman's milk is more nourishing than any other kind, the next being that of the goat; this perhaps is the origin of the story that Jupiter was nursed in this way. The sweetest milk after woman's is that of the camel, the most efficacious that of the ass. A big species or a big individual yields its milk more readily. Goat's milk is the most suited to the stomach, as the animal browses rather *Milks.*

afterwards added to the text. It should be noticed that the connection of thought is easy and natural if *maxime autem* follows immediately after *maternum*.

herba vescuntur. bubulum medicatius, ovillum dul-
cius et magis alit, stomacho minus utile, quoniam est
pinguius. omne autem vernum aquatius aestivo et
de novellis. probatissimum vero quod in ungue
haeret nec defluit. innocentius decoctum, praecipue
cum calculis marinis. alvus maxime solvitur bubulo,
minus autem inflat quodcumque decoctum. usus
125 lactis ad omnia intus exulcerata, maxime renes,
vesicam, interanea, fauces, pulmones, foris pruritum
cutis, eruptiones pituitae poti ab [1] abstinentia.[2] nam
ut in Arcadia bubulum biberent phthisici, syntectici,
cachectae, diximus in ratione herbarum. sunt inter
exempla qui asininum bibendo liberati sint podagra
126 chiragrave. medici speciem unam addidere lactis
generibus quod schiston appellavere. id fit hoc
modo: fictili novo fervet caprinum maxime, ramisque
ficulneis recentibus miscetur additis totidem cyathis
mulsi quot sint heminae lactis. cum fervet, ne [3] cir-
cumfundatur praestat cyathus argenteus cum frigida
aqua demissus ita ne quid infundat. ablatum deinde
igni refrigeratione dividitur et discedit serum a lacte.
127 quidam et ipsum serum iam multo potentissimum

[1] poti ab f : poti at F : potior d x : poscit R : post r.
[2] abstinentia Vdx *vulg.* : abstinentiam R. *In textu* poti ab
abstinentia *et Detlefsen et Mayhoff, qui addit* : "*locus nondum
sanatus.an* posci abstinentia medicaminum ut in *sqq*? *Cfr.*
XXV 94."
[3] ne *Hermolaus Barbarus, Mayhoff* : ni *codd., Detlefsen.*

[a] Dioscorides has (II. § 70) μάλιστα δὲ διαπύροις κόχλαξιν
ἐξικμασθέν ("especially when boiled down by hot pebbles ").
Pliny seems to have misunderstood his original, or to have
had different Greek before him.
[b] For a good account of modern uses of milk see W. T.
Fernie, *Animal Simples*, pp. 301–317.
[c] For *eruptiones pituitae* see List of Diseases.

than grazes. Cow's milk is more medicinal, sheep's
sweeter and more nourishing, although less useful for
the stomach because of its greater richness. All
spring milk, however, is more watery than that of
summer, as is that from new pastures. The highest
grade, however, is that of which a drop stays on the
nail without falling off. Milk is less harmful when
boiled, especially with sea pebbles.[a] Cow's milk is
the most relaxing, and any milk causes less flatulence
when boiled.[b] Milk is used for all internal ulcers,
especially those of the kidneys, bladder, intestines,
throat, and lungs, externally for irritation of the
skin, and for outbursts of phlegm,[c] but it must be
drunk after fasting.[d] And I have mentioned in my
account of herbs[e] how in Arcadia cow's milk is drunk
by consumptives, and by those in a decline or poor
state of health. Cases too are quoted of patients
who by drinking ass's milk have been freed from gout
in feet or hands. To the various kinds of milk
physicians have added another, named *schiston*, that
is, "divided." It is made in this way: milk, by
preference goat's milk, is boiled in new[f] earthenware and stirred with fresh branches of a fig-tree,
after adding as many cyathi of honey wine as there
are heminae of milk. When it boils, to prevent its
boiling over a silver cyathus of cold water is lowered
into it so that none is spilled. Then taken off the
fire it divides as it cools, and the whey separates from
the milk. Some also boil down to one-third the

---

[d] It is difficult to see why Mayhoff calls this passage *locus
nondum sanatus*. The grammar, at any rate, is no looser than
in many other places.

[e] See XXV. § 94.

[f] Why new? Probably so as to avoid contamination or
for a magical reason.

decocunt ad tertias partes et sub diu refrigerant.
bibitur autem efficacissime heminis per intervalla,
statis [1] diebus quinae; melius a potu gestari. datur
comitialibus, melancholicis, paralyticis, in lepris,
128 elephantiasi, articulariis morbis. infunditur quoque
lac contra rosiones a medicamentis factas et, si urat
dysinteria, decoctum cum marinis lapillis aut cum
tisana hordeacia. item ad intestinorum rosiones bubulum aut ovillum utilius, recens quoque dysintericis
infunditur, ad colum autem crudum, item vulvae et
propter serpentium ictus potisve pityocampis, bu-
129 presti, cantharidum aut salamandrae venenis, privatim bubulum his qui Colchicum biberint aut cicutam
aut dorycnium aut leporem marinum, sicut asininum
contra gypsum et cerussam et sulpur et argentum
vivum, item durae alvo in febri. gargarizatur quoque
faucibus exulceratis utilissime et bibitur ab imbecilli-
tate vires recolligentibus quos atrophos vocant, in
febri etiam quae careat dolore capitis. pueris ante
cibum lactis asinini heminam dari, aut si exitus cibi
rosiones sentirent, antiqui in arcanis habuerunt, si
130 hoc non esset, caprini. bubuli serum orthopnoicis
prodest ante cetera addito nasturtio. inunguntur
etiam oculi in lactis heminas additis sesamae drachmis
quattuor tritis in lippitudine. caprino lienes sanantur,
post bidui inediam tertio die hedera pastis capris,

---

[1] statis *ego* : satis *Ianus, Detlefsen, Mayhoff* : singulis
veteres edd. : salis *codd.*

---

[a] With the reading *singulis,* "separate." With *satis*
(apparently) "five heminae are enough for the days (on which
it is taken)." This is strange Latin, and exercise, or a drive,
five times a day seems excessive. It is more natural to

whey itself, which is now very vinous indeed, and cool it in the open air.   But the most efficacious way to drink it is a hemina at a time at intervals, five heminae in all on fixed [a] days; it is better to take a drive afterwards.   It is given for epilepsy, melancholia, paralysis, leprous sores, leprosy, and diseases of the joints.   Milk is also injected for smarting caused by purges, or, for the smarting of dysentery, milk boiled down with [b] sea pebbles or with barley gruel.   For smarting intestines also cow's milk or sheep's is the more effective.   Fresh milk too is injected for dysentery, and raw milk for colitis, uterus trouble, snake bite, swallowing pine-caterpillars, buprestis, the poison of Spanish fly [c] or salamander, and cow's milk is specific when there has been taken in drink Colchicum, hemlock, dorycnium, or sea hare, as ass's milk is for gypsum, white lead, sulphur, quicksilver, and constipation in fever.   It also makes a very useful gargle for ulcerated throats, is drank by convalescents from weakening illness, said to be "in a decline," [d] and also for fever which is without headache.   To give to children before food a hemina of ass's milk, or failing that of goat's milk, and if the rectum smarted at stool, the ancients held to be one of their secrets.   Better for orthopnoea than other remedies is whey of cow's milk with the addition of cress.   The eyes also are bathed for ophthalmia with a hemina of milk to which have been added four drachmae of pounded sesame.   Splenic diseases are cured by drinking goat's milk for three days without

suppose that five doses were to be taken in all, each on a fixed day, to be followed by a ride or drive. Cf. *statas febres* § 107.

   [b] This *cum* is perhaps an interpolation (dittography), but cf. § 124.

   [c] See note on § 160.   [d] Or: "undernourished."

per triduum poto sine alio cibo. lactis usus alias contrarius capitis doloribus, hepaticis, splenicis, nervorum vitio, febres habentibus, vertigini, praeterquam purgationis gratia, gravedini, tussientibus, lippis. ovillum[1] utilissimum tenesmo, dysinteriae nec non phthisicis. hoc et mulieribus[2] saluberrimum qui dicerent fuerunt.

131   XXXIV. De generibus caseorum diximus, cum de uberibus singulisque membris animalium diceremus. Sextius eosdem effectus equino quos bubulo tradit. hunc vocant hippacen. stomacho utiles qui non sunt salsi, id est recentes. veteres alvum sistunt corpusque minuunt, stomacho inutiliores[3]; et in totum
132 salsa minuunt corpus, alunt mollia. caseus recens cum melle suggillata emendat, mollis alvum sistit, sedat tormina pastillis in vino austero decoctis rursusque in patina tostis cum melle. saprum vocant qui cum sale et sorbis siccis e vino tritus potusque medetur coeliacis, genitalium carbunculis caprinus tritus inpositus. item acidus cum oxymelite maculis in balineo inlitus oleo interlinitur.

133   XXXV. E lacte fit et butyrum, barbararum gentium lautissimus cibus et qui divites a plebe discernat, plurimum e bubulo, et inde nomen, pinguissi-

---

[1] ovillum *Hard., Mayhoff, ex Dioscoride*: suillum *codd., Detlefsen.*
[2] mulieribus dTx, *Detlefsen*: mulieres VRf: mulieris *Mayhoff, qui etiam post* dysinteriae *dist.*
[3] inutiliores *Urlichs, Detlefsen, Mayhoff*: utiliores *codd.*

---

[a] With Mayhoff's reading and punctuation: "this and woman's milk are the most wholesome for consumptives."
[b] Book XI. § 240.
[c] See note on § 120.

any other food, but the goats must fast for two days and then browse on ivy the third day. Drinking milk is generally bad for headache, complaints of the liver, spleen and sinews, for fevers, for giddiness except as a purge, and for a heavy cold, cough, and ophthalmia. Sheep's milk is very beneficial for tenesmus, dysentery, and consumption; there have been some who said that this milk is also the most wholesome for women.[a]

XXXIV. The kinds of cheese I discussed when *Cheese.* speaking of udders and the separate parts of animals.[b] Sextius[c] gives to cow's-milk cheese the same properties as he gives to that from mare's milk, which is called *hippace*.[d] Beneficial to the stomach are those not salted, that is to say the fresh. Old cheeses bind the bowels and reduce flesh, being rather bad for the stomach; on the whole salty foods reduce flesh, soft foods make it. Fresh cheese with honey heals bruises, a soft cheese binds the bowels, and relieves gripes if lozenges of it are boiled in a dry wine and then roasted in a pan with honey. Coeliac affections are cured by the cheese that they call *saprum*,[e] taken in drink after being pounded in wine with salt and dried sorb apples; carbuncles of the genitals by an application of pounded goat's-milk cheese. Sour cheese also with oxymel is applied in the bath alternately with oil to remove spots.

XXXV. From milk is also made butter, among *Butter.* barbarian tribes accounted the choicest food, one that distinguishes the rich from the lower orders. Mostly cow's milk is used (hence the name [f]), but

---

[d] See note on XXV. § 83.
[e] That is, " rotten " (σαπρόν).
[f] The word means " cow cheese."

mum ex ovillo [1]—fit et ex caprino—sed hieme calefacto lacte, aestate expresso tantum crebro iactatu
in longis vasis, angusto foramine spiritum accipienti-
bus sub [2] ipso ore alias praeligato. additur paululum
134 aquae ut acescat. quod est maxime coactum in
summo fluitat, id exemptum addito sale oxygala
appellant. relicum decocunt in ollis. ibi quod
supernatat butyrum est oleosum natura. quo magis
virus resipit hoc praestantius iudicatur. pluribus
conpositionibus miscetur inveteratum. natura eius
adstringere, mollire, replere, purgare.

135   XXXVI. Oxygala fit et alio modo, acido lacte
addito in recens quod velis [3] inacescere, utilissimum
stomacho. effectus dicemus suis locis.

        XXXVII. Proxima in communibus adipi laus est,
sed maxime suillo, apud antiquos etiam religiosius.
certe novae nuptae intrantes etiamnum [4] sollemne
habent postes eo attingere. inveteratur duobus
136 modis, cum sale aut sincerus, tanto fit utilior.[5] axungiam Graeci etiam appellavere eam in voluminibus
suis. neque est occulta virium causa, quoniam id
animal herbarum radicibus vescitur—itaque etiam

---

[1] ovillo *coni. Mayhoff*: ovibus *codd.*

[2] sub *omittere velit Mayhoff.*

[3] velis *Detlefsen*: velint *Mayhoff*: inm VR: in dx:
ve - - - r: dum (acescit) *vulg. Mayhoff nonnulla verba, ut
quodve aliud cogat, excidisse putat.*

[4] etiamnum *codd.*: etiam nunc *Mayhoff.*

[5] tanto fit utilior *Mayhoff*: tanto utilior quanto sit vetustior *Detlefsen. Pro* utilior *multi codd.* vetustior (vectior R),
pro fit (dx) sit VR.

---

[a] It has been suggested that for *aqua* we should read *aceto*
(vinegar).

[b] If we omit all from *exemptum* to *supernatat*, the ancient
method of making butter is much like the modern, but then

the richest comes from sheep's—it is also made from goat's—but in winter the milk is warmed, while in summer the butter is extracted merely by shaking it rapidly in a tall vessel. This has a small hole to admit the air, made just under the mouth, which is otherwise completely stopped. There is added a little water *a* to make the milk turn sour. The part that curdles most, floating on the top, [is skimmed off, and with salt added is called oxygala; the rest they boil down in pots. What comes to the surface *b*] is butter, a fatty substance. The stronger the taste, the more highly is butter esteemed. When matured it is used as an ingredient for several mixtures. It is by nature astringent, emollient, flesh-forming, and cleansing.

XXXVI. Oxygala is made in yet another way, by adding sour milk to the fresh that it is wished to turn sour. It is very good for the stomach; of its properties I shall speak in the appropriate places. *Oxygala.*

XXXVII. Of remedies common to animals the next in repute is fat, especially pig's fat, which to the men of old was not a little sacred. At any rate brides even today touch ritually the door-posts with it on entering their homes. Lard is matured in two ways, with salt or by itself; it is so much the more beneficial when matured. The name axungia (axle-grease) is the one adopted by the Greeks also in their writings. Nor is the cause of its properties a mystery, for the pig feeds on the roots of plants, so that there are very many uses even for its dung. *Fats, especially of pigs.*

oxygala disappears, which is required because of Ch. XXXVI, and the interpolation needs to be explained. It is perhaps safer with J. Müller to regard *addito . . . relicum* as a parenthesis.

fimo innumeri usus—quamobrem non de alia loque-
mur quam e sue.[1] multo efficacior e femina est quae
non peperit, [multo vero praestantior in apris.] [2] est
igitur usus axungiae ad emollienda, excalfacienda,
137 discutienda purgandaque. medicorum aliqui admixto anseris adipe taurorumque sebo et oesypo ad
podagras uti iubent, si vero permanet dolor, cum cera,
myrto, resina, pice. sincera axungia medetur
ambustis vel nive, pernionibus autem cum hordei
cinere et galla pari modo. prodest et confricatis
membris, itinerumque lassitudines et fatigationes
levat. ad tussim veterem recens decoquitur quad-
rantis pondere in vini cyathis tribus addito melle.
138 vetus etiam phthisis pilulis sumpta sanat quae sine
sale inveterata est. omnino enim non nisi ad ea quae
purganda sint aut quae non sint exulcerata salsa reci-
pitur. quidam quadrantes axungiae et mulsi [3] in
vini cyathis ternis decocunt contra phthisis, quarto
quoque die picem liquidam in ovo sumi iubent, cir-
cumligatur et lateribus pectoribus scapulis eorum qui
phthisim sentiunt, tantaque est vis ut genibus etiam
adalligata redeat in os sapor eamque expuere

---

[1] quam e sue *Urlichs, Detlefsen* : sue *codd.* : *uncos ponit
Mayhoff.*
[2] *Uncos ego posui. In textu esse dicitur Mayhoff, qui etiam*
intellegitur *coni.* : est igitur *codd.*
[3] mulsi *vulg., Detlefsen, Mayhoff*; multis *codd.*

---

[a] The emendation of Urlichs seems to be the best solution
of the difficulty presented by the MS. reading.
[b] If we bracket, as being a scribe's or commentator's note,
from *multo* to *apris*, there is no need further to emend this
sentence.

Therefore I shall not speak of other grease than that of the pig.[a] By far the more beneficial is that from a sow that has not littered, [but much more excellent is that of boars.[b]] Axle-grease then is used for softening, warming, dispersing, and cleansing. Certain medical men recommend for gout a mixture of it with goose grease, bull suet and suint; if however the pain should persist, they add wax, myrtle berries, resin, and pitch. Unsalted axle-grease is good for burns or frost-bite; for chilblains add equal measures of barley-ash and gall nuts. It is also beneficial for chafed limbs, and relieves weariness and fatigue from a journey. Fresh axle-grease, three ounces in three cyathi of wine with honey added, is boiled down for chronic cough. Old grease taken in pills cures even consumption, but it must have matured without salt, for salt grease is not recommended at all except where cleansing is required and where there is no ulceration. Some boil down three ounces of axle-grease and of honey wine in three cyathi of wine to treat consumption, recommending that on every fourth day liquid pitch should be taken in egg. Poultices of it are applied to the sides, chest, and shoulders of consumptive patients, and so great is its power that even when fastened to the knees as an amulet the taste comes back [c] to the mouth and they seem to be spitting it out. Fat from a sow that has not littered is used with very great advantage by women as a cosmetic, but for itch any kind [d] is good, mixed with a third part

[c] In the context *redeat* is strange. May it mean : " comes to its natural place " ?

[d] With *quivis* understand *adeps* and a verb like *medetur*. So Littré : " toute espèce de graisse est bonne." Perhaps, however, it is " anybody (and not women only) can use."

97

139 videantur. e sue quae non peperit aptissime utuntur
ad cutem mulieres, contra scabiem vero quivis admixto iumentorum sebo pro parte tertia et pice, pariterque subfervefactis. sincera partus in abortum
vergentes nutriunt collyrii modo subdita. cicatrices
concolores facit cerussa admixta vel argenti spuma, at
cum sulpure unguium scabritias emendat. medetur et
capillis fluentibus et ulceribus in capite mulierum cum
gallae parte quarta et infumata pilis oculorum.
datur et phthisicis unciatim cum vini veteris hemina
decocta donec tres unciae e toto restent, aliqui et
140 mellis exiguum adiciunt. panis inlinitur cum calce,
item furunculis duritiaeque mammarum. rupta et
convulsa et spasmata et luxata sanat, clavos et rimas
callique vitia cum helleboro albo, parotidas admixta
farina salsamentariae testae, quo genere proficit et ad
strumas. pruritus et papulas in balineo perunctis
tollit, alioque etiamnum modo podagricis prodest
mixto oleo vetere, contrito una sarcophago lapide et
quinquefolio tuso in vino vel cum calce vel cum cinere.
facit et peculiare emplastrum LXXVX ponderi centum
spumae argenteae mixtis, utilissimum contra ulcerum inflammationes.[1] adipe verrino et inungui
putant utile, quaeque serpant inlinere cum resina.
141 antiqui axibus vehiculorum perunguendis maxime ad
faciliorem circumactum rotarum utebantur, unde
nomen, sic quoque utili medicina cum illa ferrugine
142 rotarum ad sedis vitia virilitatisque. [et per se
axungia][2] medici antiqui maxime probabant renibus

[1] *Hoc punctum post* verrino *ponit Mayhoff.*
[2] *Ego uncos posui ex Mayhoffii coniectura.*

---

a *spasmata* may be a gloss, for Pliny renders the Greek
σπάσματα by *convulsa.*

of beef suet and pitch, all being warmed together. Unsalted axle-grease used as a pessary nourishes the foetus when there is the threat of a miscarriage. Mixed with white lead or litharge lard gives to scars the colour of the surrounding skin, and with sulphur cleans scabrous nails. It cures too the falling-out of hair, and with a quarter of a gall nut sores on the head of women; as a fumigant it is good for eye-lashes. It is also given to consumptives, in doses of one ounce with a hemina of old wine boiled down until of the whole three ounces remain; some add also a little honey. With lime it is applied to superficial abscesses, also to boils and to indurations of the breasts. It cures ruptures, sprains, cramps,[a] and dislocations; with white hellebore corns, chaps, and callosities; and parotid swellings with pounded earthenware that has contained salted food, the same being also good for scrofulous sores. Rubbing in the bath with this fat removes irritation and pimples, and administered in yet another way it is good for gout: mixed with old oil, crushed sarcophagus[b] stone, and cinquefoil pounded in wine, or with lime, or with ash. A special plaster too is made of 75 denarii by weight of lard mixed with 100 of litharge, very useful for in-flamed ulcers. They also think it useful to treat such sores with boar's grease, and to apply it with resin to those that spread. The men of old used lard in particular for greasing the axles of their vehicles, that the wheels might revolve more easily, and in this way it received its name. So also with that rust of the wheels it made a useful medicament for affections of the anus and of the male genitals. The old physicians valued most the fat taken from

[b] See II. § 211 and XXXVI. § 161.

detractam exemptisque venis aqua caelesti fricabant
crebro decoquebantque fictili novo saepius, tum demum adservantes. convenit salsam magis mollire,
excalfacere, discutere, utilioremque esse vino lotam.
Masurius palmam lupino adipi dedisse antiquos tradidit. ideo novas nuptas illo perunguere postes
solitas ne quid mali medicamenti inferretur.

143   XXXVIII. Quae ratio adipis eadem in his quae
ruminant sebi est, aliis modis, non minoris potentiae.
perficitur omne exemptis venis aqua marina vel salsa
lotum, mox in pila tusum aspersa marina crebro.
postea coquitur donec odor omnis aboleatur, mox
adsiduo sole ad candorem reducitur. a renibus autem
144 omne laudatissimum est. si vero vetus revocetur ad
curam, liquefieri prius iubent, mox frigida aqua
lavari saepius, dein liquefacere adfuso vino quam
odoratissimo. eodemque modo iterum ac saepius
cocunt donec virus evanescat. multi privatim sic
taurorum leonumque ac pantherarum et camelorum
pinguia curari iubent. usus dicetur suis locis.

145   XXXIX. Communis et medullarum est. omnes
molliunt, explent, siccant, excalfaciunt. laudatissima e cervis, mox vitulina, dein hircina et caprina.
curantur ante autumnum recentes lotae siccataeque

ᵃ The last sentence is added as an afterthought; it differs
from a similar remark in § 135. Masurius was apparently a
jurist who lived in the reign of Tiberius and later.
ᵇ Or, "The most highly valued suet is always that from
the kidneys."

the kidneys : removing the veins they rubbed it briskly with rain water, boiled it down several times in new earthenware, and then finally stored it away. It is agreed that when salted it has increased power of softening, warming, and dispersing, and that it is more useful when washed with wine. Masurius tells us that the men of old gave the palm to wolf's fat ; that, he said, was why new brides were wont to smear with it the door-posts to keep out all evil drugs.[a]

XXXVIII. Corresponding to fat in other animals *Suet.* is suet in ruminants ; used in other ways it is of no less potency. All suet is prepared by taking out the veins, washing in sea-water or salt water, and then pounding in a mortar with frequent sprinklings of sea-water. Afterwards it is boiled until all smell disappears, and then by continual exposure to the sun it is bleached to a shining white. All suet from the kidneys is highly valued.[b] But if stale suet is being put to use, it is recommended first to melt it, then wash it several times in cold water, and then to melt it after pouring on it wine with the most fragrant bouquet. They boil it in the same way again and again, until all the rankness disappears. Many recommend that in this way should be prepared the fat in particular of bulls, lions, panthers, and camels. Their use will be given in the appropriate places.

XXXIX. The various marrows too are all in use. *Marrow.* All marrow is emollient, filling, drying, and warming. The most highly valued is that of deer, next of calves, and then of goats, male and female. Marrow is prepared before autumn ; it should be fresh, washed, dried in the shade, then passed melted through a

in umbra, per cribrum dein liquatae per lintea expri
muntur ac reponuntur in fictili locis frigidis.

146 XL. Inter omnia autem communia animalium vel
praestantissimum effectu fel est. vis eius excalfacere, mordere, scindere, extrahere, discutere.
minorum animalium subtilius intellegitur et ideo ad
oculorum medicamenta utilius existimatur. taurino
praecipua potentia etiam in aere pelvibusque aureo
colore obducendis. omne autem curatur recens
praeligato ore lino crasso, demissum in ferventem
aquam semihora, mox siccatum sine sole atque in
melle conditum. damnatur equinum tantum inter
venena. ideo flamini sacrorum equum tangere non
licet, cum Romae publicis sacris equus etiam im-
moletur.

147 XLI. Quin et sanguis eorum[1] septicam vim habet,
item equarum, praeterquam virginum; erodit, emar-
ginat ulcera. taurinus quidem recens inter venena
est excepta Aegira. ibi enim sacerdos Terrae vati-
cinatura sanguinem tauri bibit prius quam in specus
descendat. tantum potest sympathia illa de qua
loquimur, ut aliquando religione aut loco fiat.

148 Drusus tribunus plebei traditur caprinum bibisse,
cum pallore et invidia veneni sibi dati insimulare Q.
Caepionem inimicum vellet. hircorum sanguini tanta
vis est ut ferramentorum subtilitas non aliter acrius

---

[1] eorum *codd.*: equorum *Warmington.*

---

[a] A town in Achaia.
[b] See XXIV. §§ 1–3, XXIX. § 61, and Additional Note C,
p. 564. See the same note for the view that bull's blood is
poison.
[c] Tribune of the people in 91 B.C., and murdered the same
year. He was a supporter of the Italians in their claim to
Roman citizenship.

sieve, strained through a linen cloth, and then stored away in an earthenware vessel in a cool place.

XL. But of all the parts common to animals gall *Gall.* is by far the most efficacious. Its nature is warming, pungent, dissolvent, extractive, and dispersive. That of the smaller animals is understood to be more delicate, and so is thought to be more useful for eye medicaments. Bull's gall is particularly potent, staining even bronze and basins with a golden colour. All gall is prepared when fresh by tying with stout thread the mouth of the gall bladder, steeping it for half an hour in boiling water, then drying it out of the sun, and storing away in honey. That of horses alone is condemned as a poison. Therefore the sacrificial flamen is not allowed to touch a horse, although at the public sacrifices at Rome a horse is even offered as a victim.

XLI. Moreover the blood of horses has a corrosive *Blood.* power; the blood of mares also, except that of virgin animals. It cleans out ulcers and eats away their lips. Fresh bull's blood indeed is reckoned one of the poisons, except at Aegira.[a] For there the priestess of Earth, when about to prophesy, drinks bull's blood before she goes down into the caves. So strong is that famous sympathy [b] I speak of that it sometimes becomes active under the influence of religious awe or of a place. Drusus,[c] tribune of the people, is reported to have drunk goat's blood because he wished, by his pallor, to accuse his enemy Q. Caepio of having poisoned him, and so to arouse hatred against him.[d] So great is the power of he-goats' blood that iron tools cannot in any other way be hardened

---

[d] Or, "to arouse hatred against his enemy Q. Caepio, his pallor suggesting that he had been poisoned by him."

PLINY: NATURAL HISTORY

induretur, scabritia tollatur vehementius quam lima.
non igitur et sanguis animalium inter communia dici
potest et ideo suis quisque dicetur effectibus.

149   XLII. Digeremus enim in mala singula usus plurimumque [1] contra serpentes. exitio his esse cervos
nemo ignorat ut, si quae sunt,[2] extractas cavernis
mandentes. nec vero ipsi spirantesque tantum adversantur, sed membratim quoque. fugari eas nidore
cornus eorum, si uratur, dictum est, at e summo gut-
ture ustis ossibus congregari dicuntur. pelles eiusdem animalis substratae securos praestant ab eo metu
150   somnos, coagulum ex aceto potum ab ictu, et si
omnino tractatum sit, eo die non ferit serpens. testes
quoque eius inveterati vel genitale vetus [3] maris
salutariter dantur in vino, item venter quem centi-
pellionem vocant. fugiunt et omnino dentem cervi
habentes aut medulla perunctos sebove cervi aut
vituli. summis autem remediis praefertur hinnulei
coagulum matris utero execti, ut indicavimus.
151   sanguine cervino, si una urantur dracontion et cunilago et anchusa lentisci ligno, contrahi serpentes
tradunt, dissipari deinde, si sanguine detracto adiciatur pyrethrum. invenio apud auctores Graecos
animal cervo minus et pilo demum simile, quod

---

[1] plurimumque *codd.*: primumque *Pintianus, Sillig, Mayhoff.*
[2] ut, si quae sunt *codd.*: utique spiritu *Pintianus*: vesti-
gantes et *coni. Mayhoff*: ut pi credimus *Warmington.*
[3] vetus *I. Müller, Mayhoff*: eius *codd.*: *del. Detlefsen.*

---

[a] See VIII. § 118.
[b] The *centipellio* is the second stomach of ruminating animals.
[c] See VIII. § 118.

to a finer edge, and roughness is smoothed more thoroughly by it than by a file. Accordingly blood cannot be included among the remedies common to animals, and so each kind of blood will be discussed according to its effects.

XLII. For I shall arrange remedies according to each malady, serpents' bites requiring very full treatment. Nobody is unaware that deer are their deadly enemies, in that they drag any they may find from their holes and eat them. Not only, however, when whole and alive are they the enemy of serpents; the parts of their body are so also. The fumes from their horns when burnt, as I have said,[a] keep serpents away; but if the topmost bones of a stag's neck have been burnt, serpents are said to assemble. The skins of the same animal make a bed on which one may sleep without fear of snakes, and the rennet taken in vinegar prevents being bitten; if it is merely handled, in fact, on that day no serpent strikes. A stag's testicles dried, or the dried male organ, are in wine a salutary drink; so is that stomach which is called *centipellio.*[b] Serpents keep away from those who have about them merely a stag's tooth, or have been rubbed with the marrow or suet of stag or fawn. As I have already pointed out,[c] to sovereign remedies is preferred the rennet of a young stag cut from his mother's uterus. Stag's blood, if with it are burnt on a lentisk-wood fire dracontion, cunilago and anchusa, is said to collect serpents together; then they scatter, it is said, if in place of blood pyrethrum is added. In my Greek authorities I find mentioned an animal that they call *ophion,*[d] smaller than a stag and like it only

Remedies for snake bite.

[a] See XXX. § 146.

ophion vocaretur, Sardiniam tantum ferre solitam. hoc interisse arbitror et ideo medicinas ex eo omitto.

152 Apri quoque cerebrum contra eas laudatur cum sanguine, iocur etiam inveteratum cum ruta potum ex vino, item adips cum melle resinaque, simili modo verrinum iocur et fellis dumtaxat fibra X IIII pondere vel cerebrum in vino potum. caprarum cornu vel pilis accensis fugari serpentes dicunt, cineremque ex cornu potum vel inlitum contra ictus valere, item lactis haustus cum uva taminia vel urinae cum aceto scillite, caseum caprinum cum origano inpositum vel sebum cum cera. milia praeterea remediorum ex eo

153 animali demonstrantur, sicut apparebit, quod equidem miror, cum febri negetur carere. amplior potentia feris eiusdem generis, quod numerosissimum esse diximus, alia vero et hircis. Democritus etiamnum effectus auget eius qui singularis natus sit. fimo quoque caprarum in aceto decocto inlini ictus serpentium placet et recentis cinere in vino, atque in totum difficilius sese recolligentes a serpentium ictu in

154 caprilibus optime convalescunt. qui efficacius volunt mederi occisae caprae alvum dissectam cum fimo intus reperto inligant statim. alii carnem recentem haedorum cum [1] pilo suffiunt eodemque nidore fugant serpentes. utuntur et pelle eorum recente ad [2]

---

[1] cum *add. C. F. W. Müller.*
[2] *Ante* ad *comma transponit Mayhoff.*

---

[a] This seems like a vague and inaccurate reference to the goat as the cause of Malta fever.
[b] See VIII. 214.

in its hair, which is found nowhere save in Sardinia. I believe that it is extinct today, and therefore I give no remedies from it. The brain and blood of a wild boar is another approved protection against serpents, as is its liver preserved and taken in wine with rue, likewise the fat with honey and resin, and given in the same way boar's liver and the fibre only of the gall-bladder, the dose being four denarii by weight, or the brain taken in wine. The horn or hair of she-goats, when burnt, is said to keep serpents away, and the ash from the horn, whether taken in drink or applied, to be efficacious for their bites; as are also draughts of their milk with taminian grapes, or of their urine with squill vinegar; so too an application of goat cheese with marjoram, or of goat suet with wax. Thousands of remedies besides from the goat are given in prescriptions, as will be pointed out; this is surprising to me, because it is said never to be free from fever.[a] The potency of the wild-goat— goats are a very numerous species, as I have said [b]—is greater, but a he-goat too has a potency of its own. Democritus also holds that if a goat is the only one at a birth he supplies more efficacious remedies. An application also of she-goat's dung boiled down in vinegar is approved treatment for snake bite, and so is the ash of fresh dung boiled down in wine; speaking generally, slow convalescents from snake bite recover best in a goat's stable. Those who want more efficacious treatment apply immediately as a plaster a slaughtered she-goat's belly cut open, including any dung found inside. Others fumigate with fresh kid-meat, not taking away the hair, and with the same fumes drive snakes away. They also use a fresh kid-skin for the wound, or the flesh and dung

plagas, et carne et fimo equi in agro pasti, coagulo
leporis ex aceto, contraque scorpionem et murem
araneum.   aiunt non feriri leporis coagulo peruncti.
155 a scorpione percussis fimum caprae efficacius cum
aceto decoctum auxiliatur, lardum iusque decocti
potum et his qui buprestim hauserint.   quin etiam si
quis asino in aurem percussum a scorpione se dicat,
transire malum protinus tradunt, venenataque omnia
accenso pulmone eius fugere.   et fimo vituli suffiri
percussos a scorpione prodest.
156   XLIII. Canis rabiosi morsu facta volnera circumcidunt ad vivas usque partes quidam carnemque
vituli admovent—et ius ex eodem carnis decoctae
dant potui [1]—aut axungiam cum calce tusam, hirci
iocur, quo inposito ne temptari quidem aquae metu
adfirmant.   laudant et caprae fimum ex vino inlitum, melis et cuculi et hirundinis decoctum et
potum.   ad reliquos bestiarum morsus caprinum
caseum siccum cum origano inponunt et bibi iubent,
ad hominis morsus carnem bubulam coctam, efficacius
vituli, si non ante quintum diem solvant.
157   XLIV. Veneficiis rostrum lupi resistere inveteratum aiunt ob idque villarum portis praefigunt.   hoc
idem praestare et pellis e cervice solida manica existimatur, quippe tanta vis est animalis praeter ea quae
retulimus ut vestigia eius calcata equis adferant
torporem.

[1] *Parenthesim ego indicavi.*

[a] It eases the construction to take from *et ius* to *potui* as a
parenthesis, a common feature of Pliny's style.

of a horse fed by pasture and the rennet of a hare in vinegar; the same for scorpions and the shrew-mouse. It is said that rubbing with hare's rennet protects from being stung or bitten. Those stung by a scorpion are helped by she-goat's dung, more efficaciously if it is boiled down in vinegar; the fat and broth of the decoction, if drunk, helps those too who have swallowed a buprestis. Moreover, if anyone says in the ear of an ass that he has been stung by a scorpion, the mischief, it is said, at once passes over into the animal, all venomous creatures run away from an ass's burning lung, and those stung by a scorpion are benefited by fumigation with the dung of a calf.

XLIII. Wounds made by the bite of a mad dog *Remedies* some cut round into the quick and apply veal, *for bites of* giving to drink veal broth,[a] or else axle-grease *mad dogs.* pounded with lime, or he-goat's liver, an application of which is said to keep off entirely the dread of water. Approved treatment is also she-goat's dung applied in wine, and to drink a decoction of the dung of badger, cuckoo and swallow. For the other beast-bites dried goat's cheese with marjoram is applied and recommended to be taken in drink; to human bites is applied boiled beef, boiled veal being more efficacious, if it is not taken off before the fifth day.

XLIV. Sorceries are said to be counteracted by a *Sorceries.* wolf's preserved muzzle, and for this reason they hang one up on the gates of country houses. The same effect is supposed to be given by the whole fur from a wolf's neck, the legs included, for so great is the power of the animal that, besides what I have already stated, his footprints when trodden on by horses make them torpid.

158   XLV. Iis qui argentum vivum biberint lardum remedio est. asinino lacte poto venena restinguntur, peculiariter si hyoscyamum potum sit aut viscum aut cicuta aut lepus marinus aut opocarpatum aut pharicum[1] aut dorycnium et si coagulum alicui nocuerit, nam id quoque venenum est in prima lactis coagulatione. multos et alios usus eius dicemus, sed meminisse oportebit recenti utendum aut non multo postea tepefacto, nullum enim celerius evanescit. ossa quoque asini confracta et decocta contra leporis marini venenum dantur. omnia eadem onagris efficaciora.

159   de equiferis non scripserunt Graeci, quoniam terrae illae non gignebant, verum tamen fortiora omnia eadem quam in equis intellegi debent. lacte equino venena leporis marini et toxica expugnantur. nec uros aut bisontes habuerunt Graeci in experimentis, quamquam bove fero refertis Indiae silvis, portione tamen eadem efficaciora omnia ex his credi par est.

160   sic quoque lacte bubulo cuncta venena expugnari tradunt, maxime supra dicta et si ephemerum inpactum sit aut si cantharides datae, omnia vomitione egeri, sic et caprino iure cantharidas. contra ea vero quae exulceratione enecant sebum vitulinum vel bubulum auxiliatur. nam contra sanguisugas potas butyrum remedio est cum aceto ferro calefacto, quod et per se prodest contra venena. nam si oleum non

---

[1] pharicum *Hermolaus Barbarus*; cf. *Scribonius Largus CXCV*: agaricum *Detlefsen*: cerussa *Mayhoff*: carice V: tarice R: caryce d.

---

[a] Unknown.

[b] See Scribonius Largus CXCV.

[c] Ephemerum was used in a mouth-wash, and so very liable to be swallowed by accident. The word *inpactum* is curious, and probably corrupt, but the sense is clear.

XLV. Those who have swallowed quicksilver find *Remedies*
a remedy in lard. By drinking ass's milk poisons are *for poisons.*
neutralized, especially if henbane has been swallowed,
or mistletoe, hemlock, sea-hare, opocarpathum,[a]
pharicum,[b] dorycnium, or if milk has done harm by
curdling, for there is poison in the first coagulation
of it. I shall give many other uses of ass's milk, but
it should be remembered to use it when fresh, or
nearly fresh and warmed, for no milk loses its power
more rapidly. The bones too of the ass, crushed and
boiled, are given for the poison of the sea-hare. All
these remedies are more efficacious from the wild
ass. About wild horses the Greeks have not written,
because Greek lands did not breed them, but it must
be inferred that all remedies from them are more
potent than from the tame animal. By mare's milk
are neutralized the poison of the sea-hare and arrow
poisons. The Greeks had not the urus or the bison
to try out, although the Indian jungles swarm with
wild cattle. All the same remedies from them,
however, it is reasonable to believe, are proportion-
ally more efficacious. So cow's milk too is said to
neutralize all poisons, especially those mentioned
before, and if ephemerum has gone down the throat[c]
or Spanish fly[d] administered, and to expel by vomiting
all the noxious substances; goat broth also to act in the
same way on Spanish fly. Those poisons however
that cause fatal ulceration are relieved by veal-suet
or beef-suet. But for leeches swallowed in drink
butter, with vinegar warmed by hot iron, is a remedy,
butter even by itself being beneficial against poisoning, for if one has no oil butter is a good substitute.

[d] For an interesting account of Spanish fly, really a kind
of beetle, see W. T. Fernie, *Animal Simples*, pp. 176–180.

PLINY: NATURAL HISTORY

161 sit, vicem eius repraesentat. multipedae morsus cum
melle sanat. omasi quoque iure poto venena supra
dicta expugnari putant, privatim vero aconita et cicu-
tas, itemque vitulino sebo. caprinus caseus recens
his qui viscum biberint, lac contra cantharidas
remedio est et contra ephemeri potum cum taminia
uva. sanguis caprinus decoctus cum medulla contra
toxica venena sumitur, haedinus contra reliqua,
162 coagulum haedi contra viscum et chamaeleonem
album sanguinemque taurinum, contra quem et
leporis coagulum est ex aceto, contra pastinacam vero
et omnium marinorum ictus vel morsus coagulum
leporis vel haedi vel agni drachmae pondere ex vino.
leporis coagulum et contra venena additur antidotis.
papilio quoque lucernarum luminibus advolans inter
mala medicamenta numeratur. huic contrarium est
iocur caprinum, sicut fel veneficis ex mustella rustica
factis. hinc deinde praevertemur ad genera morborum.
163   XLVI. Capilli defluvia ursinus adips admixto
ladano et adianto continet alopeciasque emendat et
raritatem superciliorum cum fungis lucernarum ac
fuligine quae est in rostris earum, porriginem cum
vino. prodest ad hanc et cornus cervini cinis e vino,
utque non taedia animalium capillis increscant, item
fel caprinum cum creta Cimolia et aceto sic uti paulum
capiti inarescant, item fel scrofinum, urina tauri. si
vero vetus sit, furfures etiam adiecto sulpure emen-
164 dat. cinere genitalis asini spissari capillum putant et
a canitie vindicari, si rasis inlinatur plumboque tritus

---

a See Book XXVI. § 47, and for the plants mentioned in
this section of Pliny the *Index of Plants* in vol. VII.

112

It and honey together cure the bites of millipedes. Tripe broth and also veal suet are thought to neutralize the poisons mentioned above, especially however aconite and hemlock. Fresh goat-cheese is a remedy for those who have taken mistletoe in drink, as is goat's milk for Spanish fly, and with Taminian grapes for swallowing ephemerum. Goat's blood boiled with the marrow is taken for arrow poison, kid's for the other poisons, kid's rennet for mistletoe, white chamaeleon and bull's blood, for which another remedy is hare's rennet in vinegar; for the sting-ray however, and for the stings or bites of all sea creatures, hare's rennet or that of a kid or lamb, the dose being a drachma by weight in wine. Hare's rennet is also an ingredient of antidotes against poisons. The moth too that flutters to the lamp-light is counted among noxious drugs; an antidote is goat's liver, as is its gall for sorcerer's potions made from the field weasel. At this point I shall return to the various kinds of diseases.

XLVI. Bear's grease mixed with ladanum [a] and adiantum prevents the hair from falling out, and cures mange, and scanty eyebrows, if mixed with the lamp-black from lamp wicks and the soot that collects in their nozzles. Mixed with wine it cures dandruff. Good too for the last is the ash of deer's horn in wine, good also to prevent vermin from breeding in the hair, likewise goat's gall with Cimolian chalk and vinegar, the mixture being allowed to dry a little on the head; sow's gall too, and the urine of a bull. If indeed it should be old, with the addition of sulphur it also cures dandruff. It is thought that a thicker growth of hair and prevention of greyness are given by an ass's genital organ reduced to ash;

For complaints of the scalp, etc.

cum oleo, densari et asinini pulli illitum [1] urina; admiscent nardum fastidii gratia. alopecias felle taurino cum Aegyptio alumine tepefactis inlinunt. ulcera capitis manantia urina tauri efficaciter sanat, item hominis vetus, si cyclaminum adiciatur et sulpur, efficacius tamen vitulinum fel, quo cum aceto cale-

165 facto et lendes tolluntur. sebum vitulinum capitis ulceribus cum sale tritum utilissimum. laudatur et vulpium adips, sed praecipue felium fimum cum sinapis pari modo inlitum, caprini cornus farina vel cinis, magisque hircini, addito nitro et tamaricis semine et butyro oleoque, prius capite raso; mire continent ita fluentem capillum, sicuti carnis cinere

166 ex oleo inlita supercilia nigrescunt. lacte caprino lendes tolli tradunt, fimo cum melle [2] alopecias expleri, item ungularum cinere cum pice. fluentem capillum continet leporinus cinis cum oleo myrteo. capitis dolorem sedat pota aqua quae relicta est e bovis aut asini potu et, si credimus, vulpis masculae genitale circumligatum, cornus cervini cinis inlitus ex aceto aut rosaceo aut ex irino.

167 XLVII. Oculorum epiphoras bubulo sebo cum oleo cocto inlinunt. cervini cornus cinere scabritias ex eodem [3] inunguunt, mucrones autem ipsos efficaciores putant. lupi excrementis circumlini suffusiones

[1] illitum *Mayhoff*: cum *codd.*: del. *Detlefsen*.
[2] melle] *Coni.* oleo e *Dioscoride Mayhoff*.
[3] ex eodem *Mayhoff*: eorundem *Hard., Detlefsen*: eodem *multi codd., vulg.*

---

[a] The reading of Mayhoff is plausible and has been adopted, but the reading of the MSS., although there is a violent omission of several words understood from the preceding sentence, makes sense : " [the same part] of an ass's foal with his urine, also thickens the hair."

this should be pounded with oil in a leaden mortar, and applied after shaving the head. They also think that thicker hair is encouraged by applying [a] the urine of a young ass. Nard is mixed with it because of its nastiness. For mange is applied warmed bull's gall with Egyptian alum. Running sores on the head are healed efficaciously by bull's urine, also by stale human urine with the addition of cyclamen and sulphur, more efficaciously however by the gall of a calf, which warmed with vinegar also removes nits. For sores on the head calf's suet pounded with salt is very useful. Fox fat is also recommended, but especially cats' dung applied with an equal quantity of mustard; goat's horn, ground to powder or reduced to ash, a he-goat's being better, with the addition of soda, tamarisk seed, butter, and oil, the head being first shaved; this treatment is wonderful for preventing loss of hair, just as goat's meat, reduced to ash and applied with oil, darkens the eyebrows. Goat's milk is said to remove nits, the dung with honey to replace hair lost by mange, likewise the hoofs reduced to ash and added to pitch. Hare's flesh reduced to ash, with oil of myrtle, prevents hair from falling out. Headache is relieved by drinking the water left after an ox or ass has drunk, and also, if we care to believe it, by the genital organ of a male fox fastened round the head, and by a deer's horn reduced to ashes and applied in vinegar, rose oil, or iris oil.

XLVII. To eye fluxes is applied beef suet boiled *For complaints of the eyes.* with oil; scabrous eyes are smeared with the same and deer's horn reduced to ash, but the tips by themselves are thought to be more efficacious. Cataract is benefited by applying round the eyes the excrement of

prodest, cinere eorum cum Attico melle inungui
obscuritates, item felle ursino, epinyctidas adipe
apruno cum rosaceo.   ungulae asininae cinis inunctus
e suo lacte oculorum cicatrices et albugines tollit.
168 medulla bubula e dextro crure priore trita cum
fuligine pilis et palpebrarum vitiis angulorumque
occurrit, calliblephari modo fuligo in hoc usu tem-
peratur, optime ellychnio papyracio oleoque sesa-
mino, fuligine in novum vas pinnis detersa, efficacissime tamen evolsos ibi pilos coercet.   felle tauri
cum ovi albo collyria fiunt, aquaque dissoluta inun-
169 gunt per quadriduum.   sebum vituli cum anseris
adipe et ocimi suco genarum vitiis aptissimum est.
eiusdem medullae cum pari pondere cerae et olei vel
rosacei addito ovo duritiae genarum inlinuntur.
caseo molli caprino inposito ex aqua calida epiphorae
sedantur, si tumor sit, ex melle ; utrumque sero calido
fovendum.   sicca lippitudo lumbulis suum exustis
170 atque contritis et inpositis tollitur.   capras negant
lippire, quoniam quasdam herbas edint, item dor-
cadas ; ob id fimum earum cera circumdatum nova
luna devorari iubent.   et quoniam noctu aeque [1]
cernant, sanguine hircino lusciosos sanari putant
nyctalopas a Graecis dictos, caprae vero iocinere in
vino austero decocto.   quidam inassati iocineris sanie
inungunt aut felle caprae, carnesque vesci eas et,

---

[1] aeque *Detlefsen* : quoque aeque *Mayhoff* : aeque quoque
plerique codd. : quoque r.

---

[a] For these see List of Diseases.
[b] A possible reason for removing the eyelashes and for preventing their regrowth is revealed in § 171.
[c] A cosmetic for the eyebrows.

a wolf, dimness by smearing them with its ash and
Attic honey, also with bear's gall, and epinyctis *a* with
wild boar's fat and rose oil. The ash of an ass's hoof
smeared on the eyes with the same ass's milk removes
scars and albugo. The marrow from the right front
leg of an ox, pounded and added to soot, combats *b*
eyelashes, affections of the eyelids and of the corners
of the eyes (the soot for this purpose is prepared as
for a callamblepharum,*c* best from a papyrus wick and
sesame oil, the soot being wiped off with feathers into
a new vessel), very efficiently however it prevents the
hairs once pulled out there from growing again.
From the gall of a bull with white of egg are made
eye-salves, and dissolved in water they are applied
for four successive days. Calf suet with goose-grease
and juice of ocimum is very good for affections of the
eye-lids. Calf marrow, with equal weights of wax
and of oil or rose-oil, with an egg added, is applied to
indurations of the eye-lids. Eye fluxes are relieved by
an application in warm water of soft cheese made from
goat's milk, or, if there is swelling, in honey ; in both
cases there should be fomentation with warm whey.
Dry ophthalmia is cured by taking the small loins of
pork, burning, pounding, and then placing them on
the eyes. She-goats are said never to suffer from
ophthalmia, because of certain herbs they eat, and
likewise gazelles ; for this reason it is recommended
that at the new moon their dung should be swallowed,
coated with wax. Since they see equally well at
night, it is thought that those who have no night
vision (the Greeks call them *nyctalopes*) are cured
by the blood of a he-goat, but also by the liver of a
she-goat boiled down in a dry wine. Some smear
the eyes with the gravy from a she-goat's roasted

dum coquantur, oculos vaporari his praecipiunt. id
quoque referre arbitrantur ut rutili coloris fuerint.
171 volunt et suffiri oculos iocinere in ollis decocto, quidam inassato. fel quidem caprinum pluribus modis
adsumunt, cum melle contra caligines, cum veratri
candidi tertia parte contra glaucomata, cum vino
contra cicatrices et albugines et caligines et pterygia
et argema, ad palpebras vero evolso prius pilo cum
172 suco oleris ita ut unctio inarescat, contra ruptas tuniculas cum lacte mulieris. ad omnia inveteratum fel
efficacius putant, nec abdicant fimum ex melle inlitum epiphoris, contraque dolores medullam, item
pulmonem leporis, et ad caligines fel cum passo aut
melle. lupino quoque adipe vel medulla suum
fricari oculos contra lippitudines praecipiunt. nam
vulpinam linguam habentes in armilla lippituros
negant.
173   XLVIII. Aurium dolori et vitiis medentur urina
apri in vitreo servata, fel apri vel suis vel bubulum cum
oleo citreo [1] et rosaceo aequis portionibus, praecipue
vero taurinum cum porri suco tepidum vel cum melle,
si suppuret,[2] contraque odorem gravem per se tepefactum in malicorio. rupta in ea parte cum lacte
174 mulierum efficaciter sanat. quidam etiam in gravi-
tate aures sic perluendas putant, alii cum senecta
serpentium et aceto—includunt lana—collutas ante

---

[1] citreo *codd.*, *Detlefsen*: cedrino *Mayhoff e Marcello*:
citrino f: cicino *Caesarius*.
[2] suppuret dxr, *Detlefsen*: supperet VR: suppurent
Mayhoff, vulg.

---

[a] For these see List of Diseases.
[b] With Mayhoff's reading: "cedar."

liver, or with its gall; they prescribe its meat as a food, and fumigation of the eyes with the steam that arises from the cooking; they also consider it important for the animal to have been of a red colour. They also wish the eyes to be fumigated with the steam of the liver boiled in a clay pot; some say that it should be roasted. The gall indeed of goats is employed in many ways; with honey for dimness; with a third part of white hellebore for opaqueness of the lens; with wine for scars, albugo,[a] dimness, pterygia,[a] and argema [a]; but with cabbage juice for affections of the eyelids, the hairs being first pulled out, and the application being left to dry; with human milk for rupture of the eye-coats. For all purposes preserved gall is thought to be more efficacious. Goat's dung with honey is a not unvalued ointment for eye fluxes, or the marrow for eye pains, or a hare's lung, and for dimness its gall with raisin wine or honey. Wolf's fat also or pig's marrow is prescribed as an ointment for ophthalmia. But it is said that those who carry a fox's tongue in a bracelet will never suffer from ophthalmia.

XLVIII. Pain in the ears and ear affections are cured by the urine of a wild boar kept in a glass vessel, by the gall of a wild boar, pig, or ox, with citrus [b] oil and rose oil in equal proportions, but best of all by warm bull's gall with leek juice, or with honey should there be suppuration, and for foul odour the gall by itself warmed in a pomegranate rind. Ruptures in this region are thoroughly healed by the gall with woman's milk. Some hold that for hardness of hearing also the ears should be rinsed out with this wash, others add serpents' slough and vinegar (they insert the mixture on wool), the ears being

For complaints of the ears.

calida aqua aut, si maior sit gravitas, taurinum [1] fel
cum murra et ruta in malicorio excalfactum infundunt,
lardum quoque pingue ; item fimum asini recens cum
rosaceo instillatur, omnia tepefacta. utilior equi
spuma vel equini fimi recentis cinis cum rosaceo,
butyrum recens, sebum bubulum cum adipe anserino,
urina caprae vel tauri aut fullonia vetus, calfacta
175 vapore per lagoenae collum subeunte—admiscent
aceti tertiam partem et aliquid murrae—vituli qui
nondum herbam gustaverit fimum mixto felle eiusdem
et cute [2] quam relincunt angues, excalefactis prius
auribus ; lana autem medicamina ea includuntur.
prodest et sebum vituli cum anseris adipe et ocimi
suco, eiusdem medulla admixto cumino trito infusa,
virus verrinum e scrofa exceptum priusquam terram
176 attingat contra dolores, auribus fractis glutinum e
naturis vitulorum factum et in aqua liquatum ; aliis
vitiis adips vulpium, item fel caprinum cum rosaceo
tepido aut porri suco aut, si rupta sint aliqua ibi, e
lacte mulieris ; si gravitas audiendi, fel bubulum cum
urina caprae vel hirci, vel si pus sit. in quocumque
autem usu putant esse efficaciora haec in cornu
177 caprino per dies viginti infumata. laudant et coagulum leporis tertia denarii parte ex dimidiaque saco-
penii in Ammineo vino. parotidas ursinus adips conprimit pari pondere cerae et taurini sebi—addunt
quidam hypocisthidem—[3] et per se butyrum inlitum,

---

[1] taurinum *Urlichs, Detlefsen* : verrinum *Mayhoff e Marcello* : aurium *codd., vulg.*
[2] cute d x *Mayhoff* : cutem *multi codd., Detlefsen.*
[3] *Sic dist. Mayhoff.*

---

[a] With Mayhoff's reading : " hog's."
[b] Perhaps " taken out of " (Warmington).

first rinsed with warm water, or, if the hardness of hearing amounts to deafness, they pour in bull's gall [a] with myrrh and rue warmed in pomegranate rind, also fat bacon; or fresh ass's dung with rose oil is inserted in drops, all being warmed. More useful is the foam of a horse, or fresh horse-dung reduced to ash and mixed with rose-oil, fresh butter, beef suet with goose grease, she-goat's or bull's urine, or that used by fullers, stale, and warmed until the steam rises up the neck of the jar (a third part of vinegar is added and little myrrh), the dung, mixed with the gall, of a calf that has not tasted grass added to the slough of snakes, the ears being first warmed; these medicaments are inserted into the ears on wool. Beneficial is also veal suet, with goose grease and juice of ocimum; the marrow of a calf mixed with pounded cummin and poured into the ear; and for ear pains the seminal fluid of a hog, caught [b] as it drips from a sow before it can touch the ground; for fractures of the ears the glue made from the genitals of calves and melted in water; for other affections the fat of foxes, goat's gall with warm rose-oil or with leek juice, or, if any part of the ear has been ruptured, with woman's milk; if there is hardness of hearing, ox gall with the urine of a goat, male or female, or if there is pus. But whatever the use may be, it is thought that these remedies are more efficacious if they are smoke-dried for twenty days in a goat's horn. Another approved treatment is a third of a denarius of hare's rennet and half a denarius of sacopenium in Amminean wine. Parotid swellings are reduced by bear's grease with an equal weight of wax and bull suet (some add hypocisthis), and an application of butter by itself after previous fomentation with a decoction

si prius foveantur feni Graeci decocti suco, efficacius cum strychno. prosunt et vulpium testes et taurinus sanguis aridus tritus, urina caprae calefacta instillata auribus, fimum eiusdem cum axungia inlitum.

178 XLIX. Dentes mobiles confirmat cervini cornus cinis doloresque eorum mitigat, sive infricentur sive colluantur. quidam efficaciorem ad omnes eosdem usus crudi cornus farinam arbitrantur. dentifricia utroque modo fiunt. magnum remedium est et in luporum capitis cinere. certum est in excrementis eorum plerumque inveniri ossa; haec adalligata eundem effectum habent, item leporina coagula per aurem infusa contra dolores. et capitis eorum cinis dentifricium est adiectoque nardo mulcet graveo-
179 lentiam oris. aliqui murinorum capitum cinerem miscuisse malunt. reperitur in latere leporis os acui simile, hoc scarifare dentes in dolore suadent. talus bubulus accensus eos qui labant cum dolore admotus confirmat. eiusdem cinis cum murra dentifricium est. ossa quoque ex ungulis suum combusta eundem usum praebent, item ossa ex acetabulis pernarum
180 circa quae coxendices vertuntur. isdem sanari demissis in fauces iumentorum verminationes notum est, sed et combustis dentes confirmari, asinino quoque lacte percussu vexatos aut dentium eiusdem cinere, item lichene equi cum oleo infuso per aurem. est autem hoc non hippomanes, quod alioqui noxium
181 omitto, sed in equorum genibus ac super ungulas.

of fenugreek, more efficaciously with the addition of strychnos. Beneficial also are the testicles of foxes and bull's blood dried and pounded, she-goat's urine warmed and poured by drops into the ear, and an application of she-goat's dung with axle-grease.

XLIX. Loose teeth are made tight by the ash of *For the teeth.* deer's horn, which relieves their pain, whether used as dentifrice or in a mouth wash. Some consider more efficacious for all the same purposes the unburnt horn ground to powder. Dentifrices are made in either way. A grand remedy too is a wolf's head reduced to ash. It is certain that bones are generally found in the excrements of wolves. Used as an amulet these have the same effect, and hare's rennet relieves toothache if poured through the ear. Hare's head reduced to ash makes a dentifrice, and with nard added corrects a bad odour from the mouth. Some prefer to add as well ash from the burnt heads of mice. There is found in the flank of a hare a bone like a needle, with which they recommend aching teeth to be scraped. The ignited pastern bone of an ox, applied to teeth that are loose and aching, tightens them; the ash of the same with myrrh makes a dentifrice. The bones also of pigs' feet, when burnt, have the same effect, as have the bones from the sockets round which the hip-bones move. It is well known that by these, when inserted into the throat of draught cattle, worms are cured, that by them, when burnt, teeth are tightened, as they are, when loosened through a blow, by ass's milk, by the ash of an ass's teeth, or by the lichen of a horse poured with oil through the ear. This lichen is not the same as hippomanes, which being pernicious on several grounds I omit, but an excrescence on the knees of

praeterea in corde equorum invenitur os dentibus caninis maximis simile, hoc scarifari dolorem aut exempto
dente mortui equi maxillis ad numerum eius qui doleat demonstrant. equarum virus a coitu in ellychniis
accensum Anaxilaus prodidit[1] equinorum capitum
visus[2] repraesentare monstrifice, similiter ex asinis.
nam hippomanes tantas in veneficio vires habet ut
adfusum aeris mixturae in effigiem equae Olympiae
182 admotos mares equos ad rabiem coitus agat. medetur dentibus et fabrile glutinum in aqua decoctum
inlitum et mox paulo detractum ita ut confestim con-
luantur vino in quo decocti sunt cortices mali Punici
dulcis. efficax habetur et caprino lacte conlui dentes
vel felle taurino. talorum caprae recentium cinis
dentrifricio placet et omnium fere villaticarum
quadrupedum, ne saepius eadem dicantur.

183 L. Cutem in facie erugari et tenerescere candore[3]
lacte asinino putant, notumque est quasdam cottidie
septies genas[4] custodito numero fovere. Poppaea
hoc Neronis principis instituit, balnearum quoque
solia sic temperans, ob hoc asinarum gregibus eam
comitantibus. impetus pituitae in facie butyro inlito
tolluntur, efficacius cum cerussa, sincero vero ea vitia

---

[1] *Hic* lichenis *add. I. Müller* : *servat Mayhoff.*
[2] visus *vulg.* : usus *Detlefsen, codd.*
[3] candore *Urlichs, Detlefsen, Mayhoff, qui conicit* candore
eius aucto (*vel* lucido) : candore custodito *codd.*
[4] septies genas *Mayhoff* : septingenties *multi codd., Hard.,
Detlefsen* : septingentes VE. *Coni.* sescenties *Warmington.*

---

[a] *Candore* without an epithet or *cum* is odd, as Mayhoff felt
when he added *eius aucto*. A repeated *custodito* can hardly be
right, even in Pliny. If the *custodito* of the MSS. has replaced
a lost adjective or participle it is but guess-work to attempt
emendation.

horses and above their hoofs. Moreover, in the heart of horses is found a bone like very large canine teeth; with this they prescribe the painful tooth to be scraped, or with the tooth, corresponding to the place of the aching tooth, extracted from the jaw-bone of a dead horse. Anaxilaus has informed us that the fluid coming from mares when covered, if ignited on lamp wicks, shows weird appearances of horses' heads, and similarly with asses. But hippomanes has such virulent and magical properties that, added to the molten bronze for a figure of an Olympian mare, it maddens any stallions brought near with a raving sexual lust. Teeth are also healed by workman's glue boiled down in water, applied, and shortly after taken off, the teeth immediately to be rinsed in wine in which the rind of sweet pomegranates has been boiled. It is also thought efficacious to rinse the teeth in goat's milk or bull's gall. The ash from a freshly-killed she-goat's pastern bones makes a popular dentifrice, and, so that I need not repeat myself, the same is true of nearly all female farm quadrupeds.

L. It is thought that ass's milk removes wrinkles *For the* from the face, making the skin white [a] and soft, and *complexion.* it is well known that some women every day bathe their cheeks in it seven [b] times, keeping carefully to that number. Poppaea, wife of the Emperor Nero, began this custom, even preparing her bath-tubs with the milk, and for this purpose she was always attended by troops of she-asses. Pituitous eruptions on the face are removed by the application of butter, the addition of white-lead being an improvement, but

[b] The *septingenties* of many MSS. must surely be wrong, even as a playful exaggeration. Warmington's suggestion is happy.

quae serpunt, superinposita farina hordeacia, ulcera
184 in facie membrana e partu bovis madida. frivolum
videatur, non tamen omittendum propter desideria
mulierum, talum candidi iuvenci XL diebus nocti-
busque, donec resolvatur in liquorem, decoctum et inlitum linteolo candorem cutisque erugationem praestare. fimo taurino malas rubescere aiunt, non ut [1]
crocodileam inlini melius sit,[2] sed foveri frigida et ante
185 et postea iubent. testas et quae decolorem faciunt
cutem fimum vituli cum oleo et cummi manu subactum emendat, ulcera oris ac rimas sebum vituli vel
bovis cum adipe anserino et ocimi suco. est et alia
mixtura sebo vituli cum medulla cervi et albae spinae
foliis una tritis. idem praestat et medulla cum resina
186 vel si vaccina sit, et ius e carne vaccina. lichenas oris
praestantissime vincit glutinum factum e genitalibus
vitulorum, liquatum aceto cum sulpure vivo, ramo
ficulneo permixtum, ita ut bis die recens inlinatur,
item lepras ex melle et aceto decoctum, quas et iocur
hirci calidum inlinitum tollit, sicut elephantiasin fel
caprinum, etiamnum lepras ac furfures tauri fel addito
nitro, urina asini circa canis ortum, maculas in facie fel
utriusque per sese aqua infractum evitatisque solibus
187 ac ventis post detractam cutem. similis effectus et in
taurino vitulinove felle cum semine cunilae, cinere e

----

[1] ut *del. Gelenius.*
[2] sit *Urlichs, Mayhoff,* sed (*codd.*) *deleto.*

----

[a] See § 108. The *non ut* is curious, as the sense requires
non ut non. Gelenius would delete *ut.* Warmington suggests
ut non.
[b] Perhaps sun-burn.

spreading sores by unmixed butter with a sprinkling of barley meal on top, and ulcers on the face by the membrane, still moist, that follows the birth of a calf. The following recipe may seem a trifle, but to satisfy the women I must not omit it: the pastern bone of a white bull-calf, boiled for forty days and nights until it melts to a jelly, and applied on a linen cloth, gives whiteness to the skin and smooths away wrinkles. They say that bull's dung brings a rosy colour to the cheeks, though it is better to rub them with crocodilea,[a] but before and after they must be bathed with cold water. Brick-red spots[b] and discolorations of the skin are removed by calf dung kneaded by hand with oil and gum, sores and cracks in the mouth by veal suet or beef suet with goose grease and juice of ocimum. There is yet another compound, veal suet with deer's marrow and white-thorn leaves pounded together. The same effect is given by marrow with resin, even if it is cow marrow, and by the broth from cow beef. An excellent cure for facial lichens is the gluey substance made from the genitals of calves, dissolved in vinegar with native sulphur, stirred up with a fig branch and applied fresh twice a day, and the same boiled down in honey and vinegar for leprous sores, which are also removed by a warm application of he-goat's liver, as is leprosy by goat's gall. Moreover, leprous sores and scurf are removed by bull's gall with soda, or at the rising of the Dog-star by ass's urine; spots on the face by the gall of either animal broken up in water without addition; after the skin has come away sun and winds must be avoided. A similar effect is also obtained by bull's gall or veal gall, with the seed of cunila, and the ash of deer's horn burnt

For affections of the face.

cornu cervino, si canicula exoriente conburatur. asinino sebo cicatricibus a lichene leprisque maxime color
redditur. hirci fel et lentigines tollit admixto caseo
ac vivo sulpure spongeaeque cinere, ut sit mellis
188 crassitudo. aliqui inveterato felle maluere uti,
mixtis calidis furfuribus pondere oboli unius quattuor-
que mellis, prius defricatis maculis. efficax eiusdem
et sebum cum melanthio et sulpure et iride, labrorum
fissuris cum anserino adipe ac medulla cervina resinaque et calce. invenio aput auctores his qui lentigines
habeant negari magice sacrificiorum usum.

189   LI. Lacte bubulo aut caprino tonsillae et arteriae
exulceratae levantur. gargarizatur tepidum ut est
usus, expressum aut calefactum. caprinum utilius
cum malva decoctum et sale exiguo. linguae exul-
cerationi et arteriarum prodest ius omasi gargariza-
tum, tonsillis autem privatim renes vulpium aridi cum
melle triti inlitique, anginae fel taurinum vel capri-
190 num cum melle, iocur melis ex aqua. oris gravitatem
ulceraque butyrum emendat. spinam aliudve quid
faucibus adhaerens felis extrinsecus fimo perfricatis
aut reddi aut delabi tradunt. strumas discutit fel
aprunum vel bubulum tepidum inlitum—nam coagulum leporis e vino in linteolo exulceratis dumtaxat in-
191 ponitur—discutit et ungulae asini vel equi cinis ex
oleo vel aqua inlitus et urina calefacta et bovis ungulae cinis ex aqua, fimum quoque fervens ex aceto,
item sebum caprinum cum calce aut fimum ex aceto
decoctum testesque vulpini. prodest et sapo, Gal-

at the rising of the lesser Dog-star. By ass suet their natural colour is restored to scars, especially to those left by lichen or leprous sores. Freckles too are removed by he-goat's gall mixed with cheese, native sulphur, and sponge ash; the consistency of the mixture should be that of honey. Some have preferred to use matured gall, mixing one obolus of warm bran and four oboli of honey, the spots being first rubbed. An efficacious mixture is also he-goat's suet with melanthium, sulphur, and iris; for cracks in the lips the suet with goose grease, deer's marrow, resin, and lime. I find in my authorities that those with freckles are debarred from assisting at magic ritual.

LI. Cow's milk or goat's is helpful for ulcerated *For the* tonsils or trachea. It is used as a gargle, of the *mouth.* usual warmth, either newly milked or heated. Goat's milk is more useful, boiled down with mallow and a little salt. For ulceration of the tongue or trachea a remedy is a gargle of tripe broth, while for tonsils are specific dried fox kidneys pounded with honey and applied, and for quinsy bull's or goat's gall with honey, or badger's liver in water. Butter remedies offensive breath and ulcerated mouth. If a pointed thing or anything else sticks in the throat, external rubbing with cat's dung is said either to bring it up or to make it pass down. Scrofulous sores are dispersed by a warm application of wild-boar's gall or ox gall (but hare's rennet, on a linen cloth with wine, is applied only when there is ulceration) or by the ash of the hoof of ass or horse applied in oil or water, the urine heated, the ash of an ox's hoof in water, the hot dung in vinegar, goat suet with lime or dung boiled in vinegar, or a fox's testicles. Soap

129

liarum [1] hoc inventum rutilandis capillis. fit ex sebo et cinere, optimus fagino et caprino,[2] duobus modis, spissus ac liquidus, uterque apud Germanos maiore in usu viris quam feminis.

192 LII. Cervicium dolores butyro aut adipe ursino perfricentur, rigores bubulo sebo, quod strumis prodest cum oleo. dolorem inflexibilem—opisthotonum vocant—levat urina caprae auribus infusa aut fimum cum bulbis inlitum, ungues contusos fel cuiuscumque animalis circumligatum, pterygia digitorum fel tauri aridum aqua calida dissolutum. quidam adiciunt sulpur et alumen pari pondere omnium.

193 LIII. Tussim iocur lupi ex vino tepido sanat, ursinum fel admixto melle aut ex cornus bubuli summis partibus cinis, vel saliva equi triduo pota—ecum mori tradunt—pulmo cervinus cum gula sua arefactus in fumo, dein tusus ex melle cottidiano eligmate; efficacior est ad id subulo cervorum generis. san-
194 guinem expuentes cervini cornus cinis, coagulum leporis tertia parte denarii cum terra Samia et vino myrteo potum sanat, eiusdem fimi cinis in vino vesperi potus nocturnas tusses, pili quoque leporis suffiti extrahunt pulmonibus difficiles excreationes. purulentas autem exulcerationes pectoris pulmonisque et a pulmone graveolentiam halitus butyrum efficacissime iuvat cum pari modo mellis Attici decoctum donec

---

[1] Galliarum dT *Mayhoff*: Gallarum RE: Gallorum V, *vulg.*, *Detlefsen*.
[2] caprino *codd.*, *Mayhoff*: carpineo *Sillig*, *Detlefsen*.

---

[a] Sillig's emendation, adopted by Detlefsen, would give: "or hornbeam." It was suggested by the strange arrangement of *sebo, cinere, fagino, caprino*.

is also good, an invention of the Gallic provinces for making the hair red. It is made from suet and ash, the best from beech ash and goat suet,[a] in two kinds, thick and liquid, both being used among the Germans, more by men than by women.

LII. For pains in the neck it should be rubbed with butter or bear's grease, and for stiffness with beef suet, which with oil is good for scrofulous sores. The rigid cramp, called opisthotonus, is relieved by she-goat's urine poured into the ears or by an application of the dung with bulbs, crushed nails by binding round them the gall of any animal, and whitlows by dried bull's gall dissolved in hot water. Some add sulphur and alum, all the ingredients being of equal weight. *For pain in the neck, etc.*

LIII. Cough is cured by wolf's liver in warmed wine, by bear's gall mixed with honey, by the tips of the horns of ox or cow reduced to ash, by the saliva of a horse taken for three days (they say that the horse dies), by a deer's lung dried in smoke with the gullet, then pounded in honey and taken daily as an electuary, the species of deer more efficacious for this purpose being the subulo.[b] Spitting of blood is cured by the ash of deer's horn, and by hare's rennet, the dose being one third part of a denarius, with Samian [c] earth and myrtle wine. Hare's dung reduced to ash and taken in wine in the evening cures night coughs, and inhaling the smoke of burning hare's-fur brings up difficult expectorations. Purulent ulceration of the chest or lungs, and foul breath from the lungs, are very effectively relieved by butter boiled with an equal measure of Attic honey until it turns *For cough.*

[b] See XI. § 213.

[c] A fine clay, of which the famous Samian ware was made.

rufescat et matutinis sumptum ad mensuram lingulae.
195 quidam pro melle laricis resinam addere maluere. si
sanguis reiciatur, efficacem tradunt bubulum sanguinem, modice et cum aceto sumptum, nam de
taurino credere temerarium est. sed glutinum
taurinum tribus obolis cum calida aqua bibitur in
vetere sanguinis excreatione.
196 LIV. Stomachum exulceratum lactis asinini potus
reficit, item bubuli, rosiones eius caro bubula admixto
aceto et vino cocta, rheumatismos cornus cervini cinis,
sanguinis excreationes haedinus sanguis recens ad
cyathos ternos cum aceto acri pari modo fervens potus,
coagulum tertia parte ex aceto potum, LV. iocineris
197 dolores lupi iocur aridum ex mulso, asini iocur aridum
cum petroselini partibus duabus ac nucibus tribus ex
melle tritum et in cibo sumptum, sanguis hircinus
cibo aptatus. suspiriosis ante omnia efficax est potus
equiferorum sanguinis, proxime lactis asinini tepidi,
bubuli [1] decocti ita ut serum ex eo bibatur, addito in
tres heminas cyatho nasturtii albi perfusi aqua, deinde
melle diluti. iocur quoque vulpinum aut pulmo in
vino nigro aut fel ursinum in aqua laxat meatus
spirandi.
198 LVI. Lumborum dolores et quaecumque alia molliri opus sit ursino adipe perfricari convenit, cinerem
apruni aut suilli fimi inveterati aspergi potioni vini.
[adferunt [2] et Magi sua commenta: primum omnium
rabiem hircorum, si mulceatur barba, mitigari, eadem

---

[1] bubuli VRdT, *Detlefsen, Mayhoff*: bulbi E : bulbis r *vulg.*
[2] adferunt VRd *vulg. Mayhoff*: adiciunt *Sillig, Detlefsen.*

---

[a] It was supposed to be poison.

red, the dose being a spoonful taken in the morning; some instead of honey have preferred to add larch resin. For spitting of blood it is said to be beneficial to drink ox or cow blood, a moderate amount taken in vinegar. But to trust recommendations of bull's blood is hazardous;[a] bull glue, however, in three-oboli doses is taken with warm water for chronic spitting of blood.

LIV. An ulcerated stomach is cured by drinking *For stomach* ass's milk or cow's milk; gnawings of the stomach by *and chest.* beef boiled in a mixture of vinegar and wine; catarrhs by the ash of deer's horn; spitting of blood by fresh kid's-blood taken hot, in doses up to three cyathi, with an equal amount of strong vinegar, or by one part of kid's rennet with two parts of vinegar; LV. pains of the liver by dried wolf's liver in honey wine; by dried ass's liver, with two parts of rock parsley and three nuts, pounded in honey and taken in food, and by he-goat's blood made suitable for food. For asthma, effective above all things is to drink the blood of wild horses, next to drink warm ass's milk, or cow's milk boiled, the part drunk being the whey only, with the addition for every three heminae of a cyathus of white cress steeped in water and then tempered with honey. A fox's liver or lung also in dark wine, or bear's gall in water, loosens the breath passage.

LVI. Pains in the loins and all other complaints *For the loins.* needing emollients should be treated by rubbing with bear's grease, or the ash of wild boar's or pig's dried dung should be sprinkled in a draught of wine. [The Magi too add their usual lies: first of all, that the madness of he-goats is soothed if their beard is stroked, and if it is cut off, they do not stray

133

praecisa non abire eos in alienum gregem.[1]] ischiadicis fimum bubulum inponunt calfactum in foliis
cinere ferventi.[2] huic admiscent fimum caprinum et
subdito linteolo uncto cava manu quantum capi possit
fervens sustineri iubent ita ut, si laeva pars doleat,
haec medicina in dextera manu fiat aut e contrario.
fimum quoque ad eum usum acus aereae puncto tolli
199 iubent. modus est curationis donec vapor ad lumbos
pervenisse sentiatur, postea manum porro tuso inlinunt, item lumbos ipso fimo cum melle; suadent in
eo dolore et testes leporis devorare. in renium dolore
leporis renes crudos devorari iubent, aut certe coctos
ita ne dente contingantur. ventris quidem dolore
temptari negant talum leporis habentes.

200 LVII. Lienem sedat fel apri vel suis potum vel
cervini cornus cinis in aceto, efficacissime tamen inveteratus lien asini ita ut in triduo sentiatur utilitas.
asinini pulli fimum quod primum edidit—poleam
vocant—Syri dant in aceto mulso, datur et equi lingua
inveterata ex vino praesentaneo medicamento, ut
didicisse se ex barbaris Caecilius Bion tradidit, et lien
bubulus simili modo, recens autem assus vel elixus in
cibo. in vesica quoque bovis alii capita XX tusa cum

---

[1] *uncos add. Mayhoff.*
[2] ischiadicis . . . ferventi *transposuit Mayhoff ex § 199, ubi
post* leporis devorare *habent codd., vulg.*

---

[a] I have bracketed this sentence, following Mayhoff; where
it should be transferred is not clear.
[b] Mayhoff's transposition of *ischiadicis fimum . . . ferventi*
is not certain, although Dioscorides, II. 80, § 2, ἐπὶ ἰσχιαδικῶν
. . . καλεῖται δὲ τοιαύτη καῦσις 'Αραβική, is very similar. The
huic admiscent after *imponunt* is strange; if the transposition
is correct, *huic* must mean " the dung last mentioned," and the

to another herd.] [a] For sciatica they apply cow-dung heated in leaves over hot embers.[b] With this dung they mix goat's dung, prescribing that as much as it can contain should be held hot in the hollow of the hand, a linen cloth soaked in oil being placed underneath; if the left side aches the medicament should be held in the right hand, and *vice versa*; the dung for this purpose, they say, must be taken up with the point of a bronze needle. The treatment is continued until the warmth is felt to have reached the loins; afterwards they rub the hand with pounded leek, the loins also with the dung itself and honey. For this pain they also recommend sufferers to swallow a hare's testicles. For pain in the kidneys they prescribe the kidneys of a hare to be swallowed raw, or if boiled at least not to be touched by a tooth. Bowel pain indeed never, they say, afflicts those who carry about them the pastern bone of a hare.

LVII. The spleen is relieved by wild boar's or pig's gall taken by the mouth, by ash of deer's horn in vinegar, but most efficaciously by matured ass's spleen, with the result that benefit is felt within three days. The first dung passed by an ass's foal, called *polea*, is administered by the Syrians in oxymel. There is also administered in wine as a sovereign remedy the dried tongue of a horse, as Caecilius Bion reports that he learnt from foreigners.[c] Spleen of ox or cow is administered in a similar way; if fresh it is roasted or boiled and taken in food. There are also applied for pains in the spleen twenty crushed heads of garlic

For the spleen.

application to the hip is to be reinforced by holding some in the hand.

[c] This is interesting, for it shows how wide Pliny spread his net. The remedies given are by no means all Italian.

201 aceti sextario imponuntur ad lienis dolores. eadem
ex causa emi lienem vituli quanti indicatus sit iubent
Magi nulla pretii cunctatione, quoniam hoc quoque
religiose pertineat, divisumque per longitudinem
adnecti tunicae utrimque et induentem pati decidere
ad pedes, dein collectum arefacere in umbra. cum
hoc fiat, simul residere lienem aegri vitiatum liberari-
que eum morbo dicitur. prodest et pulmo vulpium
cinere siccatus atque in aqua potus, item haedorum
lien impositus.

202 LVIII. Alvum sistit cervi sanguis, item cornus
cinis, iocur aprunum ex vino potum citra salem
recensque, item assum, vel suillum, hircinum decoctum ad quintas [1] in vino, coagulum leporis in vino
ciceris magnitudine aut, si febris sit, ex aqua—aliqui
et gallam adiciunt, alii per se leporis sanguine contenti sunt—lac coctum, equini fimi cinis in aquae potu,
taurini cornus veteris e parte ima cinis inspersus
potioni aquae, sanguis hircinus in carbone decoctus,
corium caprinum cum suo pilo decoctum suco epoto,

203 coagulum equi et sanguis caprinus vel medulla vel
iocur. alvum solvit fel lupi cum elaterio umbilico
inlitum [2] vel lactis equini potus, item caprini cum sale
et melle, caprae fel cum cyclamini suco et aluminis
momento—aliqui et nitrum et aquam adiecisse malunt
—fel tauri cum absinthio tritum ac subditum pastillo,

---

[1] ad quintas *ego*: ad quintam heminae *Detlefsen*: ad
quintas hemina *Mayhoff*: ad quintam heminam *codd.*
[2] inlitum *vet. Dal., Mayhoff*: inligatum *codd., Detlefsen.*

---

[a] I believe that the -*s* of *quintas* was taken to be a sign for
hemina; the further change to *quinta(m) heminam* would be
inevitable. For the omission of a measure cf. *ad dimidias
partes* § 206.

in the bladder of an ox with a sextarius of vinegar. For the same purpose the Magi recommend a calf's spleen to be bought at the price asked, without any haggling, attention to this also affecting the efficacy of the ritual. This spleen should be divided lengthwise and attached to the patient's tunic on both sides. As he puts it on, the patient should allow the spleen to fall to his feet, then pick it up and dry in the shade. At the same time as this happens, the diseased spleen of the patient is said to shrink, and he himself to be freed from his complaint. Beneficial too is fox lung dried on embers and taken in water, and kids' spleen applied locally.

LVIII. Binding to the bowels are stag's blood, *For the* stag's horn reduced to ash, wild boar's liver taken in *bowels.* wine, unsalted and fresh, the same liver roasted, pig's liver, he-goat's liver boiled down to one fifth *a* in wine, hare's rennet of the size of a chick-pea in wine, or if there is fever, in water—some add a gall-nut, others are content with hare's blood by itself—boiled milk, horse dung reduced to ash in a draught of water, the root of an old horn of a bull reduced to ash and sprinkled on a draught of water, he-goat's blood boiled down over charcoal, the juice, taken by the mouth, of goat's skin boiled down with the hair on, horse rennet and goat's blood, marrow, or liver. The bowels are loosened by wolf's gall applied *b* to the navel with elaterium, or by draughts of mare's milk, or of goat's milk with salt and honey, by she-goat's gall with juice of cyclamen and a little alum—some prefer to add both soda and water—bull's gall pounded with wormwood and used in the form of a lozenge as a suppository, and by large doses of butter. Those

b Cf. § 205 *umbilico inponere.*

204 butyrum largius sumptum. coeliacis et dysintericis
medetur iocur vaccinum, cornus cervini cinis tribus
digitis captus in potione aquae, coagulum leporis
subactum in pane, si vero sanguinem detrahant, in
polenta, apruni vel suilli vel leporini fimi cinis
inspersus potioni tepidi vini. vituli quoque ius
vulgariter dari [1] inter auxilia coeliacorum et dysin-
tericorum tradunt. lactis asinini potus utilior addito
melle, nec minus efficax fimi cinis ex vino utrique
vitio, item polea supra dicta, equi coagulum, quod
205 aliqui hippacen appellant, etiam si sanguinem detrahant, vel fimi cinis dentiumque eiusdem tusorum
farina salutaris et bubuli lactis decocti potus. dysintericis addi mellis exiguum praecipiunt et, si tormina
sint, cornus cervini cinerem aut fel taurinum cumino
mixtum et cucurbitae carnes umbilico inponere.
caseus recens vaccinus inmittitur ad utrumque vitium,
item butyrum heminis quattuor cum resinae terebinthinae sextante aut cum malva decocta aut cum
rosaceo. datur et sebum vitulinum aut bubulum,
206 item medulla [2]—et cocuntur [3] cum farinae ceraeque
exiguo et oleo, ut sorberi possit; [4] medulla et in pane
subigitur—lac caprinum ad dimidias partes decoctum.
si sint et tormina, additur protropum. torminibus
satis esse remedii in leporis coagulo poto e vino tepido
vel semel arbitrantur aliqui. cautiores et sanguine

---

[1] dari *Mayhoff*: datum *Detlefsen*: datum *aut* dati *codd.*
[2] medulla VdTE *Mayhoff*: medullae R, *vulg., Detlefsen.*
[3] et coquuntur (cocuntur) VdTE: excoquuntur R, *vulg., Detlefsen*: et coquitur *Mayhoff.*
[4] possit *Mayhoff, codd.*: possint *Detlefsen, vulg.*

with coeliac disorder or dysentery are benefited by
cow's liver, a three-finger pinch of the ash of deer's
horn taken in a draught of water, by hare's rennet
kneaded in bread, but in pearl barley if blood is
brought away, and by ash of wild boar's, pig's, or
hare's dung sprinkled on a draught of warm wine.
It is also reported that veal broth is a popular remedy
to relieve sufferers from coeliac disorder or dysentery.
Ass's milk makes a more beneficial draught with the
addition of honey, the dung, reduced to ash and taken
in wine, is no less efficacious for either complaint,
polea [a] too, which I mentioned just now, horse's
rennet, that some call *hippace*, even if blood is brought
away, or the dung ash and crushed teeth of the same
animal, a health-giving powder, and taken with boiled
cow's milk.   For dysentery is prescribed the addition
of a little honey, and if there are griping pains to apply
to the navel the ash of deer's horn or bull's gall mixed
with cummin, and the fleshy parts of a gourd.   New
cheese made from cow's milk is injected for both
complaints, so also four heminae of butter with two
ounces of terebinth resin, or with a decoction of
mallows, or with rose oil.   There is administered also
veal suet, beef suet, or the marrow (they are boiled
with a little flour and wax, and with oil, so that to
drink the mixture is possible, and the marrow is also
kneaded in bread), and goat's milk boiled down to
one half; if there is also griping, *protropum* [b] is added.
It is thought by some that a sufficient remedy for
griping is even a single dose of hare's rennet taken in
warm wine; more careful people also apply as

[a] See § 200.
[b] The first wine made from grapes before pressing. See
XIV. § 75 and § 85.

caprino cum farina hordeacea et resina ventrem in-
207 linunt. ad omnes epiphoras ventris inlini caseum
mollem suadent, veterem autem in farinam tritum
coeliacis et dysintericis dari, cyatho casei in cyathis
vini cibarii tribus. sanguis caprinus decoctus cum
medulla dysintericis, iocur assum caprae coeliacis
subvenit, magisque etiam hirci, in vino austero decoctum potumque vel ex oleo myrteo umbilico inpositum.
quidam decocunt a tribus sextariis aquae ad heminam
208 addita ruta. utuntur et liene asso caprae hircive et
sebo hirci in pane qui cinere coctus sit, caprae a renibus maxime, ut per se hauriatur protinus aqua[1]
modice frigida. sorberi iubent aliqui et in aqua
decoctum sebum admixta polenta et cumino et aneto
acetoque. inlinunt et ventrem coeliacis fimo cum
209 melle decocto. utuntur ad utrumque vitium et
coagulo haedi in vino myrtite fabae magnitudine poto
et sanguine eiusdem in cibum formato quem sangui-
culum vocant. infundunt dysintericis et glutinum
taurinum aqua calida resolutum. inflationes discutit
vitulinum fimum in vino decoctum. intestinorum
vitiis magnopere prodest coagulum cervorum decoctum cum lente betaque atque in cibo ita sumptum,
leporis pilorum cinis cum melle decoctus,[2] lactis cap-
210 rini potu decocti cum malva exiguo sale addito. si
et coagulum addatur, maioribus emolumentis fiat.

---

[1] aqua *Detlefsen* : -que *Mayhoff* : que, inque, lique *codd.*
[2] decoctus d *vulg.*, *Mayhoff* : decocto *multi codd.*, *Detlefsen.*

---

[a] We should say " grated cheese."

embrocation to the belly goat's blood with barley meal and resin. For all fluxes from the belly an application of soft cheese is recommended, but matured cheese powdered [a] is used for coeliac disorders and dysentery, the dose being a cyathus of cheese in three cyathi of ordinary wine. A decoction of goat's blood with goat's marrow is beneficial for dysentery, roasted she-goat's liver for coeliac complaints, or, better still, that of a he-goat boiled down in dry wine and drunk, or applied to the navel in myrtle oil. Some boil it down from three sextarii of water to one hemina with rue added. They also use the roasted spleen of a she-goat or he-goat with the suet of a he-goat in bread baked over hot ashes, the best suet being from the kidneys of a she-goat, which should be swallowed by itself, and be immediately followed by a draught of moderately cold water. Some prescribe also a decoction of the suet in water, made into a stew with other ingredients—pearl barley, cummin, dill, and vinegar. They also rub the belly of sufferers from coeliac disorders with a decoction of honey and goat's dung. For both complaints they also use kid's rennet, of the size of a bean, taken in myrtle wine, or kid's blood made into a food, called " blood pudding." They also inject into dysentery patients bull glue dissolved in hot water. Flatulence is dispersed by calf dung boiled down in wine. Disorders of the intestines are greatly benefited by a decoction of deers' rennet with lentils and beet, and so taken in food, by the ash of hare's fur boiled down with honey, by a draught of goat's milk boiled down with mallows with the addition of a little salt; if goat's rennet too is added the beneficial effects will be much greater. The same is the effect

eadem vis est et in sebo caprino in sorbitione aliqua, uti protinus hauriatur frigida aqua. item feminum haedi cinis rupta intestina sarcire mire traditur, fimum leporis cum melle decoctum et cottidie fabae magnitudine sumptum ita ut deploratos sanaverint. laudant et caprini capitis sum suis pilis decocti sucum.

211 LIX. Tenesmos, id est crebra et inanis voluntas desurgendi,[1] tollitur poto lacte asinino, item bubulo. taenearum genera pellit cervini cornus cinis potus. quae in excrementis lupi diximus inveniri ossa, si terram non attigerint, colo medentur adalligata bracchio. polea quoque supra dicta magnopere prodest decocta in sapa, item suilli fimi farina addito cumino in aqua rutae decoctae, cornus cervini teneri cinis cocleis Africanis cum testa sua tusis mixtus in vini potione.

212 LX. Vesicae calculorumque cruciatibus auxiliatur urina apri et ipsa vesica pro cibo sumpta, efficacius, si prius fumo maceretur utrumque. vesicam elixam mandi oportet, et a muliere feminae suis. inveniuntur et in iocineribus eorum lapilli aut duritiae lapillis similes, candidae, sicut in vulgari sue, quibus contritis atque in vino potis pelli calculos aiunt. ipsi apro tam gravis urina sua est ut nisi egesta fugae non sufficiat ac velut devinctus opprimatur, exuri illa

[1] id est . . . desurgendi *in uncis ponere velit Warmington.*

---

[a] Warmington thinks that the explanation of *tenesmos* is a gloss.
[b] See § 178.
[c] See § 200.
[d] Book XIV. § 80; it was must boiled down to one third.

of goat's suet in some kind of stew, to be immediately followed by a draught of cold water. A kid's hams also reduced to ash are said to be wonderfully healing to intestinal rupture, and the dung of a hare, boiled down with honey and taken daily in doses the size of a bean, to be so beneficial as they have cured desperate cases. Highly recommended also is the broth of a goat's head with the fur still on.

LIX. Tenesmus, that is a frequent and ineffectual desire to go to stool,ᵃ is removed by drinking ass's milk, or cow's milk. Worms are expelled by ash of deer's horn, taken in drink. The bones that I have said ᵇ are found in the excrements of a wolf, tied on to the arm as an amulet without touching the earth, are a cure for colitis. Polea also, mentioned above,ᶜ is of great benefit if boiled down in sapa,ᵈ likewise too powdered pig's dung and cummin in the water of a decoction of rue, and young deer's horn reduced to ash, mixed with African snails pounded with their shells and taken in a draught of wine. *For tenesmus, etc.*

LX. The tortures of stone in the bladder are relieved by the urine of a wild boar and by his bladder itself taken as food; both remedies are more efficacious if first thoroughly smoked. The bladder should be eaten boiled, and be a sow's if the patient is a woman. There are also found in the liver of these animals little stones, or hard substances like stones, white, and like those found in the liver of the common pig. These, crushed and taken in wine, are said to expel stone. His own urine is such a burden to the boar himself that unless he has voided it he is not strong enough for flight, and is overcome as if spell-bound. It is said that the urine dissolves the stone. Stone is also expelled by a *For stone and the kidneys.*

213 tradunt eos.[1] leporis renes inveterati in vino poti
calculos pellunt. in pernae suum articulo os[2] esse
diximus quod decoctum ius facit urinae utile.
asini renes inveterati tritique ex vino mero dati
vesicae medentur. calculos expellunt lichenes equini
ex vino aut mulso poti diebus XL. prodest et ungulae equinae cinis in vino aut aqua, item fimum
caprarum in mulso, efficacius silvestrium, pili quoque
caprini cinis; verendorum carbunculis cerebrum apri
214 vel suis sanguisque. vitia vero quae in eadem parte
serpunt iocur eorum combustum, maxime iunipiri
ligno, cum charta et arrhenico sanat, fimi cinis, fel
bubulum cum alumine Aegyptio ac murra ad crassitudinem mellis subactum, insuper beta ex vino cocta
inposita, caro quoque; manantia vero ulcera sebum
cum medulla vituli in vino decoctum, fel caprinum
cum melle rubique suco, vel si serpant; fimum etiam
prodesse cum melle dicunt aut cum aceto et per se
215 butyrum. testium tumor sebo vituli addito nitro cohibetur vel fimo eiusdem ex aceto decocto. urinae
incontinentiam cohibet vesica apruna, si assa man-
datur, ungularum apri vel suis cinis potioni inspersus,
vesica feminae suis conbusta ac pota, item haedi, vel
pulmo, cerebrum leporis in vino, eiusdem testiculi
tosti vel coagulum cum anserino adipe in polenta,
renes asini in mero triti potique. Magi verrini genitalis cinere poto ex vino dulci demonstrant urinam

---

[1] ea . . . illos *coni. Mayhoff.*
[2] articulo os *Mayhoff* : articulos *codd.*

---

[a] See § 179.

hare's kidneys, dried and taken in wine. In the ham joints of pigs I have said[a] there are bones the broth from which is beneficial for urinary disorders. The kidneys of an ass, dried, pounded, and given in neat wine, cure complaints of the bladder. The excrescences on the legs of horses, taken for forty days in wine or honey wine, expel stone. Beneficial too is the ash of a horse's hoof in wine or water, the dung also in honey wine of she-goats, that of wild goats being more efficacious, the ash also of goat's hair, while for carbuncles on the privates are used the brains and blood of a wild boar or pig. Creeping sores however in the same part are cured by the burnt liver of these animals, best if the fire is of juniper wood, mixed with paper and orpiment, by their dung reduced to ash, by ox gall with Egyptian alum and myrrh, kneaded to the consistency of honey, moreover by an application of beet boiled in wine, also by beef; but running ulcers by beef suet with the marrow of a calf boiled down in wine, by goat's gall with honey and blackberry juice, even if the sores are spreading. They say that goat's dung too with honey or vinegar is beneficial, and also butter by itself. Swelling of the testicles is reduced by veal suet with the addition of soda, or by calf's dung boiled down in vinegar. Incontinence of urine is checked by a wild-boar's bladder, if eaten roasted, by the ash of a wild-boar's or pig's hoofs sprinkled on a drink, by the bladder of a sow burnt and taken in drink, of a kid also, or by its lung, by the brain of a hare in wine, by a hare's roasted testicles, or the rennet, with goose grease in pearl barley, or by the kidneys of an ass pounded in neat wine and drunk. The Magi recommend that, after drinking in sweet wine a boar's genital organ re-

facere in canis cubili ac verba adicere, ne ipse urinam
faciat ut canis in suo cubili. rursus ciet urinam vesica
suis, si terram non attigerit, inposita pubi.

216   LXI. Sedis vitiis praeclare prodest fel ursinum cum
adipe. quidam adiciunt spumam argenti ac tus.
prodest et butyrum cum adipe anserino ac rosaceo;
modum ipsae res statuunt, ut sint inlitu faciles. praeclare medetur et taurinum fel in linteolis conceptis,
rimasque perducit ad cicatricem. inflationibus in ea
parte sebum vituli, maxime ab inguinibus, cum ruta;
ceteris vitiis medetur sanguis caprinus cum polenta,
item fel caprinum condylomatis per se, item fel
217   lupinum ex vino. panos et apostemata in quacumque
parte sanguis ursinus discutit, item taurinus aridus
tritus. praecipuum tamen remedium traditur in
calculo onagri quem dicitur, cum interficiatur, reddere urina liquidiorem initio sed in terra spissantem
se. hic adalligatus femini omnes impetus discutit
omnique suppuratione liberat. est autem rarus inventu nec ex omni onagro, sed mire[1] celebrant[2]
remedio. prodest et urina asini cum melanthio et
ungulae equinae cinis cum oleo et aqua inlitus,
sanguis equi, praecipue admissarii, sanguis bubulus,
218   item fel. caro quoque eosdem effectus habet calida
inposita et ungulae cinis ex aqua aut melle, urina
caprarum, hircorum quoque carnes in aqua decoctae

[1] mire *I. Müller, Mayhoff*: medici *Brakman*: me r: ne E
om. multi codd.
[2] celebrant *I. Müller, Mayhoff*: celebrari *codd.*: celebri
vulg. Fortasse maxime celebratur.

---

[a] I. Müller's emendations, adopted by Mayhoff, have been
kept with some misgivings. Mayhoff himself suggests *maxime*,

duced to ash, the patient should make water in a dog's
bed and add a prayer, that he may not himself make
water, as a dog does, in his own bed. On the other
hand, the bladder of a pig is diuretic, if, without
touching the ground, it is applied to the pubic part.

LXI. Complaints of the anus are greatly benefited *For the*
by bear's gall and bear's fat; some add litharge and *anus.*
frankincense. Beneficial too is butter with goose
grease and rose oil; the quantities are determined by
circumstances; the mixture must be easy to apply.
Greatly beneficial too is bull's gall in scraps of linen;
it makes chaps to cicatrize. Swellings in that part
of the body are reduced by veal suet, especially by
that from the groin, with rue; other complaints are
cured by goat's blood with pearl barley, condylomata
by goat's gall by itself, or by wolf's gall in wine.
Superficial and other abscesses in any part are dispersed by bear's blood, and likewise by bull's dried
and powdered. The finest remedy, however, is said
to be the stone which the wild ass is reported to pass
in his urine when he is being killed; more fluid than
it at first, it grows thick when on the ground. This
stone fastened to the thigh as an amulet disperses all
inflamed swellings and clears away any suppuration.
It is found, however, rarely and not always in the wild
ass, but it is wonderfully famous[a] as a remedy.
Beneficial also is the urine of an ass with melanthium,
a horse's hoof reduced to ash and applied with oil and
water, the blood of a horse, especially of a stallion,
and the blood or gall of an ox or cow. Beef too has
the same effect if applied hot, the ash of the hoof in
water or honey, the urine of she-goats, the flesh too

and *celebratur* is perhaps nearer the MSS. reading than *celebrant*.
Brakman's emendation is possibly right.

aut fimum ex his cum melle decoctum, fel verrinum,
urina suum in lana inposita. femina adteri adurique
equitatu notum est. utilissimum est ad omnes inde
causas spumam equi ex ore inguinibus inlinere.
inguina et ex ulcerum causa intumescunt. remedio
sunt equi saetae tres totidem nodis alligatae intra
ulcus.

219   LXII. Podagris medetur ursinus adips taurinum-
que sebum pari pondere et cerae. addunt quidam
hypocisthidem et gallam. alii hircinum praeferunt
sebum cum fimo caprae et croco, sinapi, item[1] caulibus
hederae tritis ac perdicio vel flore cucumeris silvestris.

220 item bovis fimum cum aceti faece magnificant et
vituli qui nondum herbam gustaverit fimum aut per
se sanguinem tauri, vulpem decoctam vivam donec
ossa tantum restent, lupumve vivum oleo cerati modo
incoctum, sebum hircinum cum helxines parte aequa,
sinapis tertia, fimi caprini cinerem cum axungia.
quin et ischiadicos uri sub pollicibus pedum eo fimo
fervente utilissime tradunt, articulorumque vitiis fel
ursinum utilissimum esse et pedes leporis adalligatos,
podagras quidem mitigari pede leporis viventi absciso,

221 si quis secum adsidue habeat. perniones ursinus adips
rimasque pedum omnes sarcit, efficacius alumine addito, sebum caprinum, dentium equi farina, aprunum
vel suillum fel cum adipe, pulmo inpositus, etsi subtriti
sint contunsive offensatione, si vero adusti frigore,
leporini pili cinis, eiusdem pulmo contusis dissectus

---

[1] sinapi, item *Mayhoff e Dioscoried* : sinapive vel *Gelenius*,
Detlefsen : sinapii vel E : sinapi cum d r.

---

[a] I have adopted the emendation of Mayhoff, because he
has some confirmatory evidence in Dioscorides and Plinius
Junior. But in so amorphous a sentence any emendations
are necessarily dubious.

of he-goats boiled down in water or their dung boiled down with honey, a boar's gall, and a pigs' urine applied on wool. It is well known that riding on a horse chafes and galls the inner side of the thighs; most useful for all such troubles is to rub on the groin the foam from the mouth of a horse. The groin also swells because of sores; the remedy is to tie within the sore three horse hairs with three knots.

LXII. Gout is benefited by bear's grease and bull suet with an equal weight of wax as well; to which some add hypocisthis and gall nut. Others prefer he-goat suet with the dung of a she-goat and with saffron, mustard,[a] pounded stalks of ivy, and perdicium or the blossom of wild cucumber. Highly praised also is ox dung with lees of vinegar and the dung of a calf that has not yet tasted grass, or, by itself, the blood of a bull, a fox boiled down alive until only the bones remain, or a wolf boiled alive in oil as though to make a wax-salve, he-goat's suet with an equal quantity of helxine, a third part of mustard, calcined goat's dung and axle-grease. Moreover, to put a burning-hot poultice of this dung under the big toes is said to be excellent for sciatica, and bear's gall very useful for diseases of the joints, as are also the feet of a hare worn as an amulet, while gouty pains are alleviated by a hare's foot, cut off from the living animal, if the patient carries it about continuously on the person. Chilblains and all chaps on the feet are healed by bear's grease, more efficaciously with the addition of alum, by goat suet, by a horse's teeth ground to powder, by the gall and fat of a wild boar or pig, by the lung applied to them even if they are chafed or broken by a knock, but if they are frost bites, by a hare's fur reduced to ash; if they are broken,

For gout and other complaints.

149

222 aut pulmonis cinis. sole adusta sebo asinino aptissime curantur, item bubulo cum rosaceo. clavos et
rimas callique vitia fimum apri vel suis recens inlitum
ac tertio die solutum sanat, talorum cinis, pulmo
aprinus aut suillus aut cervinus, adtritus calciamentorum urina asini cum luto suo inlita, clavos sebum
bubulum cum turis polline, perniones vero corium
conbustum, melius si ex vetere calciamento, iniurias
223 e calceatu ex oleo corii caprini cinis. varicum
dolores sedat fimi vitulini cinis cum lilii bulbis decoctus addito melle modico, itemque omnia inflam-
mata et suppurationes minantia. eadem res et
podagris prodest et articulariis morbis, e maribus
praecipue vitulis, articulorum adtritis fel aprorum
vel suum linteo calefacto inpositum, vituli qui nondum
herbam gustaverit fimum, item caprinum cum melle
in aceto decoctum. ungues scabros sebum vituli
emendat, item caprinum admixta sandaraca. verrucas vero aufert fimi vitulini cinis ex aceto, asini urina
et lutum.

224 LXIII. Comitiali morbo testes ursinos edisse prodest vel aprunos bibisse ex lacte equino aut ex
aqua, item aprunam urinam ex aceto mulso, efficacius
quae inaruerit in vesica sua. dantur et suum testiculi inveterati tritique in suis lacte, praecedente vini
abstinentia et sequente continuis ⟨denis⟩[1] diebus,
dantur et leporis sale custoditi pulmones cum turis
225 tertia parte in vino albo per dies XXX, item coagula

---

[1] denis *coni. Mayhoff*: *om. codd.*

---

[a] It appears likely that the *d* of *diebus* has led to the
omission of a sign for *decem* or *denis.*

by the lung of the same animal cut up or reduced to
ash. Sun burns are most beneficially treated by ass
suet, and also by suet of an ox or cow with rose oil.
Corns, chaps, and calluses are cured by an application
of fresh wild-boar's dung, or pig's, taken off on the
third day, by their pastern bones reduced to ashes,
by the lung of wild boar, pig, or deer; chafing from
shoes by the application of an ass's urine with the
mud made by it; corns by beef suet with powdered
frankincense; chilblains, however, by burnt leather,
if from an old shoe so much the better, sores from
foot-wear by the ash of goat leather in oil. The pains
of varicose veins are alleviated by the ash of calf's
dung boiled down with the bulbs of a lily, with the
addition of a little honey, and so are all inflamed
places that threaten to suppurate. The same preparation is good for gout and diseases of the joints,
especially if it is taken from a male calf, for chafed
joints the gall of wild boars or of pigs applied in a
heated linen cloth, the dung of a calf that has not
tasted grass, also the dung of goats boiled down in
vinegar with honey. Scabrous nails are cured by
veal suet, also by goat suet mixed with sanderach.
Warts however are removed by the ash of calf's dung
in vinegar, or by the urine with its mud of an ass.

LXIII. For epilepsy it is beneficial to eat a bear's *For epilepsy.*
testes or to take those of a wild boar in mare's milk or
water, likewise wild-boar's urine in oxymel, with
increased efficacy if it has dried in his bladder. There
are also given the testicles of pigs dried and pounded
in sow's milk, abstinence from wine preceding and
following for ⟨ten⟩ *a* days. There are also given the
lungs of a hare preserved in salt, with a third part of
frankincense, taken in white wine for thirty days;

eiusdem, asini cerebrum ex aqua mulsa, infumatum
prius in foliis, semuncia per dies ⟨V,⟩ vel¹ ungularum
eius cinis coclearibus binis toto mense potus, item
testes sale adservati et inspersi potioni in asinarum
maxime lacte vel ex aqua. membrana partus earum,
praecipue si marem pepererint, olefactata accedente
morbo comitialium resistit. sunt qui e mare nigroque
cor edendum cum pane sub diu prima aut secunda
luna praecipiant, alii carnem, aliqui sanguinem
226 aceto dilutum per dies XL bibendum. quidam
urinam aquae ferrariae ex officinis miscent eademque
potione et lymphatis medentur. comitialibus datur
et lactis equini potus lichenque in aceto mulso biben-
dus, dantur et carnes caprinae in rogo hominis tostae,
ut volunt Magi, sebum earum cum felle taurino pari
pondere decoctum et in folliculo fellis reconditum ita
ne terram attingat, potum vero ex aqua sublime.
morbum ipsum deprehendit caprini cornus vel cervini
usti nidor. sideratis urina pulli asinini nardo admixto
perunctione prodesse dicitur.

227 LXIV. Regio morbo cornus cervini cinis, sanguis
asini ex vino, item fimum asinini pulli quod primum
edidit a partu datum fabae magnitudine e vino

---

¹ V, vel *Hard.* : vel *Detlefsen, codd.* : VII *Mayhoff.*

---

a Again, the *v* of *vel* has led to the omission of the numeral.
b In Cato (LXX and LXXI) *stare sublime* means " to stand
upright." For an epileptic to do so might be difficult.
c Neither Littré nor the Bohn translator comments on this
vague sentence. It is not clear how the presence of epilepsy
is detected by this test. Possibly a fit is diagnosed as epileptic
according as it reacts to the treatment.
d See II. § 108. Sometimes sunstroke may be referred to
by this term. Many expressions in this chapter are curious.
Why for instance both *testes* and *testiculi* ? *Morbo comitialium*

likewise a hare's rennet, an ass's brain in hydromel, first smoked on burning leaves, half an ounce a day for ⟨five⟩ [a] days, or an ass's hoofs reduced to ash and two spoonfuls taken in drink for a whole month, likewise his testes preserved in salt and sprinkled on drink, preferably on ass's milk, or on water. The odour of the after-birth of she-asses, especially if they have had a male foal, inhaled on the approach of a fit, repels it. There are some who recommend eating with bread the heart of a black jackass in the open air on the first or second day of the moon, some the flesh, others drinking for forty days the blood diluted with vinegar. Certain people mix an ass's urine with smithy water in which hot iron has been dipped, and use the same draught to treat delirious raving. To epileptics is also given mare's milk to drink, the excrescence on a horse's leg taken in oxymel; there is given too goat's flesh roasted on a funeral pyre, as the Magi would have it, goat suet boiled down with an equal weight of bull's gall stored in the gall bladder without touching the earth, and taken in water with the patient standing upright.[b] The disease itself is detected by the fumes of burnt goat's horn or deer's horn.[c] Rubbing with the urine of an ass's foal mixed with nard is said to be beneficial to the planet-struck.[d]

LXIV. Jaundice is cured within two days by deer's horn reduced to ash, by the blood of an ass, likewise by the dung of an ass's foal, the first to pass after birth,[e] of the size of a bean and taken in wine.

For jaundice.

is strange, and so is the apparent omission on two occasions of a numeral. One may add the vagueness referred to in note (c).

[e] See § 200.

medetur intra diem tertium. eadem et ex equino
pullo similiterque [1] vis est.

LXV. Fractis ossibus praesentaneus maxillarum
apri cinis vel suis, item lardum elixum atque circumligatum mira celeritate solidat. costis quidem fractis
laudatur unice caprinum fimum ex vino vetere, aperit,
extrahit, persanat.

228 LXVI. Febres arcet cervorum caro, ut diximus, eas
quidem quae certo dierum numero redeunt oculus
lupi dexter salsus adalligatusque, si credimus Magis.
est genus febrium quod amphemerinon vocant. hoc
liberari tradunt, si quis e vena auris asini tres guttas
sanguinis in duabus heminis aquae hauserit. quartanis Magi excrementa felis cum digito bubonis
adalligari iubent, et ne recidant non removeri [2] sep-
229 teno circumitu. quis hoc, quaeso, invenire potuit?
quae est ista mixtura? cur digitus potissimum bubonis
electus est? modestiores iocur felis decrescente
luna occisae inveteratum sale ex vino bibendum
ante accessiones quartanae dixere. iidem Magi fimi
bubuli cinere consperso puerorum urina inlinunt digitos pedum manuumque.[3] leporis cor adalligant. coagulum ante accessiones propinant. datur et caseus
caprinus recens cum melle diligenter sero expresso.

230 LXVII. Melancholicis fimum vituli in vino decoctum remedio est. lethargicos excitat asini lichen

---

[1] similiterque *codd. et edd.*: similiter *vel* fimi similiter dati
coni. Mayhoff.

[2] *Hic addendum* nisi *coni. Mayhoff.*

[3] manuumque *Mayhoff*: manibusque *vulg. Detlefsen*, d:
manuusque VRE: mausque r.

---

[a] Probably: removes any diseased matter before healing
takes place.

[b] See VIII. § 119.

The first dung too of a young colt, administered in a similar way, has the same effect.

LXV. For broken bones a sovereign remedy is the *For broken bones.* ash of the jaw-bone of a wild boar or of a pig ; likewise boiled bacon-fat, tied round the fracture, heals with marvellous rapidity. For broken ribs however the highest praise is given to goat's dung in old wine ; it opens, extracts,[a] and completely heals.

LXVI. Fevers are kept away by the flesh of deer, *For fevers.* as I have said,[b] those indeed which return at fixed intervals by the salted right eye of a wolf worn as an amulet, if we are to believe the Magi. There is a kind of fever called " amphemerinos." [c] It is said that he is freed from this who drinks three drops of blood from an ass's ear in two heminae of water. For quartans the Magi prescribe the excrement of a cat with the claw of a horned owl worn as an amulet, and to prevent a relapse the amulet should not be removed before the seventh periodic return. Who pray could have made this discovery ? What sort of combination is this ? Why was an owl's claw chosen rather than anything else ? Some more moderate people have prescribed the salted liver of a cat killed when the moon is on the wane, to be taken in wine before the access of a quartan. The Magi also apply to the toes and fingers ox or cow dung reduced to ash and sprinkled with children's urine. They use the heart of a hare as an amulet, and give hare's rennet before each access. There is also given with honey fresh goat's cheese with the whey carefully pressed out.

LXVII. A remedy for melancholia [d] is calf's dung *For melan-* boiled down in wine. Victims of lethargy [d] are *cholia, lethargy and consumption.*

[c] Greek for quotidian, *i.e.* returning every day.
[d] See List of Diseases.

naribus inlitus ex aceto, caprini cornus nidor aut
pilorum, iocur aprunum. itaque et veternosis datur.
phthisicis medentur iocur lupi ex vino macro, suis
feminae herbis pastae laridum, carnes asininae ex
iure sumptae. hoc genere maxime in Achaia curant
id malum. fimi quoque aridi sed pabulo viridi pasto
bove fumum harundine haustum prodesse tradunt,
bubuli cornus mucronem exustum duorum coclearium
mensura addito melle pilulis devoratis. caprae sebo
231 in pulte alicacia et phthisim et tussim sanari, vel
recenti, cum mulso liquefacto, ita ut uncia in cyathum
addatur rutaeque ramo permisceatur, non pauci tradunt. rupicaprae sebi cyatho et lactis pari mensura
deploratum phthisicum convaluisse certus auctor
adfirmat. sunt et qui suum fimi cinerem profuisse
scripserint in passo et cervi pulmonem, maxime subu-
lonis, siccatum in [1] fumo tritumque in vino.

232 LXVIII. Hydropicis auxiliatur urina e vesica capri
paulatim data in potu, efficacius quae inaruerit cum
vesica sua, fimi taurini maxime, sed et bubuli—de
armentivis loquor, quod bolbiton vocant—cinis
coclearium trium in mulsi hemina, bovis feminae in
mulieribus, ex altero sexu in viris, quod veluti myste-
rium occultarunt Magi, fimum vituli masculi inlitum,
fimi vitulini cinis cum semine staphylini, aequa

---

[1] in del. Mayhoff.

aroused by applying to the nostrils in vinegar the
excrescence on the leg of an ass, by the fumes from
goat's horns or goat's hair, and by wild boar's liver;
accordingly it is also administered to the comatose.
Consumptives are benefited by wolf's liver in thin
wine, by the lard of a sow fed on herbs, and by ass's
flesh taken in its gravy. This treatment for the
complaint is very popular in Achaia. The smoke also
from dried dung of an ox fed on green fodder, inhaled
through a reed, is said to be beneficial, or the burnt
tip of the horn of an ox, the dose being two spoonfuls, with the addition of honey, swallowed in pills.
It is held by not a few authorities that by she-goat's
suet in groat porridge consumption and cough are
cured, or by fresh suet melted with honey wine, an
ounce of suet added to a cyathus of wine and stirred
with a spray of rue. An authoritative [a] writer
assures us that a despaired-of consumptive has recovered by being treated with a cyathus of mountain-
goat suet and the same amount of the milk. Some
have written that pig's dung reduced to ash, taken in
raisin wine, has proved of value, or the lung of a stag,
especially a subulo,[b] dried in smoke and pounded in
wine.

LXVIII. Good for dropsy is urine from the bladder *For dropsy.*
of a wild boar given little by little in the drink, that
being more beneficial which has dried up with its
bladder, the ash of bull's dung especially but also
that of oxen—herd animals I mean; it is called
bolbiton—three spoonfuls in a hemina of honey
wine, cow dung for women, bull dung for men (the
Magi have made a sort of mystery of this distinction),
the dung of a bull calf applied locally, ash of calf dung
with staphylinus seed in equal proportions taken in

portione ex vino, sanguis caprinus cum medulla.
efficaciorem putant hircinum utique si lentisco
pascantur.

233 LXIX. Igni sacro ursinus adips inlinitur, maxime
qui est ad renes, vitulinum fimum recens vel bubulum,
caseus caprinus siccus cum porro, ramenta pellis
cervinae desecta pumice ex aceto trita, rubori cum
prurigine equi spuma aut ungulae cinis, eruptionibus
pituitae asinini fimi cinis cum butyro, papulis nigris
caseus caprinus siccus ex melle et aceto in balneis,
oleo remoto, pusulis suilli fimi cinis aqua inlitus vel
234 cornus cervini cinis, LXX. luxatis recens fimum
aprinum vel suillum, item vitulinum, verris spuma
recens cum aceto, fimum caprinum cum melle, bubula
caro inposita, ad tumores fimum suillum in testo
calefactum tritumque cum oleo. duritias corporum
omnes tollit optime adips e lupis inlitus. in his quae
rumpere opus est plurimum proficit fimum bubulum
in cinere calefactum aut caprinum in vino vel aceto
decoctum, in furunculis sebum bubulum cum sale aut,
si dolores sint, cum oleo liquefactum sine sale, simili
235 modo caprinum, LXXI. in ambustis ursinus adips cum
lilii radicibus, aprunum aut suillum fimum inveteratum, saetarum ex his e penicillis tectoriis cinis cum
adipe tritus, tali bubuli cinis cum cera et medulla cervina, fel tauri, fimum leporis, sed caprarum fimum [1]
236 sine cicatrice sanare dicitur. glutinum praestantissimum fit ex auribus taurorum et genitalibus, nec quic-

---

[1] fimum] " *an* fimi cinis ? " *Mayhoff.*

---

[a] The punctuation of Mayhoff is attractive. He puts a
full stop before *sine* and after *glutinum*, removing the one
after *dicitur*. It has the support of Pliny Junior, but *fimum*

wine, and goat's blood with goat's marrow. That of
a he-goat is considered more beneficial, especially if
he has browsed on lentisk.

LXIX. There is applied for erysipelas bear's fat, *For various skin diseases.*
especially that on the kidneys, fresh dung of calves
or cattle, dried goat's cheese with leek, scrapings
of deer's skin rubbed off with pumice and pounded
in vinegar. For inflamed itch the foam of a horse
or the ash of his hoof; for pituitous eruptions ass's
dung reduced to ash with butter; for black pimples
dried goat's cheese in honey and vinegar, applied
in the bath, no oil being used, for pustules pig's
dung reduced to ash and applied in water, or the
ash of deer's horn, LXX. for dislocations the fresh *For dislocations, indurations burns.*
dung of wild boar or of pig, or of calves, the fresh
foam of a boar with vinegar, the dung of a goat with
honey, an application of beef, and for swellings pig's
dung warmed in an earthen pot and beaten up
with oil. All indurations of the body are best
removed by an application of wolf's fat. In the case
of sores that need to break the most beneficial
application is ox dung warmed on hot cinders or
goat's dung boiled down in wine or vinegar, for boils
beef suet with salt, or if there is pain melted with oil
without salt, similarly with goat suet; LXXI. for
burns bear's grease with lily roots, dried dung of wild
boar or of pig, the ash of pig's bristles from plasterers'
brushes beaten up with pig fat, the ash of the pastern
bone of bull or cow with wax and deer marrow, bull's
gall, hare's dung ; but the dung of she-goats is said to
heal without a scar.[a] The finest glue is made from
the ears and genitals of bulls, and there is no better

leporis sed caprarum fimum contains a strange repetition of
fimum.

quam efficacius prodest ambustis, sed adulteratur nihil
aeque, quibusvis pellibus inveteratis calciamentisque
etiam decoctis.   Rhodiacum fidelissimum, eoque pictores et medici utuntur.   id quoque quo candidius eo
probatius, nigrum et lignosum damnatur.

237   LXXII. Nervorum doloribus fimum caprinum decoctum in aceto cum melle utilissimum putant vel
putrescente nervo.   spasmata et percussu vitiata
fimo apruno curant vere collecto et arefacto, sic et
quadrigas agentes tractos rotave vulneratos et quoquo
238 modo sanguine contuso, vel si recens inlinatur.   sunt
qui incoxisse aceto utilius putent.   quin et in potu
farinam eam ruptis convulsisque et eversis ex aceto
salutarem promittunt.   recentiores [1] cinerem eius
ex aqua bibunt, feruntque et Neronem principem hac
potione recreari solitum, cum sic quoque se trigario
adprobare vellet.   proximam suillo fimo vim putant.

239   LXXIII. Sanguinem sistit coagulum cervinum ex
aceto, item leporis, huius quidem et pilorum cinis,
item ex fimo asini cinis inlitus, efficacior vis e maribus
aceto admixto et in lana ad omne profluvium inposito,
similiter ex equino, capitis et feminum aut fimi vitulorum cinis inlitus ex aceto, item caprini cornus vel

[1] recentiores *Hard.* : reverentiores *codd.*

---

[a] With the reading of the MSS., " more cautious."

remedy for burns, but it is more adulterated than any other, a decoction being made from any old skins and even from shoes. The most reliable glue comes from Rhodes, which is used by painters and physicians. The Rhodian too is the more approved the whiter it is; the dark and wood-like is rejected.

LXXII. It is thought that for pains in the sinews, *For strains, sprains, ruptures.* even if pus is present there, the most beneficial remedy is a decoction of goat's dung in vinegar with honey. Strains and injuries from a blow are treated with wild-boar's dung collected in spring and dried; the same remedy is also good for charioteers who have been dragged along, or wounded by a wheel, or bruised in any way, even if the dung is applied while fresh. There are some who think it more beneficial to boil the dung in vinegar. Moreover, they assure us that this dung, reduced to powder and taken in drink, is curative of ruptures and sprains; for falls from vehicles it should be taken in vinegar. The more recent authorities [a] reduce it to ash and take in water, saying that even the Emperor Nero used to refresh himself with this draught, since he was ready even by this means to distinguish himself in the three-horse chariot-race. They think that the next most efficacious dung is that of pigs.

LXXIII. Bleeding is stayed by deer's rennet in *For haemorrhage.* vinegar, by hare's also, by the latter reduced to ash with the fur, also by the application of ass's dung reduced to ash—the effect is more powerful if the ass is male, vinegar mixed with the ash, and wool used for the application to any haemorrhage, horse dung being similarly used, by the head and thighs, or dung, of calves, reduced to ash and applied in vinegar, also by the ash in vinegar of goat's horn

161

240 fimi ex aceto. hircini vero iocineris dissecti sanies
efficacior, et cinis utriusque [1] ex vino potus vel naribus
ex aceto inlitus, hircini quoque utris, vinarii dumtaxat, cinis cum pari pondere resinae, quo genere
sistitur sanguis et vulnus glutinatur. haedinum quoque coagulum ex aceto et feminum eius combustorum
cinis similiter pollere traduntur.

241 LXXIV. Ulcera sanat in tibiis cruribusque ursinus
adips admixta rubrica, quae vero serpunt fel aprunum
cum resina et cerussa, maxillarum apri vel suum cinis,
fimum suum inlitum siccum, item caprinum ex aceto
subactum et subfervefactum.[2] cetera purgantur et
explentur butyro, cornus cervini cinere vel medulla
cervi, felle taurino cum cyprino aut fimo hircino.[3]
fimum recens suum vel inveterati farina inlinitur vulneribus ferro factis. phagedaenis et fistulis inmittitur
fel tauri cum suco porri aut lacte mulierum vel sanguis

242 aridus cum cotyledone herba. carcinomata curat coagulum leporis cum pari pondere capparis adspersum
vino, gangraenas ursinum fel pinna inlitum, asini ungularum cinis ea quae serpunt ulcera inspersus.
sanguis equi adrodit carnes septica vi, item fimi
equini inveterati favilla, ea vero quae phagedaenas
vocant in ulcerum genere corii bubuli cinis cum melle.
caro vituli recentia vulnera non patitur intumescere.

243 fimum bubulum cum melle, fimi vitulini cinis sordida

---

[1] *An* sexus *excidit* ?

[2] subactum et subfervefactum *Mayhoff ex Plinio Iuniore et
Marcello*: subfervefactum *codd.*

[3] aut fimo hircino *Detlefsen*: oleo aut irino *Mayhoff ex
Plinio Iuniore cum cod.* d : *varia codd.*

---

[a] For *sanies* see Celsus, V. 26, 20.
[b] Has *sexus* fallen out here ?

or dung. The sanies,[a] however, exuding from he-goat's liver when cut up is more efficacious, as is the liver of goats of either sex,[b] reduced to ash and taken in wine or applied to the nostrils in vinegar, or the leather of a he-goat, but only that of a wine bottle, reduced to ash and with an equal weight of resin, by which remedy bleeding is stayed and the wound closed. Kid's rennet also in vinegar and kid's thighs burnt to ash are reported to be similarly effective.

LXXIV. Ulcers on the shins or shanks are healed *For ulcers* by bear's grease mixed with ruddle, but spreading *and fistulae.* ulcers by wild boar's gall with resin and white lead, by the jaw-bones of wild boars or pigs reduced to ash, by the application of dried pig's-dung, also by goat's dung, kneaded in vinegar and warmed. The other kinds of sores are cleansed and filled up by butter, by the ash of deer's horn or by deer's marrow, by bull's gall with cyprus oil or he-goat's dung.[c] To wounds inflicted with iron is applied pig's dung, either fresh or dried and powdered. Injected into phagedaenic ulcers and fistulas is bull's gall with juice of leek or woman's milk, or else dried blood with the herb cotyledon. Cancerous sores are treated with hare's rennet and an equal weight of caper sprinkled in wine, gangrenes by bear's gall applied with a feather, spreading ulcers by the ash of ass's hoofs sprinkled over them. Flesh is eaten away by the corrosive action of horse's blood and by the ash of dried horse-dung, but the ulcers coming under the class they call phagedaenic by the ash of oxhide with honey. Veal prevents fresh wounds from swelling. Foul ulcers and those called malignant are healed by dung of ox or cow with

[c] With Mayhoff's reading : " cyprus oil and iris oil."

ulcera et quae cacoethe vocant e lacte mulieris sanant,
recentes plagas ferro inlatas glutinum taurinum liquefactum, tertio die solutum. caseus caprinus siccus ex
aceto ac melle purgat ulcera, quae vero serpant
cohibet sebum cum cera, item addita pice ac sulpure
percurat. similiter proficit ad cacoethe haedi feminum cinis e lacte mulieris et adversus carbunculos suis
feminae cerebrum tostum inlitumque.

244 LXXV. Scabiem hominis asininae medullae maxime
abolent et urina[1] eiusdem cum suo[2] luto inlita,[3]
butyrum etiam quod in iumentis proficit cum resina
calida, glutinum taurinum in aceto liquefactum addita
calce, fel caprinum cum aluminis cinere, bovas fimum
bubulum, unde et nomen traxere. canum scabies
sanatur bubulo sanguine recenti iterumque, cum
inarescat, inlito et postero die abluto cinere lixivo.

245 LXXVI. Spinae et similia corpori extrahuntur felis
excrementis, item caprae ex vino, coagulo quocumque, sed maxime leporis, cum turis polline et oleo aut
cum visci pari pondere aut cum propoli. cicatrices
nigras sebum asininum reducit ad colorem, fel vituli
extenuat calefactum. medici adiciunt murram et
mel et crocum aereaque puxide condunt. aliqui et
florem aeris admiscent.

246 LXXVII. Mulierum purgationes adiuvat fel tauri
in lana sucida adpositum—Olympias Thebana addidit
oesypum[4] et nitrum—cornus cervini cinis potus, item

[1] urina *Mayhoff*: urinae *codd., Detlefsen.*
[2] suo *codd.*: suillo *Urlichs, Detlefsen.*
[3] inlita *Mayhoff*: inlitae *Detlefsen*: inlito *codd.*
[4] oesypum *vet. Dal. ex Dioscoride, Mayhoff*: hysopum *Detlefsen, codd.*

---

[a] *Bovae* = " ox disease."

honey, or by the ash of calf's dung in woman's milk, fresh wounds inflicted with iron by melted bull's glue, which is· taken off on the third day. Ulcers are cleansed by dry goat's-cheese in vinegar and honey, while spreading ulcers are checked by goat suet with wax, and the addition of pitch and sulphur makes the cure complete. In a similar way malignant ulcers are improved by the ash of a kid's thighs in woman's milk, and for carbuncles are used a sow's brains, roasted and applied.

LXXV. For itch in men the best cure is the *For itch.* marrow of the ass, or ass's urine applied with its own mud, butter likewise, which with warm resin also benefits itch in draught animals, bull glue melted in vinegar and with lime added, goat gall with the ash of alum; ox or cow dung is good for *bovae,*[a] whence comes the name of the disease. Itch in dogs is cured by the fresh blood of ox or cow, applied again when it is dry, and on the following day washed off with lye ash.

LXXVI. Thorns and similar objects are extracted *For thorns,* by a cat's excrements, also by a she-goat's in wine, *etc., in the* by any kind of rennet but especially by hare's with *flesh.* powdered frankincense and oil, or else with an equal weight of mistletoe, or with bee glue. Black scars are brought back to the original colour by ass's suet, and made fainter by warmed calf's gall. Physicians add myrrh, honey and saffron, and keep in a bronze box; some add to the mixture flower of bronze.[b]

LXXVII. The purgings of women are aided by *For female* bull's gall applied as a pessary in unwashed wool— *complaints.* Olympias, a woman of Thebes added suint and soda —by ash of deer's horn taken in drink, and uterine

[b] Red oxide of copper.

vulva laborantes inlitus quoque et fel taurinum cum
opio adpositum obolis binis. vulvas et pilo cervino
suffire prodest. tradunt cervas, cum senserint se
gravidas, lapillum devorare, quem in excrementis
repertum aut in vulva—nam et ibi invenitur—custo-
247 dire partus adalligatum. inveniuntur et ossicula in
corde et in vulva perquam utilia gravidis parturienti-
busque. nam de pumice quae in vaccarum utero
simili modo invenitur diximus in natura boum.[a] lupi
adips inlitus vulvas mollit, dolores earum iocur. carnes lupi edisse parituris prodest, aut si incipientibus
parturire sit iuxta qui ederit,[b] adeo ut etiam contra in-
248 latas noxias valeat. eundem supervenire pernitiosum
est. magnus et leporis usus mulieribus. vulvas adiuvat pulmo aridus potus, profluvia iocur cum Samia
terra ex aqua potum, secundas coagulum—caventur
pridiana balnea—inlitum quoque cum croco et porri
suco, in[1] vellere adpositum abortus mortuos expellit.
si vulva leporum in cibis sumatur, mares concipi putant, hoc et testiculis eorum et coagulo profici, conceptum leporis utero exemptum his quae parere desierint
249 restibilem fecunditatem adferre. sed pro conceptu[2]
leporis saniem et viro Magi propinant, item virgini

[1] in add. Mayhoff.
[2] sed pro conceptu E r d, Detlefsen: sic conceptus Mayhoff.

[a] See XI. § 203. [b] Possibly " eat."

troubles by an application also of this, and by two-oboli pessaries of bull's gall and poppy juice. It is beneficial also to fumigate the uterus with deer's hair. It is reported that hinds, when they realise that they are pregnant, swallow a little stone which, found in their excrements or in the uterus—for it is found there also—prevents miscarriage if worn as an amulet. There are also found in the heart and in the uterus little bones that are very useful to women who are pregnant or in child-bed. But about the pumice-like stone which in a similar way is found in the uterus of cows I have spoken when dealing with the nature of oxen.[a] The uterus is softened by an application of wolf's fat, pains there by wolf's liver, but to have eaten [b] the flesh of the wolf is beneficial for women near delivery, or at the beginning of labour the near presence of one who has eaten it, so much so that sorceries put upon the woman are counteracted. But for such a person to enter during delivery is a deadly danger. The hare is also of great use to women. The uterus is benefited by the dried lung taken in drink, fluxes by the liver taken in water with Samian earth, the after-birth is eased by hare's rennet—the bath must be avoided the day before—by the rennet applied also with saffron and leek juice; a pessary of it in raw wool brings away a dead foetus. If the uterus of the hare is taken in food, it is believed that males are conceived; that the same result is obtained by eating its testicles and rennet; that the foetus of a hare, taken from its uterus, brings a renewed fertility to women who are passed child-bearing. But the sanies of a hare is given by the Magi even to the male partner that conception may occur, and likewise

viiii grana fimi ut stent perpetuo mammae. coagulo
quoque ob id cum melle inlinunt, sanguinem ubi evolsos pilos renasci nolunt. inflationi vulvae fimum
aprunum suillumve cum oleo inlini prodest. efficacius
sistit farina aridi, ut aspergatur potioni, vel si gravidae
250 aut puerperae torqueantur. lacte suis poto cum
mulso adiuvantur partus mulierum, per se vero potum
deficientia ubera puerperarum replet. eadem cir-
cumlita sanguine feminae suis minus crescent. si
dolent, lactis asinini potu mulcentur, quod addito
melle sumptum et purgationes earum adiuvat. sanat
et vulvarum exulcerationes eiusdem animalis sebum
inveteratum et in vellere adpositum duritias vulvarum
emollit. per se vero recens vel inveteratum ex aqua
251 inlitum psilotri vim optinet. eiusdem animalis lien
inveteratus ex aqua inlitus mammis abundantiam
facit, vulvas suffitu corrigit. ungulae asininae suffitio
partum maturat ut vel abortus evocetur, nec aliter
adhibentur, quoniam viventem partum necant. eiusdem animalis fimum si recens inponatur, profluvia
sanguinis mire sedare dicitur, nec non et cinis eiusdem
252 fimi, qui et vulvae prodest inpositus. equi spuma
inlita per dies XL prius quam primum nascantur pili
restinguntur, item cornus cervini decocto, melius, si
recentia sint cornua. lacte equino iuvantur vulvae
collutae. quod si mortuus partus sentiatur, lichen

---

^a Probably " fresh," " from a deer just killed."

to a maiden nine pellets of hare's droppings to make
the breasts permanently firm. They also use for this
purpose the rennet with honey as liniment, and the
blood to prevent hairs plucked out from growing
again. For inflation of the uterus it is beneficial to
make with oil a liniment of wild boar's dung or pig's.
More efficacious is the dried dung reduced to powder
to sprinkle in the drink, even if the woman is suffering
the pains of pregnancy or child-birth. By drinking
sow's milk with honey wine child-birth is eased, while
taken by itself it refills the drying breasts of nursing
mothers. These swell less if rubbed round with a
sow's blood. If they are painful they are soothed by
drinking ass's milk, which taken with the addition of
honey is also beneficial for the purgings of women.
Ulcerations also of the uterus are healed by the dried
suet of the same animal, which applied in raw wool
as a pessary softens uterine indurations, while by itself
either fresh or dried suet, applied in water, acts as a
depilatory. Dried ass's spleen, applied in water to
the breasts, produces an abundant supply of milk,
and used in fumigation corrects displacement of the
uterus. Fumigation with ass's hoofs hastens delivery, so that even a dead foetus is extracted; only
then is the treatment applied, for it kills a living
infant. Ass's dung applied fresh is said to be a
wonderful reliever of fluxes of blood, as is also the
ash of the same dung, an application which is also
beneficial to the uterus. By horse's foam, applied
for forty days before they first grow, hairs are
prevented, also by a decoction of deer's horns, which
is more beneficial if the horns are new.[a] It is
beneficial to wash out the uterus with mare's milk.
But if the foetus is felt to be dead, it is expelled by

equae e dulci potus eicit, item ungula suffitu aut
fimum aridum. vulvas procidentes butyrum infusum
sistit. induratam vulvam aperit fel bubulum rosaceo
admixto, foris vellere cum resina terebinthina in-
253 posito. aiunt et suffitu fimi e mari bove procidentes
vulvas reprimi, partus adiuvari, conceptus vero
vaccini lactis potu. sterilitatem a partus vexatione
fieri certum est. hanc emendari Olympias Thebana
adfirmat felle taurino et adipe serpentium et aerugine
ac melle medicatis locis ante coitus. vitulinum quoque fel in purgationibus sub coitu adspersum vulvae
etiam duritias ventris [1] emollit et profluvium minuit
umbilico peruncto atque in totum vulvae prodest.
254 modum statuunt fellis pondere denarii, opii tertiam
admixto amygdalino oleo quantum satis esse appareat, haec in vellere inponunt. masculi fel vituli
cum mellis dimidio tritum servatur ad vulvas. carnem vituli si cum aristolochia inassatam edant circa
conceptum, mares parituras promittunt. medulla
vituli in vino et aqua decocta cum sebo exulcerationi-
bus vulvarum inposita prodest, item adips vulpium
excrementumque felium, hoc cum resina et rosaceo
255 inpositum. caprino cornu suffiri vulvam utilissimum
putant. silvestrium caprarum sanguis cum palma
marina pilos detrahit, ceterarum vero fel callum

---

[1] ventris *codd.*, *Detlefsen* : veteres *Mayhoff*.

---

[a] Mayhoff's emendation of *ventris* to *veteres* ("chronic
indurations of the uterus") is attractive because it allows
vulvae to be taken with *duritias*, and also avoids the apparently irrelevant introduction of *ventris* in a list of female
complaints. On the other hand, with this reading one would
expect *etiam* to come immediately before *veteres*. Perhaps
ventris emphasizes the general efficacy of calf's gall as a
softener.

taking in fresh water the excrescence from the leg of a mare, also by fumigation with the hoof or the dried dung. An injection of butter stays prolapsus of the uterus. A hardened uterus is opened by ox gall mixed with rose oil, with an external application of terebinth resin on unwashed wool. They say that prolapsus of the uterus is corrected also by fumigation with the dung of an ox, that delivery is aided, and conception also, by drinking cow's milk. It is certain that sterility may result from sufferings at child-birth. This kind of barrenness, we are assured by Olympias of Thebes, is cured by bull's gall, serpents' fat, copper rust and honey, rubbed on the parts before intercourse. Calf's gall also, sprinkled on the uterus during menstruation just before intercourse, softens even indurations of the bowels,[a] checks the flow if rubbed on the navel, and is generally beneficial to the uterus. The amount of gall prescribed is a denarius by weight; this and a third part of poppy juice, with as much almond oil as seems to be called for. The mixture is laid on unwashed wool. A bull-calf's gall beaten up with half the quantity of honey is stored away for uterine complaints. If women about the time of conception eat roasted veal with aristolochia, they are assured that they will bring forth a male child. A calf's marrow, boiled down in wine and water with calf's suet and applied to an ulcerated uterus, is beneficial, as is the fat of foxes with the excrement of cats, the last being applied with resin and rose oil. It is thought that to fumigate the uterus with goat's horn is very beneficial. The blood of wild she-goats with sea palm acts as a depilatory, while of other she-goats the gall softens callus of the uterus if sprinkled on it,

vulvarum emollit inspersum et a purgatione conceptus facit. sic quoque psilotri vis efficitur, evulsis pilis triduo servatur inlitum. profluvium quamvis inmensum urina caprae pota sisti obstetrices promittunt, et si fimum inlinatur. membrana caprarum in qua partus editur inveterata potuque sumpta in vino

256 secundas pellit. haedorum pilis suffiri vulvas utile putant et in profluvio sanguinis coagulum bibi aut cum [1] hyoscyami semine inponi. e bove silvestri nigro si sanguine ricini lumbi perungantur mulieri, taedium veneris fieri dicit Osthanes, idem amoris potu hirci urinae admixto propter fastidium nardo.

257 LXXVIII. Infantibus nihil butyro utilius per se et cum melle, privatim et in dentitione et ad gingivas et ad oris ulcera. dens lupi adalligatus infantium pavores prohibet dentiendique morbos, quod et pellis lupina praestat—dentes quidem eorum maximi equis quoque adalligati infatigabilem cursum praestare

258 dicuntur. leporum coagulo ubere inlito sistitur infantium alvus. iocur asini admixta modice panace instillatum in os a comitialibus morbis et aliis infantes tuetur; hoc XL diebus fieri praecipiunt. et pellis asini iniecta inpavidos infantes facit. dentes qui equis primum cadunt facilem dentitionem praestant adalligati infantibus, efficacius, si terram non attigere.

[1] *Ante* hyoscyami *add.* cum *Mayhoff.*

and after a menstruation causes conception; such an application also acts as a depilatory; after the hairs are pulled out it is kept on for three days. Midwives assure us that a flux, however copious, is stayed by drinking the urine of a she-goat, or if an application is made of her dung. The membrane that covers the new-born offspring of she-goats, kept till dry and taken in wine, brings away the after-birth. To fumigate the uterus with the hairs of kids is thought to be beneficial, and it is so for a flux of blood if kid's rennet is taken in drink, or applied locally with seed of hyoscyamus. Osthanes says that if the loins of a woman are rubbed thoroughly with the blood of a tick from a black wild-bull, she will be disgusted with sexual intercourse, and also with her love if she drinks the urine of a he-goat, nard being added to disguise the foul taste.

LXXVIII. For babies nothing is more beneficial *Treatment* than butter, either by itself or with honey, especially *for babies.* when they are troubled with teething, sore gums, or ulcerated mouth. The tooth of a wolf tied on as an amulet keeps away childish terrors and ailments due to teething, as does also a piece of wolf's skin. Indeed the largest teeth of wolves tied as an amulet even on horses are said to give them unwearied power of speed. Hare's rennet applied to the mothers' breasts checks the diarrhoea of babies. Ass's liver mixed with a moderate amount of panaces and let drip into the mouth protects babies from epilepsy and other diseases; the treatment, it is prescribed, should continue for forty days. Ass's hide laid on babies keeps them free from fears. The first teeth of horses to fall out make the cutting of teeth easy for babies who wear them as an amulet, a more efficacious one

259 lien bubulus in melle et datur et inlinitur ad lienis
dolores, ad [1] ulcera manantia cum melle ** lien vituli
in vino decoctus tritusque et inlitus ulcuscula oris.
cerebrum caprae Magi per anulum aureum traiectum
prius quam lac detur infantibus instillant contra
comitiales ceterosque infantium morbos. caprinum
fimum inquietos infantes adalligatum panno cohibet,
maxime puellas. lacte caprino aut cerebro leporum
perunctae gingivae faciles dentitiones faciunt.

260 LXXIX. Somnos fieri lepore sumpto in cibis Cato
arbitrabatur, vulgus et gratiam corpori in VIIII dies,
frivolo quidem ioco, cui tamen aliqua debeat subesse
causa in tanta persuasione. Magi felle caprae,
sacrificatae dumtaxat, inlito oculis vel sub pulvino
posito somnum allici dicunt. sudores inhibet cornus
caprini cinis ex myrteo oleo perunctis.

261 LXXX. Coitus stimulat fel aprunum inlitum, item
medullae suum haustae, sebum asininum anseris
masculi adipe admixto inlitum, item a coitu equi a
Vergilio quoque descriptum virus et testiculi equini
aridi ut potioni interi possint dexterve asini testis in
vino potus, portione [2] vel adalligatus bracchiali, eiusdem a coitu spuma collecta russeo panno et inclusa
262 argento, ut Osthanes tradit. Salpe genitale in oleum
fervens mergi iubet septies eoque perungui perti-

---

[1] ad codd.: sedat Mayhoff: post melle lacunam indicat
Sillig.

[2] portione del. Warmington ex potioni ortum. Vide tamen
Önnerfors, Pliniana pp. 166, 167.

---

[a] With Mayhoff's reading : " running sores are soothed by
etc."

[b] The pun is on lepus " hare " and lepos " charm."

[c] See Georgics III 280.

if the teeth have not touched the ground. Ox spleen in honey is administered internally and externally for painful spleen; for running sores [a] with honey . . . a calf's spleen boiled in wine, beaten up, and applied to little sores in the mouth. The brain of a she-goat, passed through a golden ring, is given drop by drop by the Magi to babies, before they are fed with milk, to guard them from epilepsy and other diseases of babies. Restless babies, especially girls, are quietened by an amulet of goat's dung wrapped in a piece of cloth. Rubbing the gums with goat's milk or hares' brains makes easy the cutting of teeth.

LXXIX. Cato thought that to take hare as food is *Soporifics.* soporific, and a popular belief is that it also adds charm to the person for nine days, a flippant pun,[b] but so strong a belief must have some justification. According to the Magi the gall of a she-goat—she must be an animal sacrificed—induces sleep if applied to the eyes or placed under the pillow. Sweats are checked by rubbing the body with myrtle oil and ash of goat's horn.

LXXX. Aphrodisiacs are: an application of wild- *Aphrodisiacs.* boar's gall, pig's marrow swallowed, or an application of ass's suet mixed with a gander's grease; also the fluid that Virgil[c] too describes as coming from a mare after copulation, the testicles of a horse, dried so that they may be powdered into drink, the right testis of an ass taken in wine, or a portion of it worn as an amulet on a bracelet; or the foam of an ass after copulation, collected in a red cloth and enclosed, as Osthanes tells us, in silver. Salpe prescribes an ass's genital organ to be plunged seven times into hot oil, and the relevant parts to be rubbed therewith,

nentes partes, Dalion cinerem ex eodem bibi vel
tauri a coitu urinam, luto ipso inlini pubem. at e
diverso muris [1] fimo inlito cohibetur virorum venus.
ebrietatem arcet pulmo apri aut suis assus, ieiuni [2]
cibo sumptus eo die, item haedinus.

263   LXXXI. Mira praeterea traduntur in isdem
animalibus: vestigium equi excussum ungula, ut
solet plerumque, si quis collectum reponat, singultus
remedium esse recordantibus quonam loco id repo-
suerint, iocur luporum equinae ungulae simile esse et
rumpi equos qui vestigia luporum sub equite sequan-
tur, talis suum discordiae vim quandam inesse, in
incendiis, si fimi aliquid egeratur e stabulis, facilius
extrahi nec recurrere oves bovesque, hircorum carnes
264 virus non resipere, si panem hordeacium eo die quo
interficiantur ederint laserve dilutum biberint, nullas
vero teredinem sentire luna decrescente induratas
sale. adeoque nihil omissum est ut leporem surdum
celerius pinguescere reperiamus, animalium vero
265 medicinas: si sanguis profluat iumentis, suillum
fimum ex vino infundendum, boum autem morbis
sebum, sulpur vivum, alium silvestre concoctum,[3] trita
in vino danda aut vulpis adipem; carnem caballinam

---

[1] muris *vulg., Detlefsen*: tauri *Mayhoff*: muri *codd.*:
fortasse muli.
[2] ieiuni *codd., Detlefsen*: ieiunis in *C. F. W. Müller,
Mayhoff.*
[3] concoctum T, *Sillig, Detlefsen*: ovum crudum *Mayhoff,
qui* ovum non coctum *coni.*: ovum coctum *vulg.*

---

[a] With Mayhoff's reading: " bull's."
[b] The emendation of C. F. W. Müller is more normal than
the reading of the MSS., but the latter can just be construed
with the same sense.

Dalion the ash from it to be taken in drink, or the urine of a bull after copulation to be drunk, or the mud itself made by it applied to the pubic parts. On the other hand antaphrodisiac for men is an application of mouse's [a] dung. Intoxication is kept away by the roasted lung of a wild boar or pig, taken in food the same day on an empty stomach,[b] or the lung used may be that of a kid.

LXXXI. In addition, wonderful things are reported of the same animals [c]: that if a horse casts his shoe, as often happens, and some one picks it up and puts it away, it is a cure of hiccoughs in those who remember where they have put it; that a wolf's liver is like a horse's hoof; that horses burst themselves which, carrying a rider, follow the tracks of wolves; that there is a kind of quarrelsome force in the pastern bones of pigs; that if, in case of fire, a little dung is brought out of the stables, sheep and oxen are more easily pulled out and do not run back; that the flesh of he-goats does not taste strong if on the day they are killed they have eaten barley bread or drunk diluted laser [d]; that no meat, salted when the moon is on the wane, is eaten by maggots. So much care has been taken to leave nothing out, that I find that a deaf hare fattens more quickly, and that there are also medicines made for animals: it is prescribed that if draught cattle suffer from haemorrhage, there should be injected pig's dung in wine; and that for the diseases of oxen suet, native sulphur, and a decoction of wild garlic, should all be pounded and given in wine, or else fox

Beliefs about animals.

---

[c] Or, "also of animals."

[d] Or, "an infusion of laser." It depends whether the juice or the plant is meant by "laser."

discoctam potu suum morbis mederi, omnium vero
quadripedum morbis capram solidam cum corio et
rana rubeta discoctam, gallinaceos non attingi a
vulpibus qui iocur animalis eius aridum ederint, vel
si pellicula ex eo collo induta galli inierint, similiter
266 in felle mustelae, boves in Cypro contra tormina
hominum excrementis sibi mederi, non subteri pedes
boum, si prius cornua ima pice liquida perunguantur,
lupos in agrum non accedere, si capti unius pedibus
infractis cultroque adacto paulatim sanguis circa fines
agri spargatur atqne ipse defodiatur in eo loco ex quo
267 coeperit trahi, aut si vomerem quo primus sulcus eo
anno in agro ductus sit excussum aratro focus Larum
quo familia convenit[1] exurat, lupum nulli animalium
nociturum in eo agro quam diu id fiat. hinc deinde
praevertemur ad animalia sui generis quae aut
placida non sunt aut fera.

---

[1] convenit] conveniet *codd.*, *Mayhoff.*

fat; that horse flesh thoroughly boiled and taken in drink cures the diseases of pigs, while those of all quadrupeds are cured by a she-goat boiled whole with the hide and a bramble toad; that chickens are not touched by foxes if they have eaten dried fox-liver, or if the cocks have trodden the hens wearing a piece of fox skin round their necks; similarly with a weasel's gall; that the oxen in Cyprus eat human excrement to cure themselves of colic; that the hoofs of oxen are not chafed underneath if the bases of their horns are first rubbed with liquid pitch; that wolves do not enter a field if one is caught, his legs broken, a knife driven into the body, the blood sprinkled a little at a time around the boundaries of that field, and the body itself buried in that place at which the dragging of it began; or if the share, with which that year the first furrow of that field was cut, is knocked from the plough and burnt on the hearth of the Lares where the family assemble, a wolf will harm no animal in that field so long as the custom is kept up. We will now turn to animals in a peculiar class by themselves, which are not either tame or wild.

BOOK XXIX

LIBER XXIX

I. Natura remediorum atque multitudo instantium ac praeteritorum plura de ipsa medendi arte cogunt dicere, quamquam non ignarus sim, nulli ante haec Latino sermone condita ancepsque iudicium[1] esse rerum omnium novarum, talium[2] utique tam sterilis
2 gratiae tantaeque difficultatis in promendo. sed quoniam[3] occurrere verisimile est omnium qui haec noscant cogitationi, quonam modo exoleverint in medicinae usu quae iam parata atque pertinentia erant, mirumque et indignum protinus subit nullam artium inconstantiorem fuisse aut etiamnunc saepius mutari, cum sit fructuosior nulla. dis primum inven-
3 tores suos adsignavit et caelo dicavit. nec non et hodie multifariam ab oraculis medicina petitur. auxit deinde famam etiam crimine, ictum fulmine Aesculapium fabulata, quoniam Tyndareum revocavisset ad vitam. nec tamen cessavit narrare alios revixisse opera sua clara Troianis temporibus, quibus fama certior, vulnerum tamen dumtaxat remediis.
4   II. Sequentia eius, mirum dictu, in nocte densissima latuere usque ad Peloponnesiacum bellum.

---

[1] iudicium *Detlefsen* : lubricum *Mayhoff* : ac lubricum d T.
[2] talium E *Gel.*, *Detlefsen* : exordium *Mayhoff* : et talium RdTf : et alium r : et italium V : artium *coni. Warmington.*
[3] quoniam *codd.*, *Detlefsen* : quaestionem *Mayhoff.*

---

[a] Pliny seems to forget Scribonius Largus (if he knew him) and Celsus.

BOOK XXIX

I. The nature of remedies, and the great number *Early medicine.* of those already described or waiting to be described, compel me to say more about the art of medicine itself, although I am aware that no one hitherto has treated the subject in Latin,[a] and that the judgement passed on all new endeavours is uncertain, especially on such as are barren of all charm, and the difficulty of setting them forth is so great. But since it is likely to come into the minds of all students of the subject to ask why ever things ready to hand and appropriate have become obsolete in medical practice, the thought occurs at once that it is both a wonder and a shame that none of the arts has been more unstable, or even now more often changed, although none is more profitable. To its pioneers medicine assigned a place among the gods and a home in heaven, and even today medical aid is in many ways sought from the oracle. Then medicine became more famous even through sin, for legend said that Aesculapius was struck by lightning for bringing Tyndareus back to life. But medicine did not cease to give out that by its agency other men had come to life again, being famous in Trojan times, in which its renown was more assured, but only for the treatment of wounds.

II. The subsequent story of medicine, strange to say, lay hidden in darkest night down to the Pelopon-

tunc eam revocavit in lucem Hippocrates genitus in
insula Coo in primis clara ac valida et Aesculapio
dicata. is, cum fuisset mos liberatos morbis scribere
in templo eius dei quid auxiliatum esset, ut postea
similitudo proficeret, exscripsisse ea traditur, atque,
ut Varro apud nos credit, templo cremato instituisse
medicinam hanc quae clinice vocatur. nec fuit postea
quaestus modus, quoniam Prodicus[1] Selymbriae
natus, e discipulis eius, instituit quam vocant iatra-
lipticen et unctoribus quoque medicorum ac medi-
astinis vectigal invenit.

5 III. Horum placita Chrysippus ingenti garrulitate
mutavit plurimumque et ex Chrysippo discipulus eius
Erasistratus Aristotelis filia genitus. hic Antiocho
rege sanato centum talentis donatus est a rege
Ptolemaeo filio eius, ut incipiamus et praemia artis
ostendere.

IV. Alia factio ab experimentis se cognominans
empiricen coepit in Sicilia. Acrone Agragantino
6 Empedoclis physici auctoritate commendato. V.
dissederuntque hae scholae, et omnes eas damnavit
Herophilus in musicos pedes venarum pulsu discripto
per aetatum gradus. deserta deinde et haec secta

[1] Prodicus] *Coni.* Herodicus *Dal.*

---

[a] It is thought that Pliny should have said Herodicus, who
was the teacher, not the pupil, of Hippocrates.

[b] A celebrated Cnidian physician of the early third century
B.C. Perhaps Pliny, with his *ingenti garrulitate*, has confused
this physician with the Stoic philosopher, a prolific writer who
lived about the same time.

[c] Really the adopted son.

nesian War, when it was restored to the light by Hippocrates, who was born in the very famous and *Hippocrates.* powerful island of Cos, sacred to Aesculapius. It had been the custom for patients recovered from illness to inscribe in the temple of that god an account of the help that they had received, so that afterwards similar remedies might be enjoyed. Accordingly Hippocrates, it is said, wrote out these inscriptions, and, as our countryman Varro believes, after the temple had been burnt, founded that branch of medicine called " clinical." Afterwards there was no limit to the profit from medical practice, for one of the pupils of Hippocrates, Prodicus,[a] born in Selymbria, founded *The* *iatraliptice* (" ointment cure "), and so discovered *successors of Hippocrates.* revenue for the anointers even and drudges of the doctors.

III. Changes from their tenets were made, with a flood of verbiage, by Chrysippus,[b] and from Chrysippus also a violent change was made by his pupil Erasistratus, a son[c] of the daughter of Aristotle. For curing King Antiochus he received a hundred talents from King Ptolemy, his son, to begin my account of the prizes also of the profession.

IV. Another medical clique, calling themselves " Empirics " because they relied on experience, arose in Sicily, where Acron of Agrigentum received support from Empedocles, the physical scientist. V. These schools disagreed with each other, and were all condemned by Herophilus,[d] who divided pulsation into rhythmic feet for the various periods of life. Then this sect also was abandoned, because it was necessary for its members to have book-

[d] A famous physician of Alexandria, who was the first to count pulses.

est, quoniam necesse erat in ea litteras scire. mutata
et quam postea Asclepiades, ut rettulimus, invenerat.
auditor eius Themison fuit, seque inter initia adscripsit
illi, mox procedente vita[1] sua et[2] placita mutavit,
sed et illa Antonius Musa eiusdem auditor[3] auctoritate divi Augusti quem contraria medicina gravi
7 periculo exemerat. multos praetereo medicos cele-
berrimosque ex his Cassios, Calpetanos, Arruntios,
Rubrios. ducena quinquagena HS annuales[4] mer-
cedes fuere apud principes. Q. Stertinius inputavit
principibus quod sestertiis quingenis annuis contentus
esset, sescena enim sibi quaestu urbis fuisse enumera-
8 tis domibus ostendebat. par et fratri eius merces a
Claudio Caesare infusa est, censusque, quamquam
exhausti operibus Neapoli exornata, heredi HS C̅C̅C̅
reliquere, quantum aetate eadem[5] Arruntius solus.
exortus deinde est Vettius Valens adulterio Messa-
linae Claudii Caesaris nobilitatus pariterque elo-
quentia.[6] adsectatores et potentiam nanctus novam
instituit sectam. eadem aetas Neronis principatu ad
9 Thessalum transilivit delentem cuncta placita et
rabie quadam in omnis aevi medicos perorantem,
quali prudentia ingenioque aestimari vel uno argu-

[1] vita *vulg.* : vitia *codd.*
[2] sua et VRTf : ad sua E *Detlefsen* : sua d, *vulg.* : *an* et sua ?
[3] auditor] *om. codd., excidisse putat Mayhoff.*
[4] annuales dTf : annua his E *Detlefsen* : annuae iis *May-hoff.*
[5] aetate eadem *Ianus, Mayhoff* : Athenaidi *coni. Detlefsen*
Athena id est E *vulg.* : Athenade R : Athena dens d.
[6] eloquentiae adsectatores et potentiae *Mayhoff.*

---

a He used cold baths instead of hot.
b These were probably Greeks, in spite of their Roman
names.

learning, and that sect also was changed that
afterwards had been founded, as I have related, by
Asclepiades. He had a pupil called Themison, who *Asclepiades.*
at first followed his master, but then later in life he
also changed his tenets, a further change being made
by Antonius Musa, another pupil of Asclepiades,
with the support of the late Emperor Augustus,
whose life in a dangerous illness he had saved by
reversing the treatment.[a] I pass over many famous
physicians, among them men like Cassius, Calpetanus,
Arruntius and Rubrius.[b] Two hundred and fifty *Physicians'*
thousand sesterces were their annual incomes[c] from *incomes.*
the Emperors. Q. Stertinius said that the Emperors
were in his debt because he had been content with an
income of five hundred thousand sesterces a year,
proving by a counting of homes that his city practice
had brought in six hundred thousand. A like fortune
also was showered by Claudius Caesar upon his
brother, and the estates, although exhausted by
beautifying Naples with buildings, left to the heir
thirty million, Arruntius alone in the same age
leaving as much. Then there arose Vettius Valens,
celebrated for his intrigue with Messalina, wife of
Claudius Caesar, and equally so for his eloquence.
Chancing to gain followers and power he founded a
new sect. The same generation in the principate of
Nero rushed over to Thessalus, who swept away all *Thessalus.*
received doctrines, and preached against the
physicians of every age with a sort of rabid frenzy.
The wisdom and talent he showed can be fully
judged even by one piece of evidence: on his monu-

---

[c] The reading *annuales* has such strong support (R too has
anulis) that with much misgiving I retain it.

mento abunde potest, cum monumento suo, quod est
Appia via, iatronicen se inscripserit. nullius histrionum equorumque trigarii comitatior egressus in
publico erat, cum Crinas Massiliensis arte geminata,
ut cautior religiosiorque, ad siderum motus ex
ephemeride mathematica cibos dando horasque
observando auctoritate eum praecessit, nuperque
$\overline{\text{HS}}$ c̄ reliquit, muris patriae moenibusque aliis paene
10 non minore summa extructis. hi regebant fata, cum
repente civitatem Charmis ex eadem Massilia invasit
damnatis non solum prioribus medicis verum et balneis, frigidaque etiam hibernis algoribus lavari persuasit. mersit aegros in lacus. videbamus senes consulares usque in ostentationem rigentes, qua de re
11 exstat etiam Annaei Senecae adstipulatio. nec
dubium est omnes istos famam novitate aliqua aucupantes anima statim nostra negotiari. hinc illae
circa aegros miserae sententiarum concertationes,
nullo idem censente, ne videatur accessio alterius.
hinc illa infelix monumentis inscriptio, turba se
medicorum perisse. mutatur ars cottidie totiens
interpolis, et ingeniorum Graeciae flatu inpellimur,
palamque est, ut quisque inter istos loquendo polleat,

---

a See *Epistles* VI. 1, 3 and XII. 1, 5.
b Or, " ominous."
c Or, " breeze from."

188

ment on the Appian Way he described himself as
iatronices, " the conqueror of physicians." No actor,
no driver of a three-horse chariot, was attended by
greater crowds than he as he walked abroad in public,
when Crinas of Massilia united medicine with another
art, being of a rather careful and superstitious nature,
and regulated the diet of patients by the motions of
the stars according to the almanacs of the astrono-
mers, keeping watch for the proper times, and out-
stripped Thessalus in influence. Recently he left ten
millions, and the sum he spent upon building the
walls of his native city and other fortifications was
almost as much. These men were ruling our
destinies when suddenly the state was invaded by
Charmis, also from Massilia, who condemned not
only previous physicians but also hot baths, per-
suading people to bathe in cold water even during
the winter frosts. His patients he plunged into
tanks, and we used to see old men, consulars, actually
stiff with cold in order to show off. Of this we
have today a confirmation even in the writings of
Annaeus Seneca.[a] There is no doubt that all these,
in their hunt for popularity by means of some
novelty, did not hesitate to buy it with our lives.
Hence those wretched, quarrelsome consultations at
the bedside of the patient, no consultant agreeing
with another lest he should appear to acknowledge
a superior. Hence too that gloomy [b] inscription on
monuments : " It was the crowd of physicians that
killed me." Medicine changes every day, being
furbished up again and again, and we are swept
along on the puffs [c] of the clever brains of Greece.
It is obvious that anyone among them who acquires
power of speaking at once assumes supreme command

imperatorem illico vitae nostrae necisque fieri, ceu
vero non milia gentium sine medicis degant nec
tamen sine medicina, sicuti p. R. ultra sexcentesimum
annum, neque ipse in accipiendis artibus lentus, medicinae vero etiam avidus, donec expertam damnavit.

12 VI. Etenim percensere insignia priscorum in his
moribus convenit. Cassius Hemina ex antiquissimis
auctor est primum e medicis venisse Romam Peloponn-
neso Archagathum Lysaniae filium L. Aemilio M.
Livio cos. anno urbis DXXXV, eique ius Quiritium
datum et tabernam in compito Acilio emptam ob id
13 publice. vulnerarium eum fuisse tradunt,[1] mireque
gratum adventum eius initio, mox a saevitia secandi
urendique transisse nomen in carnificem et in taedium
artem omnesque medicos, quod clarissime intellegi
potest ex M. Catone, cuius auctoritati triumphus
atque censura minimum conferunt, tanto plus in ipso
est. quamobrem verba eius ipsa ponemus:

14 VII. Dicam de istis Graecis suo loco, M. fili.[2]
quid Athenis exquisitum habeam et quod bonum sit
illorum litteras inspicere, non perdiscere, vincam.
nequissimum et indocile genus illorum, et hoc puta
vatem dixisse: quandoque ista gens suas litteras

[1] tradunt *vulg.*, *Detlefsen*: egregium *Mayhoff*: credunt
codd.
[2] *Mayhoff hoc modo distinguit*: *post* fili *comma*, *post* perdiscere *punctum*; *post* vincam *punctum delet*; evincam *coni.*

---

[a] 219 B.C.
[b] With the reading of Mayhoff: "He also says that
Archagathus was an excellent surgeon, etc."

over our life and slaughter, just as if thousands of peoples do not live without physicians, though not without physic, as the Roman people have done for more than six hundred years, although not slow themselves to welcome science and art, being actually greedy for medicine until trial led them to condemn it.

VI. In fact this is the time to review the outstanding features of medical practices in the days of our fathers. Cassius Hemina, one of our earliest authorities, asserts that the first physician to come to Rome was Archagathus, son of Lysanias, who *Archagathus.* migrated from the Peloponnesus in the year of the city 535,[a] when Lucius Aemilius and Marcus Livius were consuls. He adds that citizen rights were given him, and a surgery at the cross-way of Acilius was bought with public money for his own use. They say [b] that he was a wound specialist, and that his arrival at first was wonderfully popular, but presently from his savage use of the knife and cautery he was nicknamed " Executioner," and his profession, with all physicians, became objects of loathing. The truth of this can be seen most plainly in the opinion of Marcus Cato, whose authority is very little enhanced by his triumph and censorship ; so much more comes from his personality. Therefore I will lay before my readers his very words.

VII. I shall speak about those Greek fellows in *Cato on* their proper place, son Marcus, and point out the *physicians.* result of my enquiries at Athens, and convince you what benefit comes from dipping into their literature, and not making a close study of it. They are a quite worthless people, and an intractable one, and you must consider my words prophetic. When that race gives

dabit, omnia conrumpet, tum etiam magis, si medicos
suos hoc mittet. iurarunt inter se barbaros necare
omnes medicina, et hoc ipsum mercede faciunt ut
fides is sit et facile disperdant. nos quoque dictitant
barbaros et spurcius nos quam alios opicon appella-
tione foedant. interdixi tibi de medicis.

15 VIII. Atque hic Cato sescentesimo quinto anno
urbis nostrae obiit, octogesimo quinto suo, ne quis
illi defuisse publice tempora aut privatim vitae spatia
ad experiendum arbitretur. quid ergo? damnatam
ab eo rem utilissimam credimus? minime, Hercules.
subicit enim qua medicina se et coniugem usque ad
longam senectam perduxerit, his ipsis scilicet quae
nunc nos tractamus,[1] profiteturque esse commen-
tarium sibi quo medeatur filio, servis, familiaribus,
16 quem nos per genera usus sui[2] digerimus. non rem
antiqui damnabant, sed artem, maxime vero quaes-
tum esse manipretio vitae recusabant. ideo templum
Aesculapii, etiam cum reciperetur is deus, extra
urbem fecisse iterumque in insula traduntur, et cum
Graecos Italia pellerent diu etiam post Catonem,
excepisse medicos. augebo providentiam illorum.
17 solam hanc artium Graecarum nondum exercet
Romana gravitas, in tanto fructu paucissimi Quiritium

[1] nos tractamus *Gelenius, Harduinus, Mayhoff*: nos
trademus *vulg., Detlefsen*: nostra scitamus *plerique codd.*
[2] usus sui *codd. et edd.*: ususve *coni. Mayhoff.*

[a] An uncultivated Italian tribe.
[b] Do we believe that a thing condemned by him is very
useful?
[c] A curious use of *excipio*. Yet we must either so translate
or with Sillig read *nec* for *et.*

us its literature it will corrupt all things, and even all the more if it sends hither its physicians. They have conspired together to murder all foreigners with their physic, but this very thing they do for a fee, to gain credit and to destroy us easily. They are also always dubbing us foreigners, and to fling more filth on us than on others they give us the foul nickname of Opici.[a] I have forbidden you to have dealings with physicians.

VIII. And this Cato died in the 605th year of the City and the 85th of his own life, so that nobody can think that he lacked opportunities in public life, or length of years in private life, to gather experiences. What then? Are we to believe that he condemned a very useful thing?[b] No, by heaven! For he adds the medical treatment by which he prolonged his own life and that of his wife to an advanced age, by these very remedies in fact with which I am now dealing, and he claims to have a notebook of recipes, by the aid of which he treated his son, servants, and household; these I rearrange under the diseases for which they are used. It was not medicine that our forefathers condemned, but the medical profession, chiefly because they refused to pay fees to profiteers in order to save their lives. For this reason even when Aesculapius was brought as a god to Rome, they are said to have built his temple outside the city, and on another occasion upon an island, and when, a long time too after Cato, they banished Greeks from Italy, to have expressly included [c] physicians. I will magnify yet further their wisdom. Medicine alone of the Greek arts we serious Romans have not yet practised; in spite of its great profits only a very few of our citizens have touched upon it,

193

attigere et ipsi statim ad Graecos transfugae, immo
vero auctoritas aliter quam Graece eam tractantibus
etiam apud inperitos expertesque linguae non est, ac
minus credunt quae ad salutem suam pertinent, si in-
tellegant. itaque, Hercules, in hac artium sola evenit
ut cuicumque medicum se professo statim credatur,
18 cum sit periculum in nullo mendacio maius. non
tamen illud intuemur, adeo blanda est sperandi pro
se cuique dulcedo. nulla praeterea lex quae puniat
inscitiam capitalem, nullum exemplum vindictae.
discunt periculis nostris et experimenta per mortes
agunt, medicoque tantum hominem occidisse inpuni-
tas summa est. quin immo transit convicium et
intemperantia culpatur ultroque qui periere arguun-
tur. sed decuriae pro more censuris principum
examinantur, inquisitio per parietes agitur, et qui de
nummo iudicet a Gadibus columnisque Herculis
arcessitur, de. exilio vero non nisi XLV electis viris
19 datur tabella. at de iudice ipso quales in consilium
eunt statim occisuri! merito, dum nemini nostrum
libet scire quid saluti suae opus sit. alienis pedibus
ambulamus, alienis oculis agnoscimus, aliena memoria salutamus, aliena et vivimus opera, perierunt-
que rerum naturae pretia et vitae argumenta. nihil

ᵃ This refers to the Roman custom of using slaves to carry
them in litters, or to prompt them if they forgot faces or names.
194

and even these were at once deserters to the Greeks;
nay, if medical treatises are written in a language
other than Greek they have no prestige even among
unlearned men ignorant of Greek, and if any should
understand them they have less faith in what con-
cerns their own health. Accordingly, heaven knows,
the medical profession is the only one in which anybody professing to be a physician is at once trusted,
although nowhere else is an untruth more dangerous.
We pay however no attention to the danger, so great
for each of us is the seductive sweetness of wishful
thinking. Besides this, there is no law to punish
criminal ignorance, no instance of retribution.
Physicians acquire their knowledge from our dangers,
making experiments at the cost of our lives. Only
a physician can commit homicide with complete
impunity. Nay, the victim, not the criminal, is
abused; his is the blame for want of self-control,
and it is actually the dead who are brought to account.
Panels of judges are tested according to custom by
the censorial powers of the Emperor; their examination invades the privacy of our homes; to give a
verdict on a petty sum a man is summoned from
Cadiz and the Pillars of Hercules; indeed, before the
penalty of exile can be inflicted forty-five selected
men are given power to vote on it; yet on the judge
himself what manner of men sit in consultation to
murder him out of hand! We deserve it all, so long
as not one of us cares to know what is necessary for
his own good health. We walk with the feet of
others, we recognise our acquaintances with the eyes
of others, rely on others' memory to make our
salutations,[a] and put into the hands of others our
very lives; the precious things of nature, which

20 aliud pro nostro habemus quam delicias. non
deseram Catonem tam ambitiosae artis invidiae a me
obiectum aut senatum illum qui ita censebat, idque
non criminibus artis arreptis, ut aliquis exspectaverit.
quid enim venenorum fertilius aut unde plures testa-
mentorum insidiae? iam vero et adulteria etiam in
principum domibus, ut Eudemi in Livia Drusi
Caesaris, item Valentis in qua dictum est regina
21 non sint artis ista sed hominum; non magis haec urbi
timuit Cato, ut equidem credo, quam reginas. ne
avaritiam quidem arguam rapacesque nundinas pen-
dentibus fatis et dolorum indicaturam ac mortis arram
aut arcana praecepta, squamam in oculis emovendam
potius quam extrahendam, per quae effectum est ut
nihil magis pro re videretur quam multitudo grassan-
tium; neque enim pudor sed aemuli pretia summit-
22 tunt. notum est ab eodem Charmide unum aegrum
ex provincialibus $\overline{\text{HS}}$ cc[1] reconductum Alconti vulnerum medico, $\overline{\text{HS}}$ $\overline{\text{x}}$[2] damnato ademisse Claudium
principem, eidemque in Gallia exulanti et deinde
restituto adquisitum non minus intra paucos annos.
23 et haec personis inputentur. ne faecem quidem aut
inscitiam eius turbae[3] arguamus, ipsorum intem-

---

[1] cc *Warmington*: $\overline{\text{cc}}$ *codd.*
[2] $\overline{\text{x}}$ *Warmington*: $\overline{\text{c}}$ *codd.*
[3] turbae d *vulg.*: turpem *Mayhoff*: turbam *plerique codd.*
Post ipsorum *add.* procerum *Mayhoff.*

---

[a] That a further operation may be necessary.
[b] With Mayhoff's readings: "or its disgraceful ignorance,
the irresponsibility of the leading physicians themselves."

support life, we have quite lost. We have nothing else of our own save our luxuries. I will not abandon Cato exposed by me to the hatred of so vain-glorious a profession, or yet that Senate which shared his views, and that without seizing, as one might expect, any chances of accusation against the profession. For what has been a more fertile source of poisonings? Whence more conspiracies against wills? Yes, and through it too adulteries occur even in our imperial homes, that of Eudemus with Livia, wife of Drusus Caesar, and that of Valens with the royal lady with whom his name is linked. We may grant that the blame for such sins may lie with persons, not with the medical profession; Cato, I believe, had no more fears for Rome about these matters than he had about the presence in Rome of royal ladies. Let me not even bring charges against their avarice, their greedy bargains made with those whose fate lies in the balance, the prices charged for anodynes, the earnest-money paid for death, or their mysterious instructions, that a cataract should be moved away and not pulled off.[a] The result is that the brightest side of the picture is the vast number of marauders; for it is not shame but the competition of rivals that brings down fees. It is well known that the Charmis aforesaid exchanged one sick provincial for 200,000 sesterces by a bargain with Alcon the wound-surgeon; that Charmis was condemned and fined by the Emperor Claudius the sum of 1,000,000 sesterces, yet as an exile in Gaul and on his return from banishment he amassed a like sum within a few years. Let the blame for this sort of thing also be laid on persons. I must not accuse even the dregs of that mob [b] or its ignorance: the irresponsibility of

Attack on physicians.

197

PLINY: NATURAL HISTORY

perantiam, in morbis [1] aquarum calidarum deverticulis
imperiosa inedia et ab isdem deficientibus cibo saepius
die ingesto, mille praeterea paenitentiae modis,
culinarum etiam praeceptis et unguentorum mixturis,
24 quando nullas omisere vitae inlecebras. invehi pere-
grinas merces conciliarique externa pretia displicuisse
maioribus crediderim equidem, non tamen hoc
Catonem providisse, cum damnaret artem. theriace
vocatur excogitata compositio. fit ex rebus sex-
centis,[2] cum tot remedia dederit natura quae singula
sufficerent. Mithridatium antidotum ex rebus LIIII
componitur, inter nullas [3] pondere aequali et quarun-
dam rerum sexagesima denarii unius imperata, quo
25 deorum, per Fidem, ista monstrante! hominum enim
subtilitas tanta esse non potuit, ostentatio artis et
portentosa scientiae venditatio manifesta est. ac ne
ipsi quidem illa novere, conperique volgo pro cinnabari Indica in medicamenta minium addi inscitia
nominis, quod esse venenum docebimus inter pig-
26 menta. verum haec ad singulorum salutem pertinent, illa autem quae timuit Cato atque providit,
innocentiora multo et parva opinatu quae proceres
artis eius de semet ipsi fateantur, illa perdidere imperii
mores, illa quae sani patimur, luctatus ceromata ceu
valitudinis causa instituta, balineae ardentes quibus
persuasere in corporibus cibos coqui ut nemo non

    [1] in morbis *codd.*: immodicis *Mayhoff.*
    [2] sexcentis *Sillig, Mayhoff*: externis *codd., Detlefsen.*
    [3] nullas *Mayhoff*: nullius *Detlefsen*: nullos *plerique codd.*

    [a] Celsus (V. 23, 3) gives the number of ingredients as thirty-
six. The *antidota* were stimulant, aromatic substances which,
with honey and wine, were given for falls, pains, and poisons.
    [b] Also called *cinnabaris nativa*; hence the error.
    [c] See XXXIII. § 124.

the physicians themselves, with their out-of-the-way use of hot water in sickness, their strict fasts for patients, who when in a fainting condition are stuffed with food several times a day, their thousand ways moreover of changing their minds, their orders to the kitchen, and their compound ointments; for none of life's seductive attractions have they refrained from touching. I am inclined to believe that our ancestors were displeased with imports from abroad and with the fixing of prices by foreigners, but not that Cato foresaw these things when he condemned the profession. There is an elaborate mixture called *theriace*, which is compounded of countless ingredients, although Nature has given as many remedies, anyone of which would be enough by itself. The Mithridatic antidote is composed of fifty-four [a] ingredients, no two of them having the same weight, while of some is prescribed one sixtieth part of one denarius. Which of the gods, in the name of Truth, fixed these absurd proportions? No human brain could have been sharp enough. It is plainly a showy parade of the art, and a colossal boast of science. And not even the physicians know their facts; I have discovered that instead of Indian cinnabar there is commonly added to medicines, through a confusion of names, red lead,[b] which, as I shall point out when I discuss pigments,[c] is a poison. These things however concern the health of individuals; but those other practices, which Cato feared and foresaw, much less harmful and less regarded, such as the heads of that profession themselves admit about themselves, those, I say, have ruined the morals of the Empire, I mean the practices to which we submit when in health—wrestlers' ointments, as though

minus validus exiret, oboedientissimi vero efferrentur,
potus deinde ieiunorum ac vomitiones et rursus per-
potationes ac pilorum eviratio instituta resinis
eorum, itemque pectines in feminis quidem publicati.

27 ita est profecto, lues morum, nec aliunde maior quam
e medicina, vatem prorsus cottidie facit Catonem et
oraculum: satis esse ingenia Graecorum inspicere,

28 non perdiscere. haec fuerint dicenda pro senatu illo
sescentisque p. R. annis adversus artem in qua condi-
tione insidiosissima auctoritatem pessimis boni
faciunt, simul contra attonitas quorundam persua-
siones qui prodesse nisi pretiosa non putant. neque
enim dubitaverim aliquis fastidio futura quae dicentur
animalia, at non Vergilio fuit nominare formicas nulla
necessitate et curculiones ac lucifugis congesta cubilia
blattis, non Homero inter proelia deorum inprobi-
tatem muscae describere, non naturae gignere ista,
cum gignat hominem. proinde causas quisque et
effectus, non res aestimet.

29   IX. Ordiemur autem a confessis, hoc est lanis ovis-
que, ut ¹ rebus praecipuis honos in primis perhibeatur.

¹ ut *Urlichs, Detlefsen*: ob id ut *Mayhoff*: obiter (obitur)
aut obiter ut *codd.*

ᵃ A pun on *concoquere* (and sometimes *coquere*) in the sense
of " digest."
ᵇ Or, "innumerable."
ᶜ *Georgics* I. 186 and IV. 243.

they were intended to treat ill health, broiling baths, by which they have persuaded us that food is cooked [a] in our bodies, so that everybody leaves them the weaker for the treatment, and the most submissive are carried out to be buried, the draughts taken fasting, vomitings followed by further heavy potations, effeminate depilations produced by their resins, and even the pubes of women exposed to public view. It is certainly true that our degeneracy, due to medicine more than to anything else, proves daily that Cato was a genuine prophet and oracle when he stated that it is enough to dip into the works of Greek brains without making a close study of them. Thus much must be said in defence of that Senate and those 600 [b] years of the Roman State, against a profession where the treacherous conditions allow good men to give authority to the worst, and at the same time against the stupid convictions of certain people who consider nothing beneficial unless it is costly. For I feel sure that some will be disgusted at the animals I shall treat of, although Virgil [c] did not disdain to speak quite unnecessarily of ants and weevils, and of:—

> " sleeping places heaped up by cockroaches that
>   avoid the light."

Nor did Homer [d] disdain amid the battles of the gods to tell of the greed of the fly, nor yet did Nature disdain to create them because she creates man. Therefore let each take into account, not things themselves, but causes and results.

IX. But I shall commence with admitted medical aids, that is, with wools and eggs, to give first *Wool and eggs.*

[d] *Iliad* XVII. 570.

quaedam etiam si[1] alienis locis, tamen obiter dici
necesse erit. nec deerat materia pompae, si quicquam aliud intueri liberet quam fidem operis, quippe
inter prima proditis etiam ex cinere phoenicis nidoque
medicinis, ceu vero id certum esset atque non fabulo-
sum. inridere est vitae remedia post millensimum
30 annum reditura monstrare. lanis auctoritatem
veteres Romani etiam religiosam habuere postes a
nubentibus attingi iubentes, praeterque cultum et
tutelam contra frigora sucidae plurima praestant
remedia ex oleo vinoque aut aceto, prout quaeque
mulceri morderive opus sit et adstringi laxarive,
luxatis membris dolentibusque nervis inpositae et
crebro suffusae. quidam et salem admiscent luxatis,
alii cum lana rutam tritam adipemque inponunt,
31 item contusis tumentibusque. halitus quoque oris
gratiores facere traditur confricatis dentibus atque
gingivis admixto melle. prodest et phreneticis
suffitu. sanguinem in naribus sistit cum oleo rosaceo,
et alio modo indita auribus opturatis spissius. quin
et ulceribus vetustis inponitur cum melle. vulnera
ex vino vel aceto vel aqua frigida et oleo expressa
32 sanat. arietis vellera luta frigida ex oleo madefacta
in muliebribus malis inflammationes vulvae sedant et,
si procidant, suffitu reprimunt. sucida lana inposita
subditaque mortuos partus evocat. sistit etiam pro-

---

[1] si E *vulg. Detlefsen* : sic *plerique codd., Mayhoff.*

---

[a] Or probably " chief," " best."
[b] For *phrenitis* see List of Diseases.

honours to things of the first importance. Certain matters even out of their proper place it will be necessary to discuss, at least as incidental asides. Nor would material be wanting for rhetoric if it pleased me to pay attention to anything else than to making my work trustworthy, seeing that fable even says that among the first [a] medicines was one from the ashes and nest of the phoenix, just as though the story were fact and not myth. It is to joke with mankind to point out remedies that return only after a thousand years. The old Romans assigned to wool even supernatural powers, for they bade brides touch with it the doorposts of their new homes; and besides dress and protection from cold, unwashed wool supplies very many remedies if dipped in oil and wine or vinegar, according as the particular need is for an emollient or a pungent remedy, for an astringent or a relaxing one, being applied, and frequently moistened, for dislocations and aching sinews. For dislocations some add salt also; others apply with wool pounded rue and fat, likewise for bruises and swellings. To rub too the teeth and gums with wool and honey is said to make the breath more pleasant, and to fumigate with wool benefits phrenitis.[b] Nose bleeding is checked by inserting wool and rose oil; another way is to put it into the ears and plug them rather firmly. It is applied moreover with honey to old sores. Wounds it heals if dipped in wine, or vinegar, or cold water and oil, and then squeezed out. A ram's fleece washed in cold water and soaked in oil, soothes inflammations of the uterus in women's complaints, and by fumigation reduces prolapsus. Unwashed wool applied or used as a pessary extracts a dead foetus; it also

fluvia earum, et canis rabiosi morsibus inculcata post
diem septimum solvitur. reduvias sanat ex aqua
frigida, eadem nitro, sulpure, oleo, aceto, pice liquida
fervescentibus tincta quam calidissima inposita bis die
lumborum dolores sedat. sistit et sanguinem ex
ariete sucida articulos extremitatium praeligans.
33 laudatissima omnis e collo, natione vero Galatica,
Tarentina, Attica, Milesia. sucidam inponunt et desquamatis, percussis, lividis, incussis, conlisis, contritis,
deiectis, capitis et aliis doloribus, stomachi inflammationi ex aceto et rosaceo. cinis eius inlinitur adtritis,
vulneratis, ambustis. et in oculorum medicamentis
34 additur, item in fistulas auresque suppuratas. ad
hoc detonsam eam, alii evolsam, decisis summis partibus siccant carpuntque et in fictili crudo conponunt
ac melle perfundunt uruntque. alii astulis taedae
subiectis et subinde interstratis oleo adspersam
accendunt, cineremque in labellis aqua addita con-
fricant manu et considere patiuntur, idque saepius
mutantes aquam, donec linguam adstringat leniter
nec mordeat. tunc cinerem reponunt. vis eius
septica est efficacissimeque genas purgat.
35   X. Quin ipsae sordes pecudum sudorque feminum
et alarum adhaerentes lanis—oesypum vocant—
innumeros prope usus habent. in Atticis ovibus
genito palma. fit pluribus modis, sed probatissimum

stays uterine fluxes. Plugged into the bites of a mad dog it is taken away after the seventh day. With cold water it cures hangnails. Again, dipped into a hot mixture of soda, sulphur, oil, vinegar and liquid pitch, all as hot as possible, and applied twice a day, wool relieves lumbago. Unwashed ram's wool also stays bleeding if bound round the joints of the extremities. The most highly esteemed wool is: all from the neck, and that from the districts of Galatia, Tarentum, Attica, and Miletus. Unwashed wool is applied to excoriations, blows, bruises, contusions, crushed parts, galling, falls, pains in the head and elsewhere, and with vinegar and rose oil to inflammation of the stomach. The ash of wool is applied to chafings, wounds, and burns. It is added to medicaments for the eyes, and also used for fistulas and suppurating ears. For this purpose some take shorn wool, others wool plucked out, cut off the ends, dry, card, place in a vessel of unbaked clay, steep in honey, and burn. Others place under it a layer of pitch-pine chips, make several alternate layers, sprinkle with oil, and set on fire. The ash is rubbed by the hand into little pots, with water added, and then allowed to settle. The operation is repeated several times, with changes of water, until the ash becomes slightly astringent to the tongue without stinging it; then it is stored away. It has a caustic property that makes it an excellent detergent for the eyelids.

X. Moreover, even the greasy sweat of sheep that clings to the wool under the hollows of their flanks and forelegs—it is called *oesypum* (suint)—has uses almost innumerable. The most prized is that obtained from Attic sheep. There are several ways of preparing it,

Oesypum (suint).

lana ab his partibus recenti concerpta aut quibuscumque sordibus sucidis primum collectis lento igni in aeneo subfervefactis et refrigeratis pinguique quod supernatet collecto in fictile vas iterumque decocta priore materia, quae pinguitudo utraque frigida aqua lavatur et in linteo saccatur ac sole torretur, donec candida fiat ac tralucida, tum in stagnea pyxide 36 conditur. probatio ut sordium virus oleat et manu fricante ex aqua non liquetur sed albescat ut cerussa. oculis utilissimum contra inflammationes genarumque callum. quidam in testa torrent donec pinguitudinem amittat, utilius tale existimantes erosis et duris genis, 37 angulis scabiosis et lacrimantibus. ulcera non oculorum modo sanat sed oris etiam et genitalium cum anserino adipe. medetur et vulvae inflammationibus et sedis rhagadiis et condylomatis cum meliloto ac butyro. reliquos usus eius digeremus. sordes quoque caudarum concretae in pilulas siccatae per se tusaeque in farinam et inlitae dentibus mire prosunt, etiam 38 labantibus,[1] gingivisque, si carcinoma serpat, iam vero pura vellera aut per se inposita caecis doloribus aut accepto sulpure, et cinis eorum genitalium vitiis, tantumque pollent ut medicamentis quoque superponantur. medentur ante omnia et pecori ipsi, si fastidio non pascatur. cauda enim quam artissime

---

[1] labantibus d, *vulg.*, *Mayhoff* : labantibusque VR : labantibus quae E : labantibus, uvae *coni. Detlefsen.*

---

[a] An alloy of silver and lead.
[b] Or, " sweaty grease too round the tail, if allowed to dry and congeal by itself into little balls and then etc."
[c] That is, of uncertain locality or origin. The word is used again with *dolores* in § 55.

but the most approved is to take fresh-plucked wool
from the parts mentioned, or first to gather the greasy
sweat from any part, then warm it in a bronze pot
over a slow fire, cool it again, collect in an earthen
vessel the fat that floats on the top, and boil again
the stuff originally used. Both the fats obtained
are washed in cold water, strained through linen,
heated in the sun until they become white and transparent, and then stored away in a box of *stannum*.[a]
The test of its purity is that it should retain the
strong smell of the grease, and when rubbed with
the hand in water, should not melt, but become white
like white-lead. It is very useful for inflammations
of the eyes and hard places on the eyelids. Some
bake it in an earthen jar until it is no longer fatty,
holding that in this condition it is a more useful
remedy for sores that have eaten into the eyelids, for
indurations there, and for watery itch at the corners.
It heals, not only sores of the eyes, but also with
goose grease those of the mouth and genitals.
With melilot and butter it cures inflammations of the
uterus, chaps in the anus, and condylomata. Its
other uses I shall set out in order later on. The
sweaty[b] grease too that gathers into pills about the
tail, dried by itself and ground to powder, is wonderfully beneficial if rubbed on the teeth, even when
these are loose, and on the gums when they suffer
from malignant, running sores. Furthermore, clean
pieces of fleece are applied to blind[c] pains, either
by themselves or with sulphur added, and their ash
to affections of the genitals, being so potent that they
are even placed over medicinal applications. Wool
is also the best of remedies for sheep themselves if
they lose their appetite and will not pasture. For if

praeligata, evolsa inde lana, statim vescuntur, traduntque quod extra nodum sit e cauda praemori.

39 XI. Lanae habent et cum ovis societatem simul
fronti inpositae contra epiphoras. non opus est eas
in hoc usu radicula esse curatas neque aliud quam
candidum ex ovo infundi ac pollinem turis. ova per
se infuso candido oculis epiphoras cohibent urentes-
que refrigerant—quidam cum croco praeferunt—et
pro aqua miscentur collyriis. infantibus vero contra
lippitudines ut[1] vix aliud remedio sunt[2] butyro

40 admixto recenti. eadem cum oleo trita ignes sacros
leniunt betae foliis superinligatis. candido ovorum
in oculis et pili reclinantur Hammoniaco trito
admixtoque et vari in facie cum pineis nucleis ac
melle modico. ipsa facies inlita sole non uritur.
ambusta aquis si statim ovo occupentur, pusulas non
sentiunt—quidam admiscent farinam hordeaciam et
salis parum—ulceribus vero ex ambusto cum candido
ovorum tostum hordeum et suillo adipe mire prodest.

41 eadem curatione ad sedis vitia utuntur, infantibus
quidem etiam si quid ibi procidat, ad pedum rimas
ovorum candido decocto cum cerussae denariorum

---

[1] ut vix *Mayhoff*: vix *codd.*, *Detlefsen.*
[2] sunt *Mayhoff*: est *codd.*, *Detlefsen.*

---

[a] The reading of the MSS. would mean: " scarcely anything else is a remedy except egg mixed with fresh butter,"
a startling statement even for Pliny. Mayhoff's conjectures
give the required sense, although it is hard to see how and
why corruption occurred.

their tails are tied as tightly as possible with wool plucked therefrom they at once begin to feed, and it is said that all the tail outside the knot dies off.

XI. Wool has also a close affinity with eggs, the two being laid together on the forehead for eye fluxes. There is no need for the wool, when so used, to have been treated with radicula, or for anything else except to spread on it white of egg and powdered frankincense. White of egg by itself, poured into the eyes, checks fluxes and cools inflammations, although some prefer to add saffron, and eggs can take the place of water in eye salves. But for infant ophthalmia scarcely anything else [a] is so remedial as egg mixed with fresh butter. Eggs beaten up with olive oil relieve erysipelas if beet leaves are tied on top. White of egg mixed with pounded gum ammoniac sets back eye-lashes, and removes spots on the face with pine nuts and a little honey. The face itself if smeared with egg is not burnt by the sun. If scalds are at once covered with egg they do not blister—some add barley flour and a pinch of salt—while sores from a burn are made wonderfully better by roasted barley with white of egg and pig's lard. The same treatment is used for affections of the anus, and even for procidence in the case of infants; for chaps on the feet the white of eggs is boiled down with two denarii by weight of white lead, an equal weight of litharge, a little myrrh, and then wine; for erysipelas is used the white of three eggs with starch. It is also said that white of egg closes wounds and expels stone from the bladder. The yolk of eggs, boiled hard, mixed with a little saffron and honey, and applied in woman's milk, relieves pains of the eyes; or it may be placed over

duum pondere, pari spumae argenti, murrae exiguo,
dein vino; ad ignem sacrum candido ovorum trium
cum amulo. aiunt et vulnera candido glutinari
42 calculosque pelli. lutea ovorum cocta ut indurescant,
admixto croco modice, item melle, ex lacte mulieris
inlita dolores oculorum mitigant, vel cum rosaceo et
mulso lana oculis inposita, vel cum trito apii semine
ac polenta in mulso inlita. prodest et tussientibus
per se luteum devoratum liquidum ita ut dentibus non
attingatur, thoracis destillationibus, faucium scabri-
tiae. privatim contra haemorroidis morsum inlinitur
43 sorbeturque crudum. prodest et renibus, vesicae
rosionibus exulcerationibusque.[1] cruenta excreantibus quinque ovorum lutea in vini hemina cruda
sorbentur, dysintericis cum cinere putaminis sui et
papaveris suco ac vino. dantur coeliacis cum uvae
passae pinguis pari pondere et malicorii per triduum
aequis portionibus, et alio modo lutea ovorum trium,
lardi veteris et mellis quadrantibus, vini veteris
cyathis tribus, trita ad crassitudinem mellis et, cum
44 opus sit, abellanae nucis magnitudine ex aqua
pota, item ex oleo fricta terna, totis ovis pridie
maceratis in aceto, sic et lientericis, sanguinem
autem reicientibus cum tribus cyathis musti.
utuntur isdem ad liventia, si vetustiora sint, cum
bulbis ac melle. sistunt et menses mulierum cocta
45 et e vino pota, inflationes quoque vulvae cruda
cum oleo ac vino inlita. utilia sunt et cervicis
doloribus cum anserino adipe et rosaceo, sedis etiam
vitiis indurata igni ut calore quoque prosint, et condylomatis cum rosaceo, item ambustis durata in

---

[1] *Sic dist. Mayhoff e Plinio iun.; ceteri edd. punctum post* excreantibus *ponunt.*

the eyes on wool with rose oil and honey wine, or applied in honey wine with ground celery-seed and pearl barley. Swallowed liquid, without letting it touch the teeth, the yolk by itself is good for cough, catarrh of the chest, and rough throats. Applied externally or taken internally the raw yolk is specific for the bite of the haemorrhois.[a] It is also good for the kidneys, and for irritation or ulceration of the bladder.[b] For spitting of blood five yolks of egg are swallowed raw in a hemina of wine, and for dysentery they are taken with the ash of their shells, poppy juice, and wine. With the same weight of plump raisins and pomegranate rind yolk of egg is given in equal doses for three days to sufferers from coeliac affections. Another way is to take the yolks of three eggs, three ounces of old bacon fat and of honey, and three cyathi of old wine, beat them up until they are of the consistency of honey, and take in water when required pieces of the size of a filbert. Yet another way is to fry three eggs after steeping them whole the day before in vinegar, and to use them so for spleen diseases, but to take them in three cyathi of must for the spitting of blood. Eggs are used with bulbs and honey for persistent bruises. Boiled and taken in wine they also check menstruation; inflation too of the uterus if applied raw with oil and wine. They are useful too, with goose grease and rose oil, for pains in the neck; for affections of the anus also, if hardened over fire and applied while the additional benefit of the heat is still retained; for condylomata with rose oil; for burns they are hardened in water,

[a] For this poisonous snake see Lucan IX. 709 foll.
[b] Mayhoff's punctuation avoids the awkward repetition of *in vini hemina* and *cum . . . vino* in the same prescription.

aqua, mox in pruna;[1] putaminibus exustis, tum lutea
ex rosaceo inlinuntur. fiunt et tota lutea, quae
vocant sitista; cum triduo incubita tolluntur.
stomachum dissolutum confirmant pulli ovorum cum
gallae dimidio, ita ne ante duas horas alius cibus
sumatur. dant et dysintericis pullos in ipso ovo
decoctos admixta vini austeri hemina et pari modo
46 olei polentaeque. membrana putamini detracta sive
crudo sive cocto labiorum fissuris medetur, putaminis
cinis in vino potus sanguinis eruptionibus. comburi
sine membrana oportet, sic fit et dentifricium. idem
cinis et mulierum menses cum murra inlitus sistit.
firmitas putaminum tanta est ut recta nec vi nec
pondere ullo frangantur, nec nisi paulum inflexa
47 rotunditate. tota ova adiuvant partum cum ruta et
aneto et cumino pota e vino. scabiem corporum ac
pruritum oleo et cedria mixtis tollunt, ulcera quoque
umida in capite cyclamino admixta. ad puris et
sanguinis excreationes ovum crudum cum porri sectivi
suco parique mensura mellis Graeci calefactum
hauritur. dantur et tussientibus cocta et trita cum
melle et cruda cum passo oleique pari modo. infunduntur et virilitatis vitiis singula cum ternis passi
cyathis amulique semuncia a balneis, adversus ictus
serpentium cocta tritaque adiecto nasturtio inlinun-
48 tur. cibo quot modis iuvent notum est, cum trans-

[1] *Distinxi ego.*

then over hot coals; when the shells have been burned off, finally the yolks are applied in rose oil. Eggs become entirely yolk (they are then called *sitista*) when the hen has sat upon them for three days before they are taken up. The chicks found in eggs taken with half a gall nut settle a disordered stomach, but care must be taken to eat no other food for the next two hours. There are also given to dysentery patients chicks boiled in the egg itself and added to a hemina of dry wine and the same quantity of oil and pearl barley. The membrane peeled off the shell of a raw or boiled egg heals cracks in the lips. The shell reduced to ash and taken in wine cures discharges of blood. It must be burnt without the membrane. From this ash is also made a dentifrice. It also checks menstruation if applied with myrrh. The strength of the shells is so great that no force or weight will break them when the eggs are perpendicular, but only when the oval is slightly inclined. Childbirth is made easier by whole eggs with rue, dill, and cummin, taken in wine. Itch and irritation of the skin are removed by a mixture of oil, cedar-resin, and eggs; running ulcers too on the head by eggs mixed with cyclamen. For spitting of pus or blood is swallowed a raw egg warmed with juice of cutleek and an equal amount of Greek honey. There are given to patients with a cough boiled eggs beaten up with honey, or raw eggs with raisin wine and an equal measure of oil. Eggs are also injected for complaints of the male organs, the dose being one egg with three cyathi of raisin wine and half an ounce of starch, given after the bath; for snake bite they are applied after boiling them and beating up with the addition of cress. How helpful in many

meent faucium tumorem calfactuque obiter foveant.
nullus est alius cibus qui in aegritudine alat neque
oneret simulque vim potus et cibi habeat. macera-
49 torum in aceto molliri diximus putamen. talibus
cum farina in panem subactis coeliaci recreantur.
quidam ita resoluta in patinis torrere utilius putant,
quo genere non alvos tantum sed et menses feminarum sistunt, aut si maior sit impetus, cruda cum
farina et aqua hauriuntur, et per se lutea ex his
decocta in aceto donec indurescant, iterumque cum
trito pipere torrentur [1] ad cohibendas alvos. fit et
50 dysintericis remedium singulare ovo effuso in fictile
novum eiusdemque ovi mensura, ut paria sint omnia,
melle, mox aceto, item oleo confusis crebroque per-
mixtis. quo fuerint ea excellentiora hoc praesentius
remedium erit. alii eadem mensura pro oleo et aceto
resinam adiciunt rubentem vinumque; et alio modo
temperant, olei tantum mensura pari pineique
corticis duabus sexagensimis denarii ac una eius quod
rhus diximus, mellis obolis quinque simul decoctis, ita
ut cibus alius post quattuor horas sumatur. torminibus quoque multi medentur ova bina cum alii spicis
quattuor una terendo vinique hemina calefaciendo
51 atque ita potui dando. et, ne quid desit ovorum
gratiae, candidum ex his admixtum calci vivae

---

[1] torrentur *vulg.*: *Mayhoff qui* tosta dantur *coni.*: torreantur *codd. Detlefsen.*

---

[a] Book X. § 167.   [b] See XXIV. § 91.

ways eggs are as food is well known, for they pass a
swollen throat and incidentally by their heat soothe
it.   There is no other food so nourishing in sickness
without overloading the stomach, and it has the
nature of both food and drink.   I have said *a* that
the shell is softened of eggs steeped in vinegar.
Eggs so prepared and kneaded into bread with flour
give refreshment to patients with coeliac affections.
Some think it more useful, after softening them in
this way, to bake them in shallow pans; when so prepared they check not only diarrhoea but also excessive
menstruation; or if the attack is specially severe they
are swallowed raw with flour and water, or the yolks
from these eggs by themselves are boiled hard in
vinegar, and then roasted with ground pepper to
check diarrhoea.   There is also made for dysentery
an excellent remedy by pouring an egg into a new
earthen vessel, and so that there may be equal quantities of all the ingredients, in the shell of this egg are
measured honey, then vinegar, and oil, which are
mixed, and stirred many times.   The more excellent
the quality of these ingredients the more sovereign
will the remedy be.   Others substitute for oil and
vinegar the same amounts of red resin and wine.
There is yet another method of compounding: only
the quantity of oil remains the same, and with it are
boiled down together two sixtieths of a denarius of
pine bark, one of the shrub I have called rhus,*b* and
five oboli of honey, but no other food must be taken
until four hours have passed.   Many also treat colic
by beating up two eggs together with four heads of
garlic, warming with a hemina of wine, and so giving
the mixture as a draught.   To omit no attractive
feature of eggs, white of egg mixed with quicklime

glutinat vitri fragmenta. vis vero tanta est ut lignum perfusum ovo non ardeat ac ne vestis quidem contacta aduratur. de gallinarum autem ovis tantum locuti sumus, cum et reliquarum alitum restent, magnae utilitatis,[1] sicut suis locis dicemus.

52 XII. Praeterea est ovorum genus in magna fama Galliarum, omissum Graecis. angues ea numerose convoluti salivis faucium corporumque spumis artifici conplexu glomerant. urinum appellatur;[2] Druidae sibilis id dicunt in sublime iactari sagoque oportere intercipi ne tellurem attingat, profugere raptorem equo, serpentes enim insequi donec arceantur amnis alicuius interventu; experimentum eius esse, si 53 contra aquas fluitet vel auro vinctum. atque, ut est Magorum sollertia occultandis fraudibus sagax, certa luna capiendum censent, tamquam congruere operationem eam serpentium humani sit arbitrii. vidi equidem id ovum mali orbiculati modici magnitudine, crusta cartilagineis velut acetabulis bracchiorum

---

[1] utilitatis V *Mayhoff*: utilitates *ceteri codd.*, *Detlefsen*.

[2] *Sic ego.* angues ea numero sex convoluti salivis faucium corporumque spumis artifici complexu glomerant. uranium appellatur *Detlefsen*: angues enim numerose convoluti salivis faucium corporumque spumis artifici complexu glomerant; urinum appellatur *Mayhoff*: ea VRE *vulg.*, *Detlefsen*: eo d; *del. Hermolaus Barbarus*: numero est VRd: numero est ovorum E *vulg.*: innumeri aestate *Caesarius et Hermolaus Barbarus*: inter sese *coni. Mayhoff*: glomerantur in unum d: glomerantur annum *multi codd.*: glomerantur. anguinum *vulg.*

---

[a] Or: " nor will cloth either etc."

[b] The numerous variants in the MSS. show that the scribes were as puzzled by this passage as are modern readers. It

fastens together broken glass. So great indeed is its
power that wood dipped in egg will not take fire, and
not even cloth [a] stained with it will burn. But I have
been speaking only about farmyard hen's eggs; there
remain also other birds, the eggs of which are of
great utility; about them I shall speak on the
proper occasions.

XII. There is, moreover, a kind of egg which is *The snake's*
very famous in the Gauls, but not mentioned by the *egg.*
Greeks. Snakes intertwined in great numbers in a
studied embrace make these round objects with the
saliva from their jaws and the foam from their bodies.
It is called a " wind egg." [b] The Druids say that it
is tossed aloft by the snakes' hisses, and that it ought
to be caught in a military cloak before it can touch
the earth. The catcher, they say, must flee on horse-
back, for the serpents chase him until they are
separated by some intervening river. A test of a
genuine egg is that it floats against the current, even
if it is set in gold. Such is the clever cunning of the
Magi in wrapping up their frauds that they give out as
their opinion that it must be caught at a fixed period
of the moon, as if agreement between snakes and
moon for this act depended upon the will of man. I
indeed have seen this egg, which was like a round
apple of medium size, and remarkable for its hard
covering pitted with many gristly cup-hollows, as it

seems best to keep *ea*, accept Mayhoff's *numerose* (cf. XXV.
§ 167), and take his *urinum* (cf. X. §§ 158, 166) as the best
stop-gap for the name of the egg; it is very near the reading
of the MS. d. The vulgate *anguinum* (serpent's egg) is so
obvious and easy that it is most unlikely to have been corrupted into the variants of our MSS. See A. Blanchet on
ovum anguinum in *Bulletin Archéologique du Comité des
Travaux Historiques*, 1953, pp. 555–559.

54 polypi crebris insigne.[1] Druidis ad victorias litium
ac regum aditus mire laudatur, tantae vanitatis
ut habentem id in lite in sinu equitem R. e Vocontiis
a divo Claudio principe interemptum non ob aliud
sciam. hic tamen conplexus anguium et frugifera
eorum concordia in causa videtur esse quare exterae
gentes caduceum in pacis argumentis circumdata
effigie anguium fecerint, neque enim cristatos esse in
caduceo mos est.

55 XIII. De anserum ovis magnae utilitatis ipsoque
ansere dicturi hoc in volumine debemus honorem et
commageno, clarissimae rei. fit ex adipe anserum,
alioqui celeberrimi usus, [est ad hoc in Commagene
Syriae parte] [2] cum cinnamo, casia, pipere albo, herba
quae commagene vocatur, obrutis nive vasis, odore
iucundo, utilissimum ad perfrictiones, convulsiones,
caecos aut subitos dolores omniaque quae acopis
curantur, unguentumque pariter et medicamentum

56 est. fit et in Syria alio modo, avium adipe curato ut
dicemus, additis erysisceptro, xylobalsamo, phoenice,
item tuso [3] calamo, singulorum pondere quod sit
adipis, vino bis aut ter subfervefactum. fit autem
hieme, quoniam aestate non glaciat nisi accepta cera.
multa praeterea remedia sunt ex ansere, quod miror

---

[1] insigne *codd.*, *Mayhoff* : insigni *Detlefsen*.
[2] *Uncos add. Detlefsen. Pro* est *Mayhoff* set *scribit, et* alioqui
. . . parte *in parenthesi.*
[3] item tuso *Mayhoff ex Dioscoride* : tuso item *codd.*

---

[a] The idea is that if they were crested they would be males,
and so eggless.
[b] The part in brackets seems to be inconsistent with *fit et
in Syria alio modo* (§ 56).
[c] Many *acopa* are to be found in Celsus, but they would not
be very effective. For " blind " pains see § 38.

were, like those on the tentacles of an octopus. The
Druids praise it highly as the giver of victory in the
law-courts and of easy access to potentates. Herein
they are guilty of such lying fraud that a Roman
knight of the Vocontii, for keeping one in his bosom
during a lawsuit, was executed by the late Emperor
Claudius, and for no other reason. However, this
embrace and fertile union of snakes seem to be the
reason why foreign nations, when discussing peace
terms, have made the herald's staff surrounded with
figures of snakes; and it is not the custom for the
snakes on a herald's staff to have a crest.[a]

XIII. As in this Book I am going to treat of the *The goose.*
very useful goose egg, and of the goose itself, our
respects are due to the famous preparation called
commagenum. It is made from goose grease, a
very popular medicament everywhere, [and for this
purpose especially in Commagene, a district of Syria] [b]
with cinnamon, cassia, white pepper, and the herb
called commagene. The mixture is put into vessels
and buried in snow; it has a pleasant smell, and is
very useful for chills, sprains, blind or sudden pains,
and for all the complaints treated by anodynes,[c]
being equally good as an ointment and as a medicine.
It is also prepared in Syria in another way. The
grease of the birds is treated in the manner I shall
describe,[d] and there are added to it erysisceptrum,
balsam-wood, ground palm, and also crushed reed,
the same quantity of each as of the grease, the whole
being warmed two or three times in wine. But it
must be prepared in winter, for it will not set in
summer unless wax is added. There are many other
remedies made from the goose, which surprise me as

[d] See § 134 of this book.

aeque quam in capris, namque anser corvusque ab
aestate in autumnum morbo conflictari dicuntur.

57 XIV. De anserum honore quem meruere Gallorum
in Capitolium ascensu deprehenso diximus. eadem
de causa supplicia annua canes pendunt inter aedem
Iuventatis et Summani vivi in furca sabucea armo
fixi. sed plura de hoc animali dici cogunt priscorum
58 mores. catulos lactentes adeo puros existimabant
ad cibum ut etiam placandis numinibus hostiarum
vice uterentur iis. Genitae Manae catulo res divina
fit et in cenis deum etiamnunc ponitur catulina.
aditialibus quidem epulis celebrem[1] fuisse Plauti
fabulae indicio sunt. sanguine canino contra toxica
nihil praestantius putatur, vomitiones quoque hoc
animal monstrasse homini videtur, et alios usus ex eo
mire laudatos referemus suis locis. nunc ad statutum
ordinem pergemus.

59 XV. adversus serpentium ictus efficacia habentur
fimum pecudis recens in vino decoctum inlitumque,
mures dissecti inpositi. quorum natura non est
spernenda, praecipue in adsensu siderum, ut diximus,
cum lumine lunae fibrarum numero crescente atque
decrescente. tradunt Magi iocinere muris dato

---

[1] celebrem *vulg.*, *Mayhoff*: celebres *codd.*, *Detlefsen.*

---

[a] See XXVIII. § 153.
[b] X. § 51.
[c] *I.e.*, because they had failed to give the alarm.
[d] An old divinity supposed to have presided over child-birth.
[e] Probably in the lost play *Saturio*, mentioned by Festus.

much as the many from the goat,[a] for the goose and the crow are said to be afflicted with disease from the beginning of summer well into the autumn.

XIV. I have spoken[b] of the fame won by the geese *The dog.* which detected the ascent of the Capitoline Hill by the Gauls. For the same reason[c] dogs are punished with death every year, being crucified alive on a cross of elder between the temple of Juventas and that of Summanus. But the customs of the ancients compel me to say several other things about the dog. Sucking puppies were thought to be such pure food that they even took the place of sacrificial victims to placate the divinities. Genita Mana[d] is worshipped with the sacrifice of a puppy, and at dinners in honour of the gods even now puppy flesh is put on the table. That it was commonly in fact a special dish at inaugural banquets there is evidence in the comedies of Plautus.[e] Dog's blood is supposed to be the best remedy for arrow poison, and this animal seems also to have shown mankind the use of emetics. Other highly praised remedies from the dog I shall speak of on the appropriate occasions. I will now go on with my proposed plan.[f]

XV. For snake bites efficacious remedies are con- *Snake bites.* sidered to be fresh dung of sheep boiled down in wine and applied, and mice[g] cut in two and placed on the wound. The nature of mice is not to be despised, especially in their agreement, as I have said,[h] with the heavenly bodies, for the number of their liver filaments becomes greater or less with the light of the moon. The Magi declare that if a mouse's liver

[f] Of classifying remedies according to diseases.
[g] The Latin word will include rats.
[h] See II. § 109 and XI. § 196.

porcis in fico sequi dantem id animal, in homine
quoque similiter valere, sed resolvi cyatho olei poto.

60   XVI. Mustelarum [1] duo genera, alterum silvestre;
distant magnitudine, Graeci vocant ictidas. harum
fel contra aspidas dicitur efficax, cetero venenum.
haec autem quae in domibus nostris oberrat et catulos
suos, ut auctor est Cicero, cottidie transfert mutatque
sedem, serpentes persequitur. ex ea inveterata sale
denarii pondus in cyathis tribus datur percussis aut
ventriculus coriandro fartus inveteratusque et in vino
potus, et catulus [2] mustelae etiam efficacius.

61   XVII. Quaedam pudenda dictu tanta auctorum
adseveratione commendantur ut praeterire fas non
sit, siquidem illa concordia rerum aut repugnantia
medicinae gignuntur, veluti cimicum animalis foedis-
simi et dictu quoque fastidiendi natura contra serpentium morsus et praecipue aspidum valere dicitur,
item contra venena omnia, argumento, quod dicant
gallinas quo die ederint non interfici ab aspide carnes
62 quoque earum percussis plurimum prodesse. ex his
quae tradunt humanissimum est inlinere morsibus
cum sanguine testudinis, item suffitu eorum abigere
sanguisugas adhaerentes haustasque ab animalibus
restinguere in potu datis, quamquam et oculos quidam
his inungunt tritis cum sale et lacte mulierum,

---

[1] *Warmington* genera; distant magnitudine, alterum
silvestre, *coni.*

[2] et catulus E r *vulg., Mayhoff* : et catulū *multi codd.* : ex
catulis *coni. Detlefsen.*

---

[a] In a lost work.

in a fig is offered to pigs, that animal will follow the offerer, adding that it has a similar effect on a human being also, but that the spell is broken by drinking a cyathus of oil.

XVI. Of weasels there are two kinds, one wild and *The weasel.* larger than the other, called by the Greeks *ictis.* The gall of both is said to be efficacious against asps, though otherwise poisonous. The other kind, however, which strays about our homes, and moves daily, as Cicero tells us,[a] its nest and kittens, chases away snakes. Its flesh, preserved in salt and given in doses of one denarius by weight, is given in three cyathi of drink to those who have been bitten, or its stomach stuffed with coriander seed is kept to dry and taken in wine. A kitten of the weasel is even better still for this purpose.

XVII. Certain things, revolting to speak of, are so strongly recommended by our authorities that it would not be right to pass them by, if it is indeed true that medicines are produced by that famous sympathy and antipathy between things. The nature for instance of bugs, a most foul creature and nauseating *Bugs.* even to speak of, is said to be effective against the bite of serpents, and especially of asps, as also against all poisons. As proof, they say that hens are not killed by an asp on the day they have eaten bugs, and that their flesh then is most beneficial to such as have been bitten. Of the accounts given the least disgusting is how they are applied to bites with the blood of a tortoise, how fumigation with them makes leeches loose their hold, and how they destroy leeches swallowed by animals if administered in drink. And yet some actually anoint the eyes with bugs pounded in salt and woman's milk, and the ears with

auresque cum melle et rosaceo admixtis. eos qui
agrestes sint et in malva nascantur crematos cinere
63 permixto rosaceo infundunt auribus. cetera quae
de his tradunt, vomitionum et quartanarum remedia
aliorumque morborum, quamquam ovo aut cera aut
faba inclusos censeant devorandos, falsa nec referenda
arbitror. lethargi tantum medicinae cum argumento
adhibent, quoniam vincatur aspidum somnifica vis,
septenos in cyatho aquae dantes, puerilibus annis
quaternos. et in stranguria fistulae inposuere.
64 adeo nihil parens illa rerum omnium sine ingentibus
causis genuit. quin et adalligatos laevo bracchio
binos lana subrepta pastoribus resistere nocturnis
febribus prodiderunt, diurnis in russeo panno. rursus
his adversatur scolopendra suffituque enecat.

65 XVIII. Aspides percussos torpore et somno necant
omnium serpentium minime sanabiles. sed venenum
earum si sanguinem attingit aut recens vulnus, statim
interemit, si inveteratum ulcus, tardius. de cetero
potum quantalibet copia non nocet, non enim est
tabifica vis, itaque occisa morsu earum animalia cibis
innoxia sunt. cunctarer in proferendo ex his
remedio, ni M. Varro LXXXIII vitae anno prodidisset
aspidum ictus efficacissime sanari hausta a percussis
ipsorum urina.

66 XIX. Basilisci, quem etiam serpentes ipsae fugiunt,
alias olfactu necantem, qui hominem, vel si aspiciat

bugs in honey and rose oil. Those which are field bugs and found in mallows are burnt, and the ash mixed with rose oil is poured into the ears. The other virtues attributed to bugs, that they are cures for vomiting, quartans, and other diseases, although it is prescribed that they should be swallowed in egg, wax, or a bean, I hold to be imaginary and not worth repeating. Only as a remedy for lethargy are they employed with reason, for they overcome the narcotic poison of asps, and are given in doses of seven in a cyathus of water, and for children in doses of four. For strangury bugs have been inserted into the urethra. So true it is that the Universal Mother gave birth to nothing without very good reasons. Furthermore, a couple of bugs attached to the left arm in wool stolen from shepherds have been said to keep away night fevers, and day fevers when attached in a red cloth. On the other hand, the scolopendra is their enemy, and kills them by fumigation.

XVIII. Asps kill those they strike by torpor and *Asps.* coma, inflicting of all serpents the most incurable bites. But their venom, if it comes into contact with the blood or a fresh wound, is immediately fatal, if with an old sore, its action is delayed. Apart from this, however much is drunk, it is harmless, having no corrosive property. And so the flesh of animals killed by their bite may be eaten with safety. I should hesitate to put forward a remedy obtained from these creatures, had not Marcus Varro, in the eighty-third year of his life, recorded that a sovereign remedy for asp bites is for the victim to drink his own urine.

XIX. The basilisk, which puts to flight even the *The basilisk.* very serpents, killing them sometimes by its smell,

225

tantum, dicitur interimere, sanguinem Magi miris
laudibus celebrant coeuntem picis modo et colore,
dilutum cinnabari clariorem fieri.[a] attribuunt ei suc-
cessus petitionum a potestatibus et a diis etiam pre-
cum, morborum remedia, veneficiorum amuleta.
quidam id Saturni sanguinem appellant.

67 XX. Draco[b] non habet venena. caput eius limini
ianuarum subditum propitiatis adoratione diis for-
tunatam domum facere promittitur, oculis eius inveteratis et cum melle tritis inunctos non expavescere
ad nocturnas imagines etiam pavidos, cordis pingue
in pelle dorcadum nervis cervinis adalligatum in
lacerto conferre iudiciorum victoriae, primum[1]
spondylum aditus potestatium mulcere, dentes eius
inligatos pellibus caprearum cervinis nervis mites
praestare dominos potestatesque exorabiles. sed
68 super omnia est compositio qua invictos faciunt
Magorum mendacia: cauda draconis et capite, pilis
leonis e fronte et medulla eiusdem, equi victoris
spuma, canis ungue adalligatis cervino corio nervis-
que cervi alternatis et dorcadis. quae coarguisse non
minus referet quam contra serpentes remedia demon-
strasse, quoniam et haec Magorum[2] veneficia[3] sunt.

[1] victoriae, primum *codd., Detlefsen*: victoriae plurimum,
Mayhoff.
[2] Magorum *Detlefsen*: illorum *Mayhoff*: morum VR:
morborum d E *vulg.*
[3] veneficia VRd, *Mayhoff*: beneficia E *vulg., Detlefsen*.

[a] Or, "when diluted with cinnabar."
[b] Probably the python and similar snakes.

is said to be fatal to a man if it only looks at him. Its blood the Magi praise to the skies, telling how it thickens as does pitch, and resembles pitch in colour, but becomes a brighter red than cinnabar *a* when diluted. They claim that by it petitions to potentates, and even prayers to the gods, are made successful; that it provides cures for disease and amulets against sorcery. Some call it " Saturn's blood."

XX. The dragon *b* has no venom. Its head, buried *Draco.* under the threshold of doors after the gods have been propitiated by worship, brings, we are assured, good luck to a home; those rubbed with an ointment of his eyes, dried and beaten up with honey, are not panic-stricken, however nervous, by phantoms of the night; the fat of the heart, tied in the skin of a gazelle on the upper arm by deer sinew, makes for victory in law-suits; the first *c* vertebra smooths the approach to potentates; and its teeth, wrapped in the skin of a roe and tied on with deer sinew, make masters kind and potentates gracious. But all these are nothing compared with a mixture that the lying Magi assert makes men invincible, composed of: the tail and head of a dragon, hair from the forehead of a lion and lion's marrow, foam of a victorious race-horse, and the claw of a dog, all attached in deer hide with deer sinew and gazelle sinew plaited alternately. To expose these lies will be no less worth while than to describe their remedies for snake bite, for these too are some of the sorceries *d* of the Magi. Dragon's

---

c With Mayhoff's emendation : " great success in law-suits, a vertebra smooths etc."

d With Detlefsen's reading : " for these too are among the blessings bestowed by the Magi." This, if sarcastic, makes sense.

draconum adipem venenata fugiunt, item, si uratur,
ichneumonum, fugiunt et urticis tritis in aceto
perunctos.

69 XXI. Viperae caput inpositum, vel alterius quam
quae percusserit, sine fine [1] prodest, item si quis ipsam
eam in vapore baculo sustineat, aiunt enim re-[2]
canere, item si quis exustae eiusdem cinere inlinat.
reverti autem ad percussum serpentem necessitate
naturae Nigidius auctor est. caput quidam [3] dissecant
scite [4] inter aures ad eximendum lapillum quem
aiunt ab ea devorari territa. alii ipso toto capite
70 utuntur. fiunt ex vipera pastilli qui theriaci vocantur
a Graecis, ternis digitis mensura utrimque ampu-
tatis exemptisque interaneis et livore spinae adhaerente, reliquo corpore in patina ex aqua et aneto
discocto spinisque exemptis et addita similagine
atque ita in umbra siccatis pastillis quibus ad multa
medicamenta utuntur. significandum videtur e vipera
tantum hoc fieri. quidam purgatae ut supra dictum
est adipem cum olei sextario decocunt ad dimidias.
ex eo, cum opus sit, ternis stillis additis in oleum
perunguntur ut omnes bestiae fugiant eos.

71 XXII. Praeterea constat contra omnium ictus
quamvis insanabiles ipsarum serpentium exta inposita
auxiliari, eosque qui aliquando viperae iecur coctum
hauserint numquam postea feriri a serpente. neque
anguis venenatus est nisi per mensem [5] luna instiga-

---

[1] *Warmington* percusserit, sane prodest *coni*
[2] recanere *Sillig* (cf. XXVIII. 19) : praecanere *codd.*
[3] quidam VTE: quidem *aliquot codd.*
[4] scite VTE: Scythae *aliquot codd.*
[5] per mensem R *vulg. Mayhoff, qui* primo mense *coni.*

---

[a] Mayhoff's *primo mense* would mean : "in the early part
of the month." A contraction of *primo* might easily be taken

fat is shunned by venomous creatures, and so too, when burnt, is that of the ichneumon; they shun too those rubbed with nettles pounded in vinegar.

XXI. The head of a viper, placed on the bite, even *The viper.* though the same viper did not inflict it, is infinitely beneficial, as is the snake itself, held up on a stick in steam—it is said to undo the harm done—or if the viper is burnt and the ash applied. But Nigidius asserts that a serpent instinctively comes back to the person it has bitten. Some split skilfully the head between the ears, in order to extract the pebble it is said to swallow when alarmed, but others use the entire head itself. From the viper are made the lozenges called by the Greeks *theriaci.* Lengths of three fingers are cut off from head and tail, the intestines drawn with the livid part that adheres to the spine, the rest of the body, with the vertebrae extracted and fine flour added, is thoroughly boiled in a pan of water with dill, and the mixture dried in the shade and made into lozenges, which are used in making many medicaments. We must note, it appears, that only from the viper can the preparation be made. Some take the fat from the body, cleaned as described above, boil down with a sectarius of oil to one-half, add three drops from it when necessary to oil, and use as ointment to keep off all harmful creatures.

XXII. Furthermore, it is well known that the *Snakes.* application of the entrails of a serpent itself is a help for the bites however hard to cure of any of them, and that those who once have swallowed the boiled liver of a viper are never afterwards bitten by a serpent. A snake too is venomous only when during the month [a]

for *per*, and the change of *mense* to *mensem* would naturally follow.

tus, et prodest vivus conprehensus et in aqua con-
72 tusus, si foveantur ita morsus. quin et inesse ei
remedia multa creduntur, ut digeremus, et ideo
Aesculapio dicatur. Democritus quidem monstra
quaedam ex his [1] confingit ut possint avium sermones
intellegi.[2] anguis Aesculapius Epidauro Romam
advectus est vulgoque pascitur et in domibus, ac nisi
incendiis semina exurerentur, non esset fecunditati
eorum resistere.[3] in orbe terrarum pulcherrimum
anguium genus est quod et in aqua vivit, hydri
vocantur, nullo serpentium inferiores veneno. horum
iecur servatum adversus percussos ab his auxilium
est. scorpio tritus stelionum veneno adversatur.
73 fit enim ex stelionibus malum medicamentum.
nam cum inmortuus est vino, faciem eorum qui
biberint lentigine obducit. ob hoc in unguento
necant eum insidiantes pelicum formae. remedium
est ovi luteum et mel ac nitrum. fel stelionum tritum
in aqua mustelas congregare dicitur.
74 XXIII. Inter omnia venenata salamandrae scelus
maximum est. cetera enim singulos feriunt, nec

[1] ex his *codd.*: et hic *coni. Mayhoff*: *an post* ut *ponendum?*
[2] *Post* intellegi *excidisse* angue devorato *putat Mayhoff.*
Fortasse devorato angue.
[3] *Punctum non post* resistere *sed post* terrarum *ponit Mayhoff ex cod. Dal.*: in urbe. terrestrium *coni. Mayhoff.*

[a] The words *ex his* seem in the wrong place, and Mayhoff
would change to *et hic*, " here too." A transposition to the
ut clause would be simpler.
[b] If the words *in orbe terrarum* are placed here the meaning
will be : " a plague all over the world," and *in domibus* : " in
homes everywhere."

it is angered by the moon, and it is beneficial if a
snake is caught alive, beaten up in water, and a bite
fomented with the preparation. Moreover, many
remedies are believed to be obtained from a snake, as
I shall relate in their proper order, and this is why it is
sacred to Aesculapius. Democritus indeed invents
some weird stories about snakes, how for instance they
make it *a* possible to understand the language of birds.
The Aesculapian snake was brought to Rome from
Epidaurus, and a snake is commonly kept as a pet even
in our homes; so that were not their eggs destroyed
in fires there would be an incurable plague of them.*b*
The most beautiful snake in the world is the kind,
called *hydri*, that is amphibious, no other snake being
more venomous. Its liver when preserved does good
to those who have been bitten.*c* The scorpion when
pounded up counteracts the poison of the spotted
lizard,*d* for there is made from these lizards an evil
drug: if one has been drowned in wine it covers the
face of those who drink it with an eruption of freckle-
like spots. So women, plotting to spoil the beauty
of rival courtezans, kill a spotted lizard in the ointment used by them. The remedy is yolk of egg,
honey, and soda. The gall of this kind of lizard,
beaten up in water, is said to attract weasels.

XXIII. Of all venomous creatures the salamander *The*
is the most wicked, for while the others strike *salamander.*

c In this chapter there is certainly a distinction between
serpens and *anguis.* It is especially noticeable in § 71, where
neque anguis follows immediately after *a serpente.* In this
part of Pliny, at any rate, *anguis* includes the common or grass
snake, but the proverb *latet anguis in herba* shows that it
sometimes meant a poisonous serpent. Littré is not con-
sistent; after using *couleuvre* in § 71, he later uses *serpent.*

d Often called gecko.

plures pariter interimunt, ut omittam quod perire
conscientia dicuntur homine percusso neque amplius
admitti a terra, salamandra populos pariter necare
inprovidos potest. nam si arbori inrepsit, omnia
poma inficit veneno, et eos qui ederint necat frigida
75 vi nihil aconito distans. quin immo si contacto ab
ea ligno vel lapidi[1] crusta panis inponatur, idem veneficium est, vel si in puteum cadat, quippe cum saliva
eius quacumque parte corporis vel in pede imo
respersa omnis in toto corpore defluat pilus. tamen
talis ac tanti veneni a quibusdam animalium, ut subus,
76 manditur. dominante, eadem illa rerum dissidentia
venenum eius restingui primum omnium ab his quae
vescantur illa verisimile est, ex his vero quae pro-
bantur cantharidum potu aut lacerta in cibo sumpta.
cetera adversantia diximus dicemusque suis locis.
ex ipsa quae Magi tradunt contra incendia, quoniam
ignes sola animalium extinguat, si forent vera, iam
esset experta Roma. Sextius venerem accendi cibo
earum, si· detractis interaneis et pedibus et capite in
melle serventur, tradit negatque restingui ignem ab
his.
77 XXIV. E volucribus in auxilium contra serpentes
primum vultures. adnotatum quoque minus virium

[1] vel lapidi crusta panis inponatur *Mayhoff, sed sine* vel,
quod ego servo : vel pedis crista panis incocatur *Detlefsen :* vel
pidis V′d : vel pedis E : crista V′R : invocatur R′E : incocatur *multi codd.* : " *sed locus nondum sanatus,*" *Mayhoff.*

---

[a] See §§ 92–94, where applied externally cantharides are
said to be useful, but taken in drink poisonous.
[b] The salamander of modern zoology is a timid creature,
and not venomous to man.

individuals, and do not kill several together, to say nothing (according to report) of their dying of remorse when they have bitten a man, and of earth's refusal to grant them further admission, the salamander can kill whole tribes unawares. For if it has crawled into a tree, it infects with its venom all the fruit, killing like aconite by its freezing property those who have eaten of it. Nay, moreover, if a slice of bread is placed upon wood or stone that has been touched by a salamander, or if one falls into a well, the bread and the water, like the fruit, are poisoned, while all the hair on the whole body falls off if its saliva has sprinkled any part whatever of the body, even the sole of the foot. Nevertheless, although it is so venomous a creature, some animals, such as pigs, eat it. Under the sway of that same antipathy between things it is likely that his venom is neutralized best of all by those who eat the salamander; but among approved remedies are cantharides [a] taken in drink or a lizard taken in food. The other antidotes I have spoken of, and shall speak of, in the appropriate places. As to the power to protect against fires, which the Magi attribute to the animal, since according to them no other can put fire out, could the salamander really do so, Rome by trial would have already found out. Sextius tells us that as food the salamander, preserved in honey after entrails, feet, and head have been cut away, is aphrodisiac, but he denies its power to put fire out.[b]

XXIV. Of birds, the chief protection against *The* serpents is the vulture, and it has been noticed *vulture.* that there is less power [c] in the black vulture.

---

[c] Pliny uses the plural (*virium*) because Latin has no genitive singular of *vis*. The phrase can hardly mean that a black vulture is a weaker bird than other vultures.

esse nigris. pinnarum ex his[1] nidore, si urantur,
fugari eas dicunt, item cor eius alitis habentes tutos
esse ab impetu non solum serpentium sed etiam
ferarum latronumque et regum ira.

78 XXV. Carnibus gallinaceorum ita ut tepebunt
avulsae adpositis venena serpentium domantur,
item cerebro in vino poto. Parthi gallinae malunt
cerebrum plagis inponere. ius quoque ex his potum
praeclare medetur et in multis aliis usibus mirabile.
Pantherae, leones non attingunt peruncti eo, prae-
79 cipue si et alium fuerit incoctum. alvum solvit
validius e vetere gallinaceo, prodest et contra longin-
quas febres et torpentibus membris tremulisque et
articulariis morbis et capitis doloribus, epiphoris,
inflationibus, fastidiis, incipiente tenesmo, iocineri,
80 renibus, vesicae, contra cruditates, suspiria. itaque
etiam faciendi eius extant praecepta: efficacius coci
cum olere marino aut cybio aut cappari aut apio aut
herba Mercuriali, polypodio aut aneto, utilissime
autem in congiis tribus aquae ad tres heminas cum
supra dictis herbis et refrigeratum sub diu dari
tempestivis antecedente vomitione. non praeteribo
miraculum quamquam ad medicinam non pertinens:
si auro liquescenti gallinarum membra misceantur,
consumunt id in se. ita hoc venenum auri est. at
gallinacei ipsi circulo e ramentis addito in collum
non canunt.

---

[1] his *codd.* : alis *Mayhoff, e Sereno.*

---

[a] Mayhoff's correction, *alis* for *his*, would give " burning
wing-feathers."

They say that the fumes of their *a* burning feathers chase serpents away, and that those who carry about them a vulture's heart are protected not only from the attacks of serpents, but also from those of wild beasts, bandits, and angry potentates.

XXV. The flesh of chickens, torn away and applied *Chickens.* warm to the bite, overcomes the venom of serpents, as will also a chicken's brain taken in wine. The Parthians prefer to put on the wound the brain of a hen. Chicken broth also, taken by the mouth, is a splendid remedy, being wonderfully good for many other purposes. Panthers and lions do not touch those rubbed over with this broth, especially if garlic has been boiled in it. A rather powerful purge is the broth of an old cock, which is also good for prolonged fevers, paralysed and palsied limbs, diseases of the joints, headaches, eye-fluxes, flatulence, loss of appetite, incipient tenesmus, complaints of liver, kidneys, and bladder, indigestion and asthma. And so instructions even are current for making it: they tell us that it is more effective boiled with sea-cabbage, or tunny-fish, or caper, or celery, or the herb mercury, with polypodium or dill, but most beneficial when three congii of water are boiled down to three heminae, with the above-mentioned herbs, cooled in the open air and administered, the best time being when an emetic has preceded. I will not pass over a marvel, though it has nothing to do with medicine: if the limbs of hens are stirred up in melted gold they absorb it all into themselves, so violent a poison of gold is chicken. But cocks themselves do not crow if they have a collar of gold shavings round their necks.

81   XXVI. Auxiliatur contra serpentes et columbarum
caro recens concerpta et hirundinum, bubonis pedes
usti cum plumbagine herba. nec omittam in hac
quoque alite exemplum magicae vanitatis, quippe
praeter reliqua portentosa mendacia cor eius inpositum mammae mulieris dormientis sinistrae
tradunt efficere ut omnia secreta pronuntiet, prae-
82 terea in pugnam ferentes id fortiores fieri. eiusdem
ovo ad capillum remedia demonstrant. quis enim,
quaeso, ovum bubonis umquam visere potuit, cum
ipsam avem vidisse prodigium sit? quis utique
experiri et praecipue in capillo? sanguine quidem
pulli bubonis etiam crispari capillum promittunt.
83 cuius generis prope videri possint quae tradunt et
de vespertilione, si ter circumlatus domui vivus per [1]
fenestram inverso capite infigatur, amuletum esse,
privatimque ovilibus circumlatum totiens et pedibus
suspensum susum super limine. sanguinem quoque
eius cum carduo contra serpentium ictus inter
praecipua laudant.
84   XXVII. Phalangium est Italiae ignotum et plurium
generum: unum simile formicae, sed multo maius,
rufo capite, reliqua parte corporis nigra, albis guttis.
acerbior huius quam vespae ictus. vivit maxime
circa furnos et molas. in remedio est, si quis eiusdem
generis alterum percusso ostendat, et ad hoc servantur mortui. inveniuntur et cortices eorum qui
triti et poti medentur; mustelae catuli ut supra.[2]

[1] per *codd.*: super Mayhoff.
[2] mustelae catuli ut supra.] *Omittunt Urlichs et Detlefsen.*

[a] With Mayhoff's reading " over."
[b] Why *mortui* (masculine) when *phalangium* is neuter?
Perhaps *aranei* was in Pliny's mind.
[c] See § 60 of this Book.

XXVI. A help against snake-bite is also flesh of
doves or swallows freshly torn away, and the feet of a
horned owl burnt with the herb plumbago. Speaking
of this bird I will not omit a specimen of Magian
fraud, for besides their other monstrous lies they
declare that an horned owl's heart, placed on the left
breast of a sleeping woman, makes her tell all her
secrets, and that men carrying it into battle are made
braver by it. From the horned owl's egg they
prescribe recipes for the hair. Now who, I ask,
could have ever looked at an horned owl's egg, when
it is a portent to have seen the bird itself? Who in
any case could have tried it, particularly on the hair?
The blood, indeed, of a horned owl's chick is
guaranteed even to curl the hair. Of much the same
kind would seem to be also their stories about the
bat: that if carried alive three times round the house
and then fastened head downwards through *a* the
window, it acts as a talisman, and is specifically such
to sheepfolds if carried round them three times and
hung up by the feet over the threshold. Its blood
also with thistle the Magi praise as one of the
sovereign remedies for snake-bite.

XXVII. The phalangium is unknown to Italy and
of several kinds. One is like the ant, but much
larger, having a red head and the rest of the body
black with white spots. Its wound is more painful
than that of the wasp, and it lives especially near
furnaces and mills. One remedy is to show to the
bitten person another phalangium of the same kind;
for this purpose are kept dead *b* specimens. Their
dry bodies are also found, which are pounded and
taken as a remedy, as are a weasel's young prepared
as I have described.*c* Among classes of spiders the

85 aeque phalangion Graeci vocant inter genera araneorum, sed distingunt lupi nomine. tertium genus
est eodem phalangi nomine araneus lanuginosus
grandissimo capite, quo dissecto inveniri intus
dicuntur vermiculi duo adalligatique mulieribus pelle
cervina ante solis ortum praestare ne concipiant, ut
Caecilius in commentariis reliquit. vis ea annua est,
quam solam ex omni atocio dixisse fas sit, quoniam
aliquarum fecunditas plena liberis tali venia indiget.

86 vocatur et rhox acino nigro similis, ore minimo sub
alvo, pedibus brevissimis tamquam inperfectis.
dolor a morsu eius qualis a scorpione, urina similis
araneis textis. idem erat asterion, nisi distingueretur
virgulis albis. huius morsus genua labefactat. peior
utroque est caeruleus, lanugine nigra, caliginem
concitans et vomitus araneosos. etiamnum deterior
a crabrone pinna tantum differens. hic et ad

87 maciem perducit. myrmecion formicae similis capite,
alvo nigra, guttis albis distinguentibus, vesparum
dolore torquet. tetragnathii duo genera habent:
peior medium caput distinguente linea alba et trans-
versum altera; hic oris tumorem facit. at cinereus
posteriore parte candicans lentior, minime autem
noxius eodem colore qui telas muscis in parietibus

88 latissime pandit. contra omnium morsus remedio est
gallinaceorum cerebrum cum piperis exiguo potum in

---

a Or: "and then the urine looks like spider's web."
b I.e., "four-jawed."

Greeks also include a phalangion which they distinguish by the name of "wolf." There is also a third kind of phalangium, a hairy spider with an enormous head. When this is cut open, there are said to be found inside two little worms, which, tied in deer skin as an amulet on women before sunrise, act as a contraceptive, as Caecilius has told us in his *Commentarii*. They retain this property for a year. Of all such preventives this only would it be right for me to mention, to help those women who are so prolific that they stand in need of such a respite. There is another phalangium called *rhox*, like a black grape, with a very small mouth under the abdomen, and very short legs as though not fully grown. Its bite is as painful as a scorpion's sting, forming in the urine as it were spider's web.[a] The asterion is exactly like it, except that it is marked with white streaks. Its bite makes the knees weak. Worse than either is the blue spider; it is covered with black hair, and causes dimness of vision and vomit like spider's web. There is an even worse phalangium, which differs from the hornet only in having no wings. The bite from one of this kind also makes the body thin. The myrmecion in its head resembles the ant, with a black body marked by white spots, and a bite as painful as a wasp. There are two kinds of the phalangium called tetragnathius,[b] the worse of which has two white lines crossed on the middle of its head, and its bite makes the mouth swell; but the ash-coloured kind, which is whitish in its hind part, is less vicious. Least dangerous of all is the ash-coloured spider which spins its web all over our walls to catch flies. For the bites of all spiders remedial is a cock's brain with a little pepper taken in vinegar and water,

posca, item formicae quinque potae, pecudum fimi
cinis inlitus ex aceto et ipsi aranei quicumque in oleo
putrefacti. muris aranei morsus sanatur coagulo agni
e vino[1] poto, ungulae arietinae cinere cum melle,
mustelae catulo ut in serpentibus dictum est. si
iumenta momorderit, mus recens cum sale inponitur
89 aut fel vespertilionis ex aceto. et ipse mus araneus
contra se remedio est divulsus inpositus. nam si
praegnas momordit, protinus dissilit. optimum, si
is inponatur qui momorderit, sed et alios ad hunc
usum servant in oleo aut luto circumlitos. est et
contra morsum eius remedio terra ex orbita, ferunt
enim non transiri ab eo orbitam torpore quodam
naturae.

90 XXVIII. Scorpionibus contrarius maxime invicem
stelio traditur, ut visu quoque pavorem his adferat et
torporem frigidi sudoris. itaque in oleo putrefaciunt
eum et ita ea vulnera perungunt. quidam oleo illo
spumam argenteam decocunt ad emplastri genus
atque ita inlinunt. hunc Graeci coloten vocant et
ascalaboten et galeoten. in Italia non nascitur.
est enim hic plenus lentigine, stridoris acerbi, et
vescitur araneis,[2] quae omnia a nostris stelionibus
aliena sunt.

91 XXIX. Prodest et gallinarum fimi cinis inlitus,
draconis iocur, lacerta divulsa, mus divulsus, scorpio

---

[1] agni e vino *ex Plinio Iuniore Mayhoff* : agnino *Detlefsen* :
agne vino r : anguino Vd.
[2] araneis *add. Urlichs ex Arist. Hist. Anim. IX.* 1 : herba
vet. Dal. : vermibus *coni. Ianus* : illis (*sc.* scorpionibus) *coni.
Mayhoff.*

---

[a] See § 60 of this Book.
[b] Possibly *invicem* here means " mutually."

five ants also taken in drink, the ash of sheep's dung applied in vinegar, or spiders themselves of any sort that have rotted in oil.

The bite of the shrew-mouse is healed by lamb's *The shrew-mouse.* rennet taken in wine, by the ash of a ram's hoof with honey, and by a young weasel, as I have prescribed for snake-bite.[a] If it has bitten draught-animals, a freshly killed mouse is applied with salt, or a bat's gall in vinegar. The shrew-mouse itself, torn asunder and applied, is a remedy for its own bite; but if a pregnant shrew-mouse has bitten, it bursts open at once. It is best if the mouse applied is the one which gave the bite, but they preserve them for this purpose in oil, or enclosed in clay. Another remedy for its bite is earth from a wheel rut. For they say that it will not cross a wheel rut owing to a sort of natural torpor.

XXVIII. The stelio is said in its turn[b] to be such a *Lizards.* great enemy to scorpions that the mere sight of one strikes them with panic, and torpor with cold sweat. Accordingly they let it rot in oil and so smear on scorpion wounds. Some boil down that oil with litharge to make a sort of ointment which they thus apply. This lizard the Greeks call colotes, ascalabotes, or, galeotes. This kind is not found in Italy, for it is covered with spots, has a shrill cry, and feeds on spiders, all which characteristics are lacking in our stelios.[c]

XXIX. Beneficial too is ash of hen's dung applied, the liver of a python,[d] a lizard or a mouse torn open,

[c] Pliny has just said that the stelio is not native to Italy, but now speaks of " our stelios." Littré translates " nos lézards," taking *nostris stelionibus* to be used loosely.

[d] See XXIX. § 67, 68.

ipse suae plagae inpositus aut assus in cibo sumptus
aut potus in meri cyathis duobus. proprium est
scorpionum quod manus palmam non feriunt nec
nisi ¹ pilosa ² attingere. lapillus qualiscumque ab ea
parte quae in terra erat adpositus plagae levat
dolorem, item testa terra operta ex aliqua parte sicut
erat inposita liberare dicitur—non debent respicere
qui inponunt et cavere ne sol aspiciat—vermes terreni
92 triti inpositi. multa et alia ex his remedia sunt
propter quae in melle servantur. noctua apibus
contraria et vespis crabronibusque et sanguisugis,
pici quoque Martii rostrum secum habentes non
feriuntur ab his. adversantur et locustarum minimae
sine pinnis, quas attelebos vocant. est et formicarum
genus venenatum, non fere in Italia. solipugas
Cicero appellat, salpugas Baetica, his cor vespertilionis contrarium omnibusque formicis. salamandris
cantharidas diximus resistere.

93 XXX. Sed in his magna quaestio, quoniam ipsae
venena sunt potae vesicae cum cruciatu praecipuo.
Cossinum equitem Romanum amicitia Neronis
principis notum, cum is lichene correptus esset,
vocatus Aegypto medicus ob hanc valetudinem eius
a Caesare, cum cantharidum potum praeparare
voluisset, interemit. verum inlitas prodesse non
dubium est cum suco taminiae uvae et sebo ovis vel

---

¹ nisi *codd.*: visi *Io. Müller, Mayhoff.*
² pilosa *Ianus, Detlefsen*: pilos *Mayhoff, codd.*

---

a With Mayhoff's reading: " nor have they been seen to
touch hairs." The change from *quod* with the indicative to
the infinitive *attingere* is strange, and the emendation *visi* may
be right.
b See § 76 of this Book.

the scorpion laid on the wound it has itself inflicted, or roasted and taken in food or in two cyathi of neat wine. Scorpions are peculiar in that they do not sting the palm of the hand or touch any but hairy parts.[a] A pebble of any kind, if the part next the ground is laid on the wound, relieves the pain, and a potsherd too is said to be a cure if a part covered with earth is applied just as it was taken up—those making the application must not look back, and must take care that the sun does not behold them—and another cure is an application of pounded earth-worms. Many other remedies are obtained from earth-worms, so they are kept in honey for this purpose. The night owl is an enemy of bees, wasps, hornets, and leeches, and those are not stung by them who carry about their person a beak of the woodpecker of Mars. Hostile to them are also the smallest of the locusts, which are wingless and called *attelebi*. There is also a venomous kind of ant, not generally found in Italy. Cicero calls it *solipuga* and in Baetica it is called *salpuga*. A bat's heart is hostile to these, as it is to all ants. I have said [b] that cantharides are hostile to salamanders.

XXX. But herein arises a much-disputed question, *Spanish* for the fly taken in drink is a poison, causing excru- *fly.* ciating pain in the bladder. Cossinus, a Roman knight, well known for his friendship with the Emperor Nero, fell a victim to lichen.[c] Caesar called in a specialist physician from Egypt, who decided on preliminary treatment with Spanish fly taken in drink, and the patient died. But there is no doubt that, with juice of taminian grapes, sheep suet, or that of a she-goat, an external application is beneficial.

[c] See List of Diseases.

94 caprae. ipsarum cantharidum venenum in qua parte sit non constat inter auctores. alii in pedibus et in capite existimant esse, alii negant. convenit tantum pinnas earum auxiliari, in quacumque parte sit venenum. ipsae nascuntur ex vermiculo, in spongea maxime cynorrhodi quae fit in caule, sed fecundissime in fraxino; ceterae in alba rosa, minus efficaces. potentissimae inter omnes variae, luteis lineis quas in pinnis transversas habent, pingues; multum [1] inertiores minutae, latae, pilosae, inutilissimae vero

95 unius coloris macrae. conduntur in calice fictili non picato et linteo conligato, contectae [2] rosa matura, et suspenduntur super acetum cum sale fervens donec per linteolum vaporentur, postea reponuntur. vis earum adurere corpus, crustas obducere. eadem pityocampis in picea nascentibus, eadem bupresti, similiterque praeparantur. efficacissimae omnes ad lepras, lichenas, dicuntur et menses ciere et urinam, ideo

96 Hippocrates et hydropicis dabat. cantharides obiectae sunt Catoni Uticensi, ceu venenum vendidisset in auctione regia, quoniam eas HS $\overline{\text{LX}}$ addixerat. et sebum autem struthocamelinum tunc venisse HS $\overline{\text{XXX}}$ obiter dictum sit, efficacioris ad omnia usus quam est anserinus adips.

---

[1] pingues; multum *Urlichs* : multum pingues *codd.*
[2] contectae *Ianus* : coniecta et *Mayhoff* : coniectae *codd.*

---

[a] This description suggests " Robin's pin-cushions," caused by the gall-wasp, and not a beetle. There were probably several kinds of cantharides.

In what part of the Spanish fly [a] itself the poison lies authorities disagree; some think in the feet and in the head, but others say not. The only point agreed upon is that, wherever the poison lies, their wings help.[b] The fly itself is bred from a grub found in the sponge-like substance on the stalk of the wild rose especially, but also very plentifully on the ash. The third kind breeds on the white rose, but is less efficacious. The most potent flies of all are marked with yellow lines across their wings and are plump; much less potent are those that are small, broad and hairy; the least useful however are of one colour, and thin. They are stored away in an earthen pot, not lined with pitch, but the mouth closed with a cloth. They are covered with full-blown roses and hung over boiling vinegar and salt until the steam, passing through the cloth, suffocates them. Then they are stored away. Their property is to cauterise the flesh and to form scabs. Of the same character is the pine-caterpillar, which is found on the pitch-pine, and the buprestis, and they are prepared in a similar way. All these are very efficacious for leprous sores and lichen. They are also said to be emmenagogue and diuretic, and so Hippocrates [c] used them also for dropsy. Spanish fly was the subject of a charge against Cato Uticensis that he had sold poison at an auction of royal property, for he had knocked some down for 60,000 sesterces. And I may remark in passing that at this sale there was sold for 30,000 sesterces ostrich suet, a far more useful fat for all purposes than goose-grease.

[b] A mysterious sentence, that might mean either that the wings increase the poison, or that they are remedial.

[c] *Regimen in Acute Diseases*, 104.

97 XXXI. Diximus et mellis venenati genera; contra utuntur melle in quo apes sint mortuae. idem potum in vino remedium est vitiorum quae e cibo piscium gignuntur.

98 XXXII. In canis rabidi morsu tuetur a pavore aquae canini capitis cinis inlitus vulneri, oportet autem comburi omnia eodem modo, ut semel dicamus, in vase fictili novo argilla circumlito atque ita in furnum indito. idem et in potione proficit. quidam ob id edendum dederunt. aliqui et vermem e cadavere canino adalligavere menstruave canis in panno subdidere calici aut intus¹ ipsius caudae pilos com-
99 bustos inseruere vulneri. cor caninum habentem fugiunt canes, non latrant vero lingua canina in calciamento subdita pollici aut caudam mustelae quae abscissa ea dimissa sit habentes. est limus salivae sub lingua rabiosi canis qui datus in potu hydrophobos fieri non patitur, multo tamen utilissime iocur eius qui in rabie momorderit datur, si fieri possit, crudum mandendum, sin minus, quoquo modo coctum, aut
100 ius coctis carnibus. est vermiculus in lingua canum qui vocatur a Graecis lytta, quo exempto infantibus catulis nec rabidi fiunt nec fastidium sentiunt. idem ter igni circumlatus datur morsis a rabioso ne rabidi

---

¹ intus] "*an* imos (*vel potius* calciamentis *pro* caliciautintus)?" *Mayhoff.*

---

a Book XXI, § 74.
b Mayhoff's clever emendation of *calciamentis* for *caliciaut intus* would give: "placed the fluid in a cloth at the bottom (*sub-*) of the shoes." But it gives rather a strange meaning to *subdidere*, and *intus* is just possible as indicating the under part between the tail and the body.

XXXI. I have also mentioned ^a kinds of poisonous *Honey.*
honey. To counteract it honey is used in which bees
have died. The same honey is also a remedy for illness caused by eating fish.

XXXII. If a person has been bitten by a mad dog, *Mad dogs*
protection from hydrophobia is given by an applica- *and hydro-*
tion to the wound of ash from the burnt head of a dog. *phobia.*
Now all reduction to ash (that I may describe it once
for all) should be carried out in the following way:
a new earthen vessel is covered all over with clay and
so put into a furnace. The same method is also good
when the ash is to be taken in drink. Some have
prescribed as a cure eating a dog's head. Others too
have used as an amulet a worm from a dead dog, or
placed in a cloth under the cup the sexual fluid of a
bitch, or have rubbed into the wound the ash from
the hair under ^b the tail of the mad dog itself. Dogs
run away from one who carries a dog's heart, and
indeed do not bark if a dog's tongue is placed in the
shoe under the big toe, or at those who carry the
severed tail of a weasel which has afterwards been
set free. Under the tongue of a mad dog is a slimy
saliva, which given in drink prevents hydrophobia,
but much the most useful remedy is the liver of the
dog that bit in his madness to be eaten raw, if that
can be done, if it cannot, cooked in any way, or a
broth must be made from the boiled flesh. There is
a little worm ^c on the tongue of dogs which the
Greeks call *lytta* (madness), and if this is taken away
when they are baby puppies they neither go mad
nor lose their appetite. It is also carried three times
round fire and given to those bitten by a mad dog to

^c Really white pustules under the tongue, which break of
their own accord when the puppies are twelve days old.

fiant. et cerebello gallinaceo occurritur, sed id devoratum anno tantum eo prodest. aiunt et cristam
galli contritam efficaciter inponi et anseris adipem
cum melle. saliuntur et carnes eorum qui rabidi
101 fuerunt ad eadem remedia in cibo dandae. quin et
necantur catuli statim in aqua ad sexum eius qui
momorderit, ut iocur crudum devoretur ex iis. prodest et fimum gallinaceum, dumtaxat rufum, ex
aceto inpositum et muris aranei caudae cinis, ita
ut ipse cui abscissa sit vivus dimittatur, glaebula ex
hirundinum nido inlita ex aceto, vel pulli hirundinis
combusti, membrana sive senectus anguium vernatione
exuta cum cancro masculo ex vino trita, (nam hac[1]
etiam per se reposita in arcis armariisque tineas necant)
102 mali tanta vis est ut urina quoque calcata rabiosi
canis noceat, maxime ulcus habentibus. remedio est
fimum caballinum adspersum aceto et calefactum in
fico inpositum. minus hoc miretur qui cogitet lapidem a cane morsum usque in proverbium discordiae
venisse. qui in urinam canis suam egesserit torporem lumborum sentire dicunt. lacerta, quam sepa,
alii chalcidem vocant, in vino pota morsus suos sanat.
103   XXXIII. Veneficiis ex mustela silvestri factis contrarium est ius gallinacei veteris large haustum,
peculiariter contra aconita, addi parum salis oporteat;
gallinarum fimum, dumtaxat candidum, in hysopo
decoctum aut mulso, contra venena fungorum boletor-

---

[1] nam hac *ego*: nam *codd.*: hac *Mayhoff.*

---

[a] A Plinian parenthesis.
[b] The last sentence, bracketed by Mayhoff, has obviously
been misplaced, but its proper place is not clear. Some
other sentences seem to be careless.

prevent their going mad. The brains of poultry are an antidote, but to swallow them gives protection for that year only. They say that it is also efficacious to apply to the wound a cock's comb pounded up, or goose grease with honey. The flesh of dogs that have gone mad is also preserved in salt to be used for the same purposes given in food. Puppies too of the same sex as the bitten patient are immediately drowned and their livers swallowed raw. An application in vinegar of poultry dung, if it is red, is also of advantage, or the ash of a shrew-mouse's tail (but the mutilated animal must be set free alive), an application in vinegar of a bit of earth from a swallow's nest, of the chicks of a swallow reduced to ash, or the skin or cast slough of snakes, pounded in wine with a male crab; for by it even when put away by itself in chests and cupboards they kill moths.[a] So great is the virulence of this plague that even the urine of a mad dog does harm if trodden on, especially to those who are suffering from sores. A remedy is an application of horse dung sprinkled with vinegar and warmed in a fig. Less surprised at all this will be one who remembers that " a dog will bite a stone thrown at him " has become a proverb to describe quarrelsomeness. It is said that he who voids his own urine on that of a dog will suffer numbness in his loins. The lizard called *seps* by some and *chalcis* by others, if taken in wine is a cure for its own bites.[b]

XXXIII. For sorcerers' poisons obtained from the wild weasel a remedy is a copious draught of chicken broth made from an old bird; it is specific for aconite poisoning, and there should be added a dash of salt. Hens' dung, provided it is white, boiled down in hyssop or honey wine, is used for poisonous fungi and *Antidotes for various poisons.*

umque, item inflationes ac strangulationes, quod
miremur, cum, si aliud animal gustaverit id fimum,
104 torminibus et inflationibus adficiatur. sanguis anserinus contra lepores marinos valet cum olei aequa portione, item[1] contra mala medicamenta omnia—adservatur cum Lemnia rubrica et spinae albae suco, ut[2]
pastillorum drachmis quinque in cyathis ternis aquae
bibatur—item mustelae catulus ut supra diximus
praeparatus. coagulum quoque agninum adversus
omnia mala medicamenta pollet, item sanguis anatum
Ponticarum. itaque et spissatus servatur vinoque
diluitur. quidam feminae anatis efficaciorem putant.
105 simili modo contra venena omnia ciconiarum ventriculus valet, coagulum pecoris, ius ex carne arietum,
privatim adversus cantharidas, item lac ovium calidum
praeterque iis qui buprestim aut aconitum biberint,
columbarum silvestrium fimum privatim contra
argenti vivi potum, contra toxica mustela vulgaris inveterata drachmis binis pota.
106   XXXIV. Alopecias replet fimi pecudum cinis cum
oleo cyprio et melle, item ungularum muli vel mulae
ex oleo myrteo, praeterea, ut Varro noster tradit,
murinum fimum, quod item[3] muscerdas appellat, aut
muscarum capita recentia prius folio ficulneo aspera-
tas. alii sanguine muscarum utuntur, alii decem diebus cinerem earum inlinunt cum cinere chartae vel
nucum ita ut sit tertia pars e muscis, alii lacte
mulierum cum brassica cinerem muscarum subigunt,

---

[1] item *codd.*: idem *Mayhoff.*
[2] *Post* suco *add.* ut *Mayhoff.*
[3] item R *vulg.*: rite *Detlefsen*: ille *Mayhoff*: lite VE:
linthe d.

---

[a] See § 60.

mushrooms, as well as for flatulence and suffocations—
a matter for wonder, because if any animal save man
should taste this dung, it will suffer from colic and
flatulence. Goose blood, with the same quantity of
oil, is good for the poison of sea hares, also for all
sorcerers' poisons—it is kept with red Lemnian earth
and the sap of white thorn, and five drachmae of the
lozenges should be taken as a dose in three cyathi of
water—also a baby weasel prepared as I have
described.[a] Lamb's rennet too is a powerful antidote
to all sorcerers' poisons, as is the blood of Pontic
ducks; and so when thickened it is also stored away
and dissolved in wine. Some are of opinion that the
blood of a female duck is more efficacious. In like
manner general remedies for all poisons are the crop
of storks, sheep's rennet, the broth of ram's flesh
(which is specific for cantharides), likewise warmed
sheeps' milk, which is also good for those who have
swallowed buprestis or aconite, the dung of wild
doves (specific if quicksilver has been swallowed),
and for arrow poisons the common weasel, preserved
and taken in drink, two drachmae at a time.

XXXIV. Bald patches through mange are covered *Mange.*
again with hair by an application of ash of sheeps'
dung with cyprus oil and honey, by the hooves,
reduced to ash, of a mule of either sex, applied in
myrtle oil; moreover, as our countryman Varro
relates, by mouse dung, which he calls also *muscerdae*,
or by the fresh heads of flies, but the patches must first
be roughened with a fig leaf. Some use the blood
of flies, others for ten days apply their ash with that
of paper or nuts, but a third of the whole must be
that of flies; others make a paste of fly ash, woman's
milk, and cabbage, while some add honey only. No

quidam melle tantum. nullum animal minus docile
existimatur minorisve intellectus; eo mirabilius est
Olympiae sacro certamine nubes earum immolato
tauro deo quem Myioden vocant extra territorium id
107 abire. alopecias cinis ex murium capitibus caudisque
et totius muris emendat, praecipue si veneficio acciderit haec iniuria, item irenacei cinis cum melle aut
corium combustum cum pice liquida. caput quidem
eius ustum per se etiam cicatricibus pilos reddit. alopecias autem in ea curatione praeparari oportet novacula. et sinapi quidam ex aceto uti maluerunt.
quae de irenaceo dicentur omnia tanto magis vale-
108 bunt in hystrice. lacertae quoque ut docuimus combustae cum radice recentis harundinis, quae ut una
cremari possit, minutim findenda est, ita myrteo oleo
permixto cineres [1] capillorum defluvia continent.
efficacius virides lacertae omnia eadem praestant,
etiamnum utilius admixto sale et adipe ursino et cepa
tusa. quidam denas virides in decem sextariis olei
veteris discocunt, contenti semel in mense unguere.
109 pellium viperinarum cinis alopecias celerrime explet,
item gallinarum fimum recens inlitum. corvi ovum
in aereo vase permixtum inlitumque deraso capite
nigritiam capilli adfert, sed donec inarescat oleum in
ore habendum est ne et dentes simul nigrescant,
idque in umbra faciendum neque ante quadriduum

---

[1] cineres *codd.*: cinere *Mayhoff*.

---

[a] The Fly-catcher, who protected his worshippers from flies.
[b] See § 98.

creature is thought to be less teachable or less intelligent than the fly; it is all the more wonderful that at the Olympic sacred games, after the bull has been sacrificed to the god they call Myiodes,[a] clouds of flies depart from out Olympic territory. Hair lost by mange is restored by the ash of mice, their heads and tails, or their whole bodies, especially when this affliction is the result of sorcery; it is restored too by the ash of a hedge-hog mixed with honey, or by its burnt skin with liquid pitch. The head indeed of this animal, reduced to ash, by itself restores the hair even to scars. But for this treatment the patches must first be prepared by shaving with a razor. Some too have preferred to use mustard in vinegar. All that will be said about the hedgehog will apply even more to the porcupine. Hair is also prevented from falling out by the ash of a lizard that, in the way I have described,[b] has been burnt with the root of a fresh-cut reed, which must be chopped up fine so that the two may be consumed together, an ointment being made by the admixture of myrtle-oil. All the same results are given more efficaciously by green lizards, and with even greater benefit if there are added salt, bear's grease, and crushed onion. Some thoroughly boil ten green lizards at a time in ten sextarii of old oil, being content with one application a month. Vipers' skins reduced to ashes very quickly restore hair lost through mange, as does also an application of fresh hens' dung. A raven's egg, beaten up in a copper vessel and applied to the head after shaving it, imparts a black colour to the hair, but until it dries oil must be kept in the mouth lest the teeth too turn black at the same time; the application too must be made in the shade, and not

253

110 abluendum. alii sanguine et cerebro eius utuntur
cum vino nigro, alii excocunt ipsum et nocte concubia
in plumbeum vas condunt. aliqui alopecias can-
tharide trita inlinunt cum pice liquida, nitro praeparata cute—caustica vis earum, cavendumque ne
exulcerent alte—postea ad ulcera ita facta capita
murium et fel murium et fimum cum helleboro et
pipere inlini iubent.

111   XXXV. Lendes tolluntur adipe canino vel anguibus
in cibo sumptis anguillarum modo aut eorum vernatione quam exuunt pota, porrigines felle ovillo cum
creta Cimolia inlito capite donec inarescat.

112   XXXVI. Capitis doloribus remedio sunt coclearum quae nudae inveniuntur nondum peractae,
ablato capite, et his duritia lapidea exempta—est
autem ea calculi latitudine—quae [1] adalligantur et
minutae fronti inlinuntur tritae, item oesypum, ossa
e capite vulturis adalligata aut cerebrum cum oleo et

113 cedria, peruncto capite et intus naribus inlitis, cornicis cerebrum coctum in cibo sumptum vel noctuae,
gallinaceus si inclusus abstineatur die ac nocte, pari
inedia eius qui [2] doleat, evulsis collo plumis circum-
ligatisque vel cristis, mustelae cinis inlitus, surculus
ex nido milvi pulvino subiectus, murina pellis cremata
ex aceto inlito cinere, limacis inter duas orbitas in-
ventae ossiculum per aurum argentum ebur traiectum

---

    [1] quae *codd.* : eaque *Mayhoff.*
    [2] qui *fere omnes codd.* : cuius E, *Mayhoff.*

---

    [a] Perhaps a reference to slugs.
    [b] Or, " of the size of a bit of gravel." Perhaps, " as big
as a calculus."

washed off before three days have passed. Some use a raven's blood and brains added to dark wine; others thoroughly boil the raven itself and store it away at bed time in a vessel of lead. Some apply to patches of mange Spanish fly pounded with liquid pitch, first preparing the skin with soda—the application is caustic, and care must be taken not to cause deep sores—and prescribe that afterwards to the sores so formed be applied the heads, gall, and dung of mice with hellebore and pepper.

XXXV. Nits are removed by dog fat, snakes taken in food like eels, or by the cast slough of snakes taken in drink; dandruff by sheeps' gall with Cimolian chalk rubbed on the head until it dries off. *Cures for nits.*

XXXVI. Headaches have a remedy in the heads of snails, cut off from those that are found without shells, being not yet complete,[a] and the hard stony substance taken from them—it is of the width of a pebble [b]—which are used as an amulet, while the small snails are crushed, and rubbed on the forehead; there is also wool grease; the bones from the head of a vulture attached as an amulet, or its brain with oil and cedar resin, the head being rubbed all over and the inner part of the nostrils smeared with the ointment; the brain of a crow or owl boiled and taken in food; a cock penned up without food for a day and a night, the sufferer fasting with him at the same time, feathers plucked from the neck, or the comb, being tied round the head; the application of a weasel reduced to ash; a twig from a kite's nest placed under the pillow; a mouse's skin burnt and the ash applied in vinegar; the little bone of a slug found between two wheel ruts, passed through gold, silver and ivory, and attached in dog skin as an *For head-aches.*

in pellicula canina adalligatum, quod remedium pluri-
114 bus semper prodest. fracto capiti aranei tela ex oleo
et aceto inposita non nisi vulnere sanato abscedit.
haec et vulneribus tonstrinarum sanguinem sistit, a
cerebro vero profluentem anseris sanguis aut anatis
infusus, adeps earundem alitum cum rosaceo. cocleae
matutino[1] pascentis harundine caput praecisum,
maxime luna plena lineo panno adalligant capitis
doloribus liceo, aut cera alba[2] fronti inlinunt et pilos
caninos panno adalligant.

115   XXXVII. Cerebrum cornicis in cibo sumptum
palpebras gignere dicitur, oesypum cum murra calido
penicillo inlitum. idem praestare muscarum fimique
murini cinerem aequis portionibus ut efficiatur dimi-
dium pondus denarii promittitur, additis duabus sextis
denarii e stibi, ut omnia oesypo inlinantur, item
murini catuli triti in vino vetere ad crassitudinem
116 acopi. pilos in his incommodos evulsos renasci non
patitur fel irenacei, ovorum stelionis liquor, salamandrae cinis, lacertae viridis fel in vino albo sole coactum
ad crassitudinem mellis in aereo vase, hirundinis
pullorum cinis cum lacte tithymalli, spuma coclearum.

117   XXXVIII. Glaucomata dicunt Magi cerebro catuli
septem dierum emendari specillo demisso in dex-

----

[1] cocleae matutino *Harduinus*: coctae (cocta) matutina
codd.
[2] *Post* alba *add.* addita *Mayhoff*: *nolit Brakman.*

amulet, a remedy that always does good to most. Applied in oil and vinegar to a fractured skull, cobweb does not come away until the wound is healed. Cobweb also stops bleeding from a razor cut, but haemorrhage from the brain is stayed by pouring into the wound the blood of goose or duck, or the grease of these birds with rose oil. The head of a snail cut off with a reed as he feeds in the morning, by preference when the moon is full, is attached in a linen cloth by a thread to the head of a sufferer from headaches, or else made into an ointment for the forehead with white wax, and an amulet attached of dog's hair in a cloth.

XXXVII. A crow's brain taken in food is said to make eyelashes grow, and also wool grease and myrrh applied with a warmed probe. We are assured that the same result is obtained by taking the ash of flies and of mouse dung in equal quantities, so that the weight of the whole amounts to half a denarius, then adding two-sixths of a denarius of antimony and applying all with wool grease; or one may use baby mice beaten up in old wine to the consistency of an anodyne salve. When inconvenient hairs in the eyelashes have been plucked out they are prevented from growing again by the gall of a hedgehog, the fluid part of a spotted lizard's eggs, the ash of a salamander, the gall of a green lizard in white wine condensed by sunshine to the consistency of honey in a copper vessel, the ash of a swallow's young added to the milky juice of tithymallus and the slime of snails. *Eyelashes.*

XXXVIII. Opaqueness of the eye-lens is cured, say the Magi, by the brain of a seven-day-old puppy, the probe being inserted into the right side of the eye to *Cures for eye diseases.*

257

teram partem, si dexter oculus curetur, in sinistram,
si sinister, aut felle recenti axionis. noctuarum est id
genus, quibus pluma aurium modo micat. suffu-
sionem oculorum canino felle malebat quam hyaenae
curari Apollonius Pitanaeus cum melle, item albugines
118 oculorum. murium capitum caudaeque cinere ex
melle inunctis claritatem visus restitui dicunt, multoque magis gliris aut muris silvestris cinere aut
aquilae cerebro vel felle cum Attico melle. cinis
et adips [1] soricis cum stibi tritus lacrimosis oculis
plurimum confert—stibi quid sit dicemus in metallis
—mustelae cinis suffusionibus, item lacertae, hirundinis cerebrum. cocleae tritae fronti inlitae epiphoras sedant sive per se sive cum polline sive cum
119 ture. sic et solatis [id est sole correptis] [2] prosunt.
vivas quoque cremare et cinere earum cum melle
Cretico inunguere caligines utilissimum est. iumentorum oculis membrana aspidis quam exuit vere, cum
adipe eiusdem claritatem inunctis facit. viperam
vivam in fictili novo comburere addito feniculi suco ad
cyathum unum et turis manna [3] una, atque ita suffusiones oculorum et caligines inunguere utilissimum
120 est. medicamentum id echeon vocatur. fit et
collyrium vipera in olla putrefacta vermiculisque
enatis cum croco tritis. et uritur [4] in olla cum sale

---

[1] et adips d, *vulg. Detlefsen* : e capite *Mayhoff* : et alipe,
alipe, et adipe *ceteri codd.*

[2] id est sole correptis] *uncos ego addidi.*

[3] manna *Hermolaus Barbarus, Mayhoff* : mina E, *Detlefsen* :
mammam Vdf : mamma *vulg.*

[4] et uritur *Mayhoff* : exuritur *codd.*

---

[a] Mayhoff would omit " or . . . honey " as a gloss.
[b] A strange phrase, and Mayhoff's " ash from the head "
may be right, but some sort of grease would be needed.

treat the right eye and into the left side to treat the
left eye; or by the fresh gall of the *axio*, a kind of
owl whose feathers twitch like ears. Apollonius of
Pitane preferred to treat cataract with honey and
dog's gall rather than using hyaena's, as he did also
to treat white eye ulcers. The heads and tails of mice,
reduced to ash and made into an ointment with
honey, restore, they say, clearness of vision; much
better the ash of a dormouse or wild mouse, or the
brain of an eagle or the gall with Attic honey.[a] The
ash and fat [b] of the shrew-mouse, beaten up with
antimony, is very good for watery eyes—what antimony is I shall say when I speak [c] of metals—the ash
of the weasel for cataract, likewise of the lizard, or
the brain of the swallow.[d] Pounded snails applied to
the forehead relieve eye fluxes, either by themselves
or with fine flour or with frankincense; so applied
they are also good for sunstroke.[e] To burn them
alive also, and to use as ointment the ash with Cretan
honey is very good for dimness of vision. For the
eyes of draught animals the slough cast in spring by
the asp makes with asp fat an ointment that improves
their vision. To burn a viper alive in new earthenware, with addition of fennel juice up to one cyathus,
and of one grain of frankincense, makes an ointment
very good for cataract and dimness of vision; this
prescription is called *echeon*. An eye salve is also
made by letting a viper rot in a jar, and pounding with
saffron the grubs that breed in it. A viper is also

[c] XXXIII. § 101.
[d] Or, " likewise the brain of lizard or swallow."
[e] On the whole it seems better to omit *id est sole correptis* as
a gloss. Although a colloquial word of the countryside,
solatus would scarcely require explanation to a Roman ear.

quem lingendo claritatem oculorum consecuntur et
stomachi totiusque corporis tempestivitates. hic sal
et pecori datur salubritatis causa et in antidotum
contra serpentes additur. quidam † et adtollitur [1] †
121 viperis utuntur in cibis. primum omnium occisae
statim salem in os addi iubent donec liquescat, mox
quattuor digitorum mensura utrimque praecisa exemptisque interaneis discoquunt in aqua, oleo,[2] sale,
aneto, et aut statim vescuntur aut pane colligunt, ut
saepius utantur. ius praeter supra dicta pediculos e
toto corpore expellit pruritusque etiam summae cutis.
effectum ostendit et per se capitis viperini cinis;
utilissime eo [3] oculos inunguit, itemque adeps viper-
122 inus. de felle non audaciter suaserim quae praecipiunt, quoniam, ut suo loco docuimus, non aliud est
serpentium venenum. anguium adeps aerugini
mixtus ruptas oculorum partes sanat, et membrana
sive senectus vernatione eorum exuta si adfricetur,
claritatem facit. boae quoque fel praedicatur ad
albugines, suffusiones, caligines, adeps similiter ad
123 claritatem. aquilae, quam diximus pullos ad con-
tuendum solem experiri, felle mixto cum melle Attico
inunguntur nubeculae et caligationes suffusionesque

---

[1] et adtollitur *codd.* : ex Athoitis *Detlefsen* (VII § 27 *coll.*) :
et ad oculos *Mayhoff*, " *locus nondum sanatus.*" *Fortasse* ad
tollendos pruritus *Warmington*.

[2] discoquunt in aqua, oleo] *Mayhoff coni.* discoquunt cum
vino atque oleo.

[3] eo *add. Mayhoff*.

---

[a] The reading of the MSS. is obviously wrong, and although
the *ad* of *adtollitur* seems to be a preposition, the name of the
complaint to follow it is a mystery; Mayhoff's *oculos* would
scarcely have been misunderstood and suffered corruption.
There is a late word *tolles*, meaning goitre. Palaeographically
an easy correction, it scarcely suits the sense of the passage.

burned in a jar with salt, to lick which gives clearness
of vision, and is a tonic to the stomach and to the
whole body. This salt is also given to sheep to keep
them in health, and is an ingredient of an antidote to
snake bite. Some use vipers [a] . . . as food. They prescribe that, first of all, as soon as the viper has been
killed, salt should be placed in its mouth until it melts ;
then at both ends a length of four fingers is cut off
and the intestines taken out ; the rest they thoroughly
boil in water,[b] oil, salt and dill, and either eat at once,
or mix in bread so that it can be used several times.
In addition to what has been said above, the broth
removes lice from any part of the body, as well as
itching from the surface of the skin. Even by itself,
the ash of a viper's head shows results ; as ointment
for the eyes it is very effective, and the same is true
of viper's fat. I would not confidently recommend
what is prescribed about a viper's gall, because, as I
have pointed out in the appropriate place,[c] a serpent's
poison is nothing but gall. The fat of snakes mixed
with bronze rust heals ruptured parts of the eyes, and
rubbing with their skin, or slough, cast in spring,
gives clear vision. The gall of the boa also is recommended for white ulcers, cataract, and dimness, and
its fat similarly for clear vision.

The gall of the eagle, which, as I have said,[d] tests
its chicks for gazing at the sun, makes, when mixed
with Attic honey, an ointment for film on the eyes,
dimness of vision, and cataract. There is the same

---

[b] Mayhoff (Appendix to vol. IV, p. 495) points out that
water and oil will not mix, and proposes an emendation that
would give : " boil with wine and oil etc."

[c] II. § 163.

[d] X. § 10.

oculorum. eadem vis est et in vulturino felle cum
porri suco et melle exiguo, item in gallinacei felle ad
argema et albugines ex aqua diluto, item in suffusiones oculorum, maxime candidi gallinacei. fimum
quoque gallinaceorum, dumtaxat rubrum, lusciosis
124 inlini monstrant. laudant et gallinae fel, et praecipue adipem, contra pusulas in pupillis; nec scilicet
eius rei gratia saginant. adiuvat mirifice et ruptas
oculorum tuniculas admixtis schisto et haematite
lapidibus. fimum quoque earum, dumtaxat candidum, in oleo vetere corneisque pyxidibus adservant
ad pupillarum albugines, qua in mentione significandum est pavones fimum suum resorbere tradi invi-
125 dentes hominum utilitatibus. accipiter decoctus in
rosaceo efficacissimus ad inunctiones omnium vitiorum
putatur, item fimi eius cinis cum Attico melle.
laudatur et milvi iocur, fimum quoque columbarum,
ex aceto ad aegilopia, similiter ad albugines et cicatrices, fel anserinum, sanguis anatum contusis oculis
ita ut postea oesypo et melle inunguantur, fel perdicum cum mellis aequo pondere, per se vero ad
claritatem. ex Hippocratis putant auctoritate adici
126 quod in argentea pyxide id servari iubent. ova perdicum in vase aereo decocta cum melle ulceribus
oculorum et glaucomatis medentur. columbarum,
palumbium, turturum, perdicum sanguis oculis
cruore suffusis eximie prodest. in columbis masculae efficaciorem putant, vena autem sub ala

---

[a] I place the phrase here, instead of at the end of the
sentence, to show the *similia similibus*.

[b] The phrases in this part of the chapter are difficult to
join correctly.

property also in vulture's gall with leek juice and a little honey, likewise in the gall of a cock, especially of a white cock,[a] diluted with water and used for white specks, white ulcers, and cataract. The dung of poultry also, provided that it is red, is prescribed as an ointment for night blindness. The gall of a hen also, and in particular the fat, is recommended for pustules on the pupils, but of course hens are not fattened specially for this purpose. It is a wonderful help, combined with the stones schistos and haematites, for the coats of the eye when torn. The dung also of hens, provided it is white, is kept in old oil and horn boxes for white ulcers on the pupil; while on the subject I must mention the tradition that peacocks swallow back their own dung, begrudging men its benefits. A hawk boiled down in rose oil is thought to make a very efficacious liniment for all eye complaints, as is its dung reduced to ash and added to Attic honey. A kite's liver too is recommended, and also pigeons' dung, applied in vinegar for fistulas, similarly for white ulcers and for scars, goose's gall and duck's blood for bruised eyes, provided that afterwards they are treated with wool grease and honey; partridge gall can be used with an equal weight of honey, but by itself for clear vision.[b] It is on the supposed authority of Hippocrates that the further instruction is given to keep this gall in a silver box. Partridge eggs boiled down with honey in a bronze vessel cure ulcers on the eyes and opaqueness of the lens. The blood of pigeons, doves, turtle doves, or partridges, makes an excellent application for blood-shot eyes. Among pigeons, male birds are supposed to have the more efficacious blood, and a vein under a wing is cut for this purpose, because its

ad hunc usum inciditur, quoniam suo calore utilior est. superinponi oportet splenium e melle decoctum
127 lanamque sucidam ex oleo aut[1] vino. earundem avium sanguis nyctalopas sanat et iocur ovium, ut in capris diximus, efficacius fulvae. decocto quoque eius oculos abluere suadent et medulla dolores tumoresque inlinere. bubonis oculorum cinis collyrio mixtus claritatem oculis facere promittitur. turturis fimum albugines extenuat, item coclearum cinis, fimum cenchridis. accipitrum generis hanc Graeci
128 faciunt. argema ex melle omnibus quae supra scripta sunt sanatur. mel utilissimum oculis in quo apes sint inmortuae. ciconiae pullum qui ederit negatur annis[2] continuis lippiturus, item qui draconis caput habeat. huius adipe et melle cum oleo vetere incipientes caligines discuti tradunt. hirundinum pullos plena luna excaecant, restitutaque eorum acie capita comburuntur, cinere cum melle utuntur ad
129 claritatem et dolores ac lippitudines et ictus. lacertas quoque pluribus modis ad oculorum remedia adsumunt. alii viridem includunt novo fictili, et lapillos qui vocantur cinaedia, quae et inguinum tumoribus adalligari solent, novem signis signant et singulos detrahunt per dies. nono emittunt lacer-
130 tam, lapillos servant ad oculorum dolores. alii terram substernunt lacertae viridi excaecatae et una in vitreo vase anulos includunt e ferro solido vel auro. cum

[1] aut E, *Pl. Iun.*, *Mayhoff*: ac *Detlefsen*.
[2] *Inter* annis *et* continuis *add.* multis *Mayhoff*: V(=quinque) Brakman.

[a] See XXVIII. § 170.
[b] Here Mayhoff would add " many " and Brakman " five."
[c] I.e. dazzled.

natural heat makes it more useful. Over the application should be placed a plaster boiled in honey and greasy wool boiled in oil or wine. Night blindness is cured by the blood of the same birds and by the liver of sheep, as I said [a] when speaking of goats, with greater benefit if the sheep are tawny. With a decoction also of the liver it is recommended to bathe the eyes and to apply the marrow to those that are painful or swollen. We are assured that the eyes of the horned owl, reduced to ash and mixed with a salve, improves the vision. White ulcers are made better by the dung of a turtle dove, by snails reduced to ash, and by the dung of the cenchris, a bird considered by the Greeks to be a species of hawk. White specks are cured by all the above remedies applied with honey. The honey most beneficial for the eyes is that in which bees have died. He who has eaten the chick of a stork, it is said, will not suffer from ophthalmia for [b] years on end, likewise he who carries about the head of a python. Its fat with honey and old oil is said to disperse incipient dimness. The chicks of swallows are blinded [c] by the full moon, and when their sight is restored their heads are burnt and the ash used with honey to improve the vision and for pains, ophthalmia, and blows. Lizards too are employed in several ways for eye remedies. Some shut up a green lizard in new earthenware, and with them the pebbles called *cinaedia*, which are used as amulets for swellings on the groin, mark them with nine marks and take away one daily; on the ninth day they set the lizard free, but keep the pebbles for pains in the eyes. Others put earth under a green lizard after blinding it, and shut it in a glass vessel with rings of solid iron or gold. When they can see

recepisse visum lacertam apparuit per vitrum,
emissa ea anulis contra lippitudinem utuntur, alii
capitis cinere pro stibi ad scabritias. quidam viridem
collo longo in sabulosis nascentem comburunt et
incipientem epiphoram inungunt, item glaucomata.
131 mustelae etiam oculis punctu erutis aiunt visum
reverti, eademque quae in lacertis et anulis faciunt,
serpentis oculum dextrum adalligatum contra epiphoras prodesse, si serpens viva dimittatur. lacrimantibus sine fine oculis cinis stelionum capitis cum
stibi eximie medetur. aranei muscarii tela et praecipue spelunca ipsa inposita per frontem ad duo tempora in splenio aliquo, ita ut a puero inpube et
capiatur et inponatur nec is triduo se ostendat ei cui
medebitur, neve alter nudis pedibus terram attingat
132 his diebus, mirabiliter epiphoris mederi dicitur,
albugines quoque tollere inunctione araneus candidus
longissimis ac tenuissimis pedibus contritus in oleo
vetere. is etiam cuius crassissimum textum est in
contignationibus fere adalligatus panno epiphoras
sanare traditur. scarabaei viridis natura contuen-
tium visum exacuit, itaque gemmarum scalptores
contuitu eorum adquiescunt.
133 XXXIX. Aures purgat fel pecudis cum melle,
canini lactis instillatio sedat dolorem, gravitatem
adeps cum absinthio et oleo vetere, item adeps anserinus. quidam adiciunt sucum cepae, alii pari

through the glass that the lizard has recovered its sight, they let it out, and use the rings for ophthalmia; others use the ash of the head instead of antimony for scabrous eyes. Some burn the green lizard with a long neck that is found in sandy places, and use it as ointment for incipient fluxes, as well as for opaqueness of the lens. They also say that when a weasel's eyes have been gouged out with a pointed tool, the sight is restored, and they use the animal as they used the lizards and rings, saying also that a serpent's right eye worn as an amulet, is good for eye fluxes, if the serpent is set free alive. The ash of a spotted lizard's head makes with antimony an excellent remedy for continually streaming eyes. The web of a fly-spider, particularly its very lair, is said to be a marvellous cure for fluxes if laid in a plaster across the forehead from temple to temple; but it must be collected and applied by a boy before puberty, who waits three days before showing himself to the patient needing cure, during which days the latter must not touch the earth with bare feet. White ulcers also are said to be removed by the white spider with very long and very thin legs, which is pounded in old oil and used as ointment. The spider too, whose very coarse web is generally found in rafters, is said to cure fluxes if worn in cloth as an amulet. The green beetle has the property of sharpening the sight of those who gaze at it, and so the carvers of jewels gaze on one to rest their eyes.

XXXIX. The ears are cleaned by sheep's gall with honey; pain is relieved by drops of bitch's milk; hardness of hearing by her fat with wormwood and old oil, also by goose grease. Some add the juice of onion and a like measure of garlic. They also use *Cures for the ears.*

modo. utuntur et per se ovis formicarum, namque et huic animali est medicina, constatque ursos aegros hoc
134 cibo sanari. anserum omniumque avium adeps praeparatur,[1] exemptisque omnibus venis patina nova fictili operta in sole subdita aqua ferventi liquatur, saccatusque lineis saccis et in fictili novo repositus loco frigido; minus putrescit addito melle. murium cinis cum melle instillatus aut cum rosaceo decoctus aurium dolores sedat. si aliquod animal intraverit, praecipuum remedium est murium fel aceto dilutum, si aqua intraverit, adeps anserinus cum cepae suco.
135 gliris detracta pelle intestinisque exemptis discoquitur melle in vase novo. medici malunt e nardo decoqui usque ad tertias partes atque ita adservari, dein, cum opus sit, strigili tepefacta infundere. constat deplorata aurium vitia eo remedio sanari aut si terreni vermes cum adipe anseris decocti infundantur, item ex arboribus rubri cum oleo triti exulceratis et ruptis auribus praeclare medentur. lacerti
136 inveterati,[2] in os pendentium addito sale, contusas et ab ictu miseras aures sanant, efficacissime autem ferrugineas maculas habentes, lineis etiam per caudam distincti.[3] milipeda ab aliis centipeda aut multipeda dicta animal est e vermibus terrae pilosum,

---

[1] *Post* praeparatur *lacunam indicat Mayhoff.*
[2] lacerti inveterati *codd.* : lacertae inveteratae *Mayhoff.*
[3] distincti *Caesarius* : distinctae (-te) *codd.*

---

[a] Some words appear to have dropped out. Perhaps "washed."
[b] The MSS. have *distinctae* (or *distincte*). Hence Mayhoff would emend *lacerti* (above) to *lacertae*. It is perhaps more likely that a scribe unconsciously slipped into the more usual feminine. One should note in this chapter the many references

without addition ants' eggs, for this creature also has its use in medicine, and it is well known that bears when sick cure themselves by eating these eggs. The fat of geese and of all birds is prepared [a] . . . all the veins are taken out, and in a new earthenware pan with a lid it is melted in the sun with boiling hot water underneath, strained through linen strainers and set aside in new earthenware in a cool place; if honey is added the fat is less likely to go rancid. The ash of mice, either added to honey or boiled with rose oil, if dropped into the ears relieves pain. If some creature has crept into the ear, the sovereign remedy is mouse gall diluted with vinegar; if it is water that has got in, goose grease with the juice of an onion. A dormouse, skinned and the intestines taken out, is thoroughly boiled in honey in a new vessel. Physicians prefer it to be boiled down to one-third in nard, and so stored away, and then when needed poured into the ear in a warmed strigil. It is well ascertained that desperate ear complaints are cured by this remedy, or if a decoction of earthworms and goose grease is injected. The red worms also that are taken off trees, if pounded with oil, make excellent treatment for ulcerated or ruptured ears. Preserved lizards, with salt put into their mouths as they hang suspended, heal bruised ears that are suffering from a blow, most efficaciously those covered with spots of the colour of iron rust and also marked [b] by streaks along the tail. The millipede, by some called centipede or multipede, is one of the earth worms; it is hairy, with many feet, moving sinuously

to broken ears, owing perhaps to the head wounds common in war and gladiatorial fights, and to the heavy *caestus* used by boxers.

multis pedibus arcuatim repens tactuque contrahens
se, oniscon Graeci vocant, alii iulon.[1] efficacem
narrant ad aurium dolores in cortice Punici mali
decoctum vel [2] porri suco. addunt et rosaceum et in
alteram aurem infundunt, illam autem quae non
arcuatur sepa Graeci vocant, alii scolopendram,
137 minorem perniciosamque. cocleae quae sunt in usu
cibi cum murra aut turis polline adpositae, item
minutae latae fracturis aurium inlinuntur cum melle.
senectus serpentium fervente testa usta instillatur
rosaceo admixto, contra omnia quidem vitia efficax,
sed contra graveolentiam praecipue, ac si purulenta
sint, ex aceto, melius cum felle caprino vel bubulo aut
testudinis marinae—vetustior anno eadem membrana
non prodest, nec imbre perfusa, ut aliqui putant—
138 aranei sanies cum rosaceo aut per se in lana vel cum
croco, gryllus cum sua terra effossus et inlitus.
magnam auctoritatem huic animali perhibet Nigidius,
maiorem Magi, quoniam retro ambulet terramque
terebret, stridat noctibus. venantur eum formica
circumligata capillo in cavernam eius coniecta, efflato
prius pulvere ne sese condat, ita formicae conplexu
139 extrahitur. ventris gallinaceorum membrana quae
abici solet inveterata et in vino trita auribus purulentis calida infunditur, item [3] gallinarum adeps et
quaedam pinguitudo blattae, si caput avellatur. hanc
tritam una cum rosaceo auribus mire prodesse dicunt,

[1] iulon *Detlefsen ex Indice* : tulion, tullon, tollen, tollon
codd.

[2] vel *Urlichs, Detlefsen* : et *Mayhoff* : mel VdT.

[3] item *ego addidi* : *Mayhoff* est *pro* et.

---

[a] The ailment is supposed to be driven *out* by the remedy
inserted into the other ear.

its back as it crawls, drawing itself together when touched, and called by the Greeks *oniscos* or *iulos*. It is said to be a good cure for ear pains if boiled down in pomegranate rind or leek juice. They add also rose oil, and pour it into the ear that is not painful.[a] The kind however that does not move sinuously its back the Greeks call *seps* or scolopendra; it is smaller and very venomous. The snails that are edible are applied with myrrh or powdered frankincense, and the small, broad snails are made into an ointment with honey for fractured ears. The slough of serpents, burnt in a heated pot, is mixed with rose oil and dropped into the ears, efficacious indeed for all affections, but especially for offensive smell; if pus is present, vinegar is used, and it is better if there be added gall of goat, ox, or turtle—the slough, as some think, loses power if older than a year, or if soaked with rain —the gore of a spider on wool with rose oil, by itself, or with saffron; a cricket dug out with its earth and applied.[b] Great efficacy is attributed to this creature by Nigidius, greater still by the Magi, just because it walks backwards, bores into the earth, and chirrups at night. They hunt it with an ant tied to a hair and put into the cricket's hole, first blowing the dust away lest it bury itself, and so when the ant has embraced it the cricket is pulled out. The lining of the crop of poultry, usually thrown away, if dried and pounded in wine, is poured warm into suppurating ears, likewise hens' fat and a kind of greasy substance coming from the black beetle if its head is pulled off. This, pounded with rose oil, is said to be

---

[b] A formless sentence. Some verbal expression, such as " benefits pus in the ears," must be understood with the last clause.

sed lanam qua incluserint post paulum extrahendam,
celerrime enim id pingue transire in animal fierique
vermiculum. alii binas ternasve in oleo decoctas
efficacissime auribus mederi scribunt et tritas in
140 linteolo inponi contusis. hoc quoque animal inter
pudenda est, sed propter admirationem naturae
priscorumque curae totum in hoc loco explicandum.
plura earum genera fecerunt: molles, quas in oleo
decoctas verrucis efficaciter inlini experti sunt.
141 alterum genus myloecon appellavere circa molas fere
nascens. his capite detracto adtritas lepras sanasse
Musaeum[1] pycten in exemplis reliquerunt. tertium
genus et odoris taedio invisum, exacuta clune, cum
pisselaeo sanare ulcera alias insanabilia, strumas,
panos diebus xxi inpositas, percussa, contusa et
cacoethe, scabiem furunculosque detractis pedibus
142 et pinnis. nos haec etiam audita fastidimus. at,
Hercules, Diodorus et in morbo regio et orthopnoicis
se dedisse tradit cum resina et melle. tantum
potestatis habit ars ea pro medicamento dandi quidquid velit. humanissimi eorum cinerem crematarum
servandum ad hos usus in cornea pyxide censuere aut
tritas clysteribus infundendas orthopnoicis aut

---

[1] Musaeum *Ianus*: muscum *aut* muscum *codd.*

wonderfully good for the ears, but the wool on which
it is inserted must be taken out after a short time,
for this grease very quickly turns into something
alive, forming a grub. Some write that a dose of
two or three of these beetles, boiled down in oil,
make very good treatment for the ears, and that
when these are bruised crushed beetles are placed
in them in a piece of linen. This insect is one of the
things that arouse disgust, but because Nature and
the research of the ancients are so wonderful I must
go fully into the matter here. They have made
several classes of them: first the soft kind which,
boiled down in oil, they found to make a good ointment for warts. The second kind they called
myloecos, because they are found commonly about
mills. The instances they quoted include Musaeus
the boxer, who cured leprous sores by this kind
rubbed on without their heads. A third kind,
one with a loathsome smell and a sharp-pronged
tail-end, they say will cure, if applied with pis-
selaeum for twenty-one days, ulcers otherwise
incurable, scrofulous sores and superficial abscesses;
and without legs and wings bruises, contusions,
even malignant sores, itch scab, and boils. Even
to hear these remedies mentioned makes me feel
sick; but, heaven help us! Diodorus says that he had
given these beetles with resin and honey even in cases
of jaundice and orthopnoea. So much power has the
art of medicine to prescribe any medicament it
may wish. The kindliest among physicians have
thought that the ash of burnt black beetles should be
kept for the purposes mentioned in a horn box, or
that crushed they should be given in enemas to
sufferers from orthopnoea or catarrh. It is a known

rheumaticis. infixa utique corpori inlitas extrahere
143 constat. mel utilissimum auribus quoque est in quo
apes inmortuae sint. parotidas comprimit columbinum stercus vel per se vel cum farina hordeacea aut
avenacea, noctuae cerebrum vel iocur cum oleo infusum auriculae a parotide,[1] multipeda cum resinae
parte tertia inlita, grylli sive inliti sive adalligati. ad
reliqua morborum genera medicinam ex isdem
animalibus aut eiusdem generis sequenti dicemus
volumine.

[1] a parotide *in uncis Mayhoff.*

fact at any rate that an application brings away things embedded in the flesh. The most suitable honey for the ears also is that in which bees have died. Parotid swellings are reduced by pigeon's dung either by itself or with barley meal or oatmeal, by the brain or liver of an owl, poured with oil into the ear on the side of the swelling, by a multipede with a third part of resin used as ointment, and by crickets, used as ointment or as amulets. Medicine for the remaining kinds of disease from the same animals or from animals of the same kind, I shall speak of in the next Book.

BOOK XXX

LIBER XXX

1 I. Magicas vanitates saepius quidem antecedente operis parte, ubicumque causae locusque poscebant, coarguimus detegemusque etiamnum. in paucis tamen digna res est de qua plura dicantur, vel eo ipso quod fraudulentissima artium plurimum in toto terrarum orbe plurimisque saeculis valuit. auctoritatem ei maximam fuisse nemo miretur, quandoquidem sola artium tres alias imperiosissimas humanae

2 mentis complexa in unam se redegit. natam primum e medicina nemo dubitabit ac specie salutari inrepsisse velut altiorem sanctioremque medicinam, ita blandissimis desideratissimisque promissis addidisse vires religionis, ad quas maxime etiamnunc caligat humanum genus, atque, ut hoc quoque successerit,[1] miscuisse artes mathematicas, nullo non avido futura de sese sciendi atque ea e caelo verissime peti credente. ita possessis hominum sensibus triplici vinculo in tantum fastigii adolevit ut hodieque etiam in magna parte gentium praevaleat et in oriente regum regibus imperet.

3 II. Sine dubio illic orta in Perside a Zoroastre, ut inter auctores convenit. sed unus hic fuerit an

---

[1] successerit *C. F. W. Müller*: suggesserit *aut* suggerit *codd.*

---

a Or, "Few themes deserve more to receive fuller treatment."

BOOK XXX

I. In the previous part of my work I have often *Origin of* *magic.*
indeed refuted the fraudulent lies of the Magi, whenever the subject and the occasion required it, and I
shall continue to expose them. In a few respects,
however, the theme deserves [a] to be enlarged upon,
were it only because the most fraudulent of arts has
held complete sway throughout the world for many
ages. Nobody should be surprised at the greatness
of its influence, since alone of the arts it has embraced
three others that hold supreme dominion over the
human mind, and made them subject to itself alone.
Nobody will doubt that it first arose from medicine,
and that professing to promote health it insidiously
advanced under the disguise of a higher and holier
system; that to the most seductive and welcome
promises it added the powers of religion, about which
even today the human race is quite in the dark;
that again meeting with success it made a further
addition of astrology, because there is nobody who
is not eager to learn his destiny, or who does not
believe that the truest account of it is that gained by
watching the skies. Accordingly, holding men's
emotions in a three-fold bond, magic rose to such a
height that even today it has sway over a great part
of mankind, and in the East commands the Kings of
Kings.

II. Without doubt magic arose in Persia with
Zoroaster. On this our authorities are agreed, but

postea et alius non satis constat. Eudoxus, qui inter sapientiae sectas clarissimam utilissimamque eam intellegi voluit, Zoroastrem hunc sex milibus annorum ante Platonis mortem fuisse prodidit, sic et Aristoteles.
4 Hermippus qui de tota ea arte diligentissime scripsit et viciens centum milia versuum a Zoroastre condita indicibus quoque voluminum eius positis explanavit, praeceptorem a quo institutum diceret tradidit Agonacen, ipsum vero quinque milibus annorum ante Troianum bellum fuisse. mirum hoc in primis, durasse memoriam artemque tam longo aevo commentariis intercidentibus,[1] praeterea nec claris nec
5 continuis successionibus custoditam. quotus enim quisque [2] auditu saltem cognitos habet, qui soli nominantur, Apusorum et Zaratum Medos, Babyloniosque Marmarum et Arabantiphocum, Assyrium Tarmoendam, quorum nulla exstant monumenta? maxime tamen mirum est in bello Troiano tantum de arte ea silentium fuisse Homero tantumque operis ex eadem in Ulixis erroribus, adeo ut totum [3] opus non aliunde
6 constet, siquidem Protea et Sirenum cantus apud eum non aliter intellegi volunt, Circe [4] utique et inferum evocatione hoc solum agi. nec postea quisquam dixit quonam modo venisset Telmesum religiosissimam [5]

---

[1] intercidentibus VGd *Sillig.* : non intercedentibus R? *Detlefsen* : non *ante* commentariis *ponit Mayhoff.*
[2] *Ante* auditu *in codd.* communi *aut* commi, *om.* Er: hominum *Mayhoff.*
[3] *Ante* totum *in codd. multis* de : *om. Detlefsen* : vel *Mayhoff.*
[4] *Ante* Circe *coni. in Mayhoff.*
[5] *Post* religiosissimam *coni. in Mayhoff.*

---

[a] An *index* might be a mere title or a brief list of contents (or both).

whether he was the only one of that name, or whether there was also another afterwards, is not clear. Eudoxus, who wished magic to be acknowledged as the noblest and most useful of the schools of philosophy, declared that this Zoroaster lived six thousand years before Plato's death, and Aristotle agrees with him. Hermippus, a most studious writer about every aspect of magic, and an exponent of two million verses composed by Zoroaster, added summaries *a* too to his rolls, and gave Agonaces as the teacher by whom he *b* said that he had been instructed, assigning to the man himself a date five thousand years before the Trojan War. What especially is surprising is the survival, through so long a period, of the craft and its tradition; treatises are wanting, and besides there is no line of distinguished or continuous successors to keep alive their memory. For how few know anything, even by hearsay, of those who alone have left their names but without other memorial—Apusorus and Zaratus of Media, Marmarus and Arabantiphocus of Babylon, or Tarmoendas of Assyria? The most surprising thing, however, is the complete silence of Homer about magic in his poem on the Trojan War, and yet so much of his work in the wanderings of Ulysses is so occupied with it that it alone forms the backbone of the whole work, if indeed they put a magical interpretation upon the Proteus episode in Homer and the songs of the Sirens, and especially upon the episode of Circe and of the calling up of the dead from Hades, of which magic is the sole theme. And in later times nobody has explained how ever it reached Telmesus,

b The omission of the pronouns makes the subject of *diceret* uncertain—Zoroaster or Hermippus.

urbem, quando transisset ad Thessalas matres,
quarum cognomen diu optinuit in nostro orbe, aliena
genti Troianis utique temporibus Chironis medicinis
7 contentae et solo Marte fulminante.[1] miror equidem
Achillis populis famam eius in tantum adhaesisse, ut
Menander quoque litterarum subtilitati sine aemulo
genitus Thessalam cognominaret fabulam [2] com-
plexam ambages feminarum detrahentium lunam.
Orphea putarem e propinquo eam [3] primum pertulisse
ad vicina eiusque [4] superstitionem a medicina [5] provectam,[6] si non expers sedes eius tota Thrace magices
8 fuisset. primus, quod exstet, ut equidem invenio,
commentatus est de ea Osthanes Xerxen regem
Persarum bello quod is Graeciae intulit comitatus,
ac velut semina artis portentosae sparsit obiter in-
fecto quacumque commeaverant mundo. diligen-
tiores paulo ante hunc ponunt Zoroastrem alium
Proconnensium. quod certum est, hic maxime
Osthanes ad rabiem, non aviditatem modo scientiae
eius Graecorum populos egit, quamquam anim-
adverto summam litterarum claritatem gloriamque
ex ea scientia antiquitus et paene semper petitam.

---

[1] fulminante *multi codd.*, *Detlefsen* : fulminanti̊ *Mayhoff* :
fulminati V^{1}GR^{1}d.

[2] fabulam *Detlefsen* : famulam *Mayhoff. Neuter editor
alias indicat lectiones.*

[3] propinquo eam *Gronovius, Ianus* : propinquo artem
Mayhoff : propinquo R(?) E *vulg. Detlefsen* : propinquorum
VGd : propinquum *coni. Warmington.*

[4] eiusque *P. Green* : usque *codd.*

[5] a medicina *Gronovius, Sillig* : ac medicinae (*et* super-
stitionis) *Mayhoff* : ac medicinae (superstitiones E, superstitionem R) ER.

[6] provectam *coni. Mayhoff* : provectum *aut* profectum *codd.*

a city given up to superstition, or when it passed over
to the Thessalian matrons, whose surname [a] was long
proverbial in our part of the world, although magic
was a craft repugnant to the Thessalian people, who
were content, at any rate in the Trojan period, with
the medicines of Chiron, and with the War God as
the only wielder of the thunderbolt.[b] I am indeed
surprised that the people over whom Achilles once
ruled had a reputation for magic so lasting that
actually Menander, a man with an unrivalled gift for
sound literary taste, gave the name " Thessala " to
his comedy, which deals fully with the tricks of the
women for calling down the moon. I would believe
that Orpheus was the first to carry the craft to his
near neighbours, and that his superstition grew from
medicine, if the whole of Thrace, the home of
Orpheus, had not been untainted by magic. The
first man, so far as I can discover, to write a still-
extant treatise on magic was Osthanes, who accompanied the Persian King Xerxes in his invasion
of Greece, and sowed what I may call the seeds of this
monstrous craft, infecting the whole world by the
way at every stage of their travels. A little before
Osthanes, the more careful inquirers place another
Zoroaster, a native of Proconnesus. One thing
is certain; it was this Osthanes who chiefly roused
among the Greek peoples not so much an eager
appetite for his science as a sheer mania. And
yet I notice that of old, in fact almost always,
the highest literary distinction and renown have
been sought from that science. Certainly Pytha-

---

[a] I.e. " Thessalian." The word suggested witchcraft.
[b] With the reading *fulminanti* : " whose only thunder was
that of their War God."

9 certe Pythagoras, Empedocles, Democritus, Plato ad hanc discendam navigavere exiliis verius quam peregrinationibus susceptis, hanc reversi praedicavere, hanc in arcanis habuere. Democritus Apollobechem Coptitem et Dardanum e Phoenice inlustravit, voluminibus Dardani in sepulchrum eius petitis, suis vero ex disciplina eorum editis, quae recepta ab ullis hominum atque transisse per memoriam aeque ac nihil in vita mirandum est. in tantum fides istis fasque omne deest, adeo ut qui

10 cetera in viro probant haec opera eius esse infitientur.[1] sed frustra, hunc enim maxime adfixisse animis eam dulcedinem constat. plenumque miraculi et hoc, pariter utrasque artes effloruisse, medicinam dico magicenque, eadem aetate illam Hippocrate, hanc Democrito inlustrantibus, circa Peloponnensiacum Graeciae bellum quod gestum est a trecentesimo

11 urbis nostrae anno. est et alia magices factio a Mose et Janne et Lotape [2] ac Iudaeis pendens, sed multis milibus annorum post Zoroastrem. tanto recentior est Cypria. non levem et Alexandri Magni temporibus auctoritatem addidit professioni secundus Osthanes comitatu eius exornatus, planeque, quod nemo dubitet, orbem terrarum peragravit.

12 III. Extant certe et apud Italas gentes vestigia eius in XII tabulis nostris aliisque argumentis quae

---

[1] infitientur *Mayhoff*: inficientur *codd.*
[2] Lotape *codd.*: Iotape *Gelenius.*

---

[a] See Torrey, *The Magic of Lotapes* (Journal of Biblical Literature, 1949, 325–327). Pliny should have written *Iotape* = ἰῶτα πῆ = Yahweh. Jannes was not a Hebrew

goras, Empedocles, Democritus and Plato went overseas to learn it, going into exile rather than on a journey, taught it openly on their return, and considered it one of their most treasured secrets. Democritus expounded Apollobex the Copt and Dardanus the Phoenician, entering the latter's tomb to obtain his works and basing his own on their doctrines. That these were accepted by any human beings and transmitted by memory is the most extraordinary phenomenon in history; so utterly are they lacking in credibility and decency that those who like the other works of Democritus deny that the magical books are his. But it is all to no purpose, for it is certain that Democritus especially instilled into men's minds the sweets of magic. Another extraordinary thing is that both these arts, medicine I mean and magic, flourished together, Democritus expounding magic in the same age as Hippocrates expounded medicine, about the time of the Peloponnesian War, which was waged in Greece from the three-hundredth year of our city. There is yet another branch of magic, derived from Moses, Jannes, Lotapes,[a] and the Jews, but living many thousand years after Zoroaster. So much more recent is the branch in Cyprus. In the time too of Alexander the Great, no slight addition was made to the influence of the profession by a second Osthanes, who, honoured by his attendance on Alexander, travelled certainly without the slightest doubt all over the world.

III. Among Italian tribes also there still certainly exist traces of magic in the Twelve Tables, as is

but an Egyptian magician, who competed with Moses. See *Epistle to Timothy*, II. 3, 8.

285

priore volumine exposui. DCLVII demum anno
urbis Cn. Cornelio Lentulo P. Licinio Crasso cos.
senatusconsultum factum est ne homo immolaretur,
palamque in tempus illut sacra prodigiosa celebrata.

13   IV. Gallias utique possedit, et quidem ad nostram
memoriam. namque Tiberii Caesaris principatus
sustulit Druidas eorum et hoc genus vatum medi-
corumque. sed[1] quid ego haec commemorem in arte
oceanum quoque transgressa et ad naturae inane per-
vecta? Britannia hodieque eam adtonita celebrat
tantis caerimoniis ut dedisse Persis videri possit.
adeo ista toto mundo consensere quamquam discordi
et sibi ignoto. nec satis aestimari potest quantum
Romanis debeatur, qui sustulere monstra, in quibus
hominem occidere religiosissimum erat, mandi vero
etiam saluberrimum.

14   V. Ut narravit Osthanes, species eius plures sunt.
namque et aqua et sphaeris et aëre et stellis et
lucernis ac pelvibus securibusque et multis aliis
modis divina promittit, praeterea umbrarum in-
ferorumque colloquia. quae omnia aetate nostra
princeps Nero vana falsaque comperit. quippe non
citharae tragicique cantus libido illi maior fuit,
fortuna rerum humanarum summa gestiente[2] in
profundis animi vitiis, primumque imperare dis con-

---

[1] sed *Gelenius, Mayhoff* : ipse *codd.*
[2] gestiente *codd.* : gestienti *coni. Mayhoff.*

---

[a] XXVIII. § 17.
[b] 97 B.C.
[c] Or: "agreement in that subject of magic."

proved by my own and the other evidence set forth in an earlier Book.[a] It was not until the 657th year of the City [b] that in the consulship of Gnaeus Cornelius Lentulus and Publius Licinius Crassus there was passed a resolution of the Senate forbidding human sacrifice; so that down to that date it is manifest that such abominable rites were practised.

IV. Magic certainly found a home in the two Gallic provinces, and that down to living memory. For the principate of Tiberius Caesar did away with their Druids and this tribe of seers and medicine men. But why should I speak of these things when the craft has even crossed the Ocean and reached the empty voids of Nature? Even today Britain practises magic in awe, with such grand ritual that it might seem that she gave it to the Persians. So universal is the cult of magic [c] throughout the world, although its nations disagree or are unknown to each other. It is beyond calculation how great is the debt owed to the Romans, who swept away the monstrous rites, in which to kill a man was the highest religious duty and for him to be eaten a passport to health.

V. As Osthanes said, there are several forms of magic; he professes to divine from water, globes, air, stars, lamps, basins and axes, and by many other methods, and besides to converse with ghosts and those in the underworld. All of these in our generation the Emperor Nero discovered to be lies and frauds. In fact his passion for the lyre and tragic song was no greater than his passion for magic; his elevation to the greatest height of human fortune aroused desire in the vicious depths of his mind; his greatest wish was to issue commands to the gods,

cupivit, nec quicquam generosius voluit. nemo um-
15 quam ulli artium validius favit. ad hoc non opes
defuere, non vires, non discentis ingenium, quae non
alia patiente mundo! inmensum, indubitatum exemplum est falsae artis quam dereliquit Nero, utinamque inferos potius et quoscumque de suspitioni-
bus suis deos consuluisset quam lupanaribus atque
prostitutis mandasset inquisitiones eas! nulla profecto sacra, barbari licet ferique ritus, non mitiora
quam cogitationes eius fuissent. saevius sic[1] nos
replevit umbris.

16 VI. Sunt quaedam Magis perfugia, veluti lenti-
ginem habentibus non obsequi numina aut cerni.
num obstitit[2] forte hoc in illo? nihil membris defuit.
nam dies eligere certos liberum erat, pecudes vero
quibus non nisi ater colos esset facile. nam homines
immolare etiam gratissimum. Magus ad eum Tiridates venerat Armeniacum de se triumphum adferens
17 et ideo provinciis gravis. navigare noluerat, quoniam
expuere in maria aliisque mortalium necessitatibus
violare naturam eam fas non putant. Magos secum
adduxerat, magicis etiam cenis eum initiaverat, non
tamen, cum regnum ei daret, hanc ab eo artem accipere valuit.[3] proinde ita persuasum sit, intestabilem,
inritam, inanem esse, habentem tamen quasdam
veritatis umbras, sed in his veneficas artes pollere,
18 non magicas. quaerat aliquis, quae sint mentiti

---

[1] hic *vel* is sic *coni. Warmington.*

[2] num obstitit *ego coni. post Pintianum*: an obstitit *May-hoff*: non (*pro* num) dTE: obstet *aliquot codd., Detlefsen.*

[3] valuit d(?) *vulg., Detlefsen, Mayhoff*: voluit *paene omnes codd. et Mayhoff in Appendice.*

and he could rise to no nobler ambition. No other of the arts ever had a more enthusiastic patron. Every means were his to gratify his desire—wealth, strength, aptitude for learning—and what else did the world not allow! That the craft is a fraud there could be no greater or more indisputable proof than that Nero abandoned it; but would that he had consulted about his suspicions the powers of Hell and any other gods whatsoever, instead of entrusting these researches to pimps and harlots. Of a surety no ceremony, outlandish and savage though the rites may be, would not have been gentler than Nero's thoughts; more cruelly behaving than any did Nero thus fill our Rome with ghosts.

VI. The Magi have certain means of evasion; for example that the gods neither obey those with freckles nor are seen by them. Was this perhaps their objection to Nero? But his body was without blemish; he was free to choose the fixed days, could easily obtain perfectly black sheep, and as for human sacrifice, he took the greatest delight in it. Tiridates the Magus had come to him bringing a retinue for the Armenian triumph over himself, thereby laying a heavy burden on the provinces. He had refused to travel by sea, for the Magi hold it sin to spit into the sea or wrong that element by other necessary functions of mortal creatures. He had brought Magi with him, had initiated Nero into their banquets; yet the man giving him a kingdom was unable to acquire from him the magic art. Therefore let us be convinced by this that magic is detestable, vain, and idle; and though it has what I might call shadows of truth, their power comes from the art of the poisoner, not of the Magi. One might well ask what were the

289

veteres Magi, cum adulescentibus nobis visus Apion grammaticae artis prodiderit cynocephalian herbam, quae in Aegypto vocaretur osiritis, divinam et contra omnia veneficia, sed si tota erueretur, statim eum qui eruisset mori, seque evocasse umbras ad percunctandum Homerum quanam patria quibusque parentibus genitus esset, non tamen ausus profiteri quid sibi respondisse diceret.

19 VII. Peculiare vanitatis sit argumentum quod animalium cunctorum talpas maxime mirantur tot modis a rerum natura damnatas, caecitate perpetua, tenebris etiamnum aliis [1] defossis sepultisque similes. nullis aeque credunt extis, nullum religionum capacius iudicant animal, ut si quis cor eius recens palpitansque devoret,[2] divinationis et rerum efficien-
20 darum eventus promittant. dente talpae vivae exempto sanari dentium dolores adalligato adfirmant. cetera ex eo animali placita eorum suis reddemus locis. nec quicquam probabilius invenietur quam muris aranei morsibus adversari eas, quoniam et terra orbitis, ut diximus, depressa adversatur.

21 VIII. Cetero dentium doloribus, ut idem narrant, medetur canum qui rabie perierunt capitum cinis crematorum sine carnibus instillatus ex oleo cyprio per aurem cuius e parte doleant, caninus dens sinister maximus circumscarifato qui doleat aut draconis os

---

[1] aliis *aut* alis *codd.*, *Mayhoff*: altis *Detlefsen*.
[2] devoret V¹GRdTf : devoraret V²E¹ : devorarit E² *vulg.*, *Detlefsen*.

---

[a] See XXIX. § 89.

lies of the old Magi, when as a youth I saw Apion the grammarian, who told me that the herb cynocephalia, called in Egypt osiritis, was an instrument of divination and a protection from all kinds of sorcery, but if it were uprooted altogether the digger would die at once, and that he had called up ghosts to inquire from Homer his native country and the name of his parents, but did not dare to repeat the answers which he said were given.

VII. It should be unique evidence of fraud that *The mole.* they look upon the mole of all living creatures with the greatest awe, although it is cursed by Nature with so many defects, being permanently blind, sunk in other darkness also, and resembling the buried dead. In no entrails is placed such faith; to no creature do they attribute more supernatural properties; so that if anyone eats its heart, fresh and still beating, they promise powers of divination and of foretelling the issue of matters in hand. They declare that a tooth, extracted from a living mole and attached as an amulet, cures toothache. The rest of their beliefs about this animal I will relate in the appropriate places. But of all they say nothing will be found more likely than that the mole is an antidote for the bite of the shrewmouse, seeing that an antidote for it, as I have said,[a] is even earth that has been depressed by cart wheels.

VIII. Toothache is also cured, the Magi tell us, *Remedies for* by the ash of the burnt heads without any flesh of *the teeth.* dogs that have died of madness, which must be dropped in cyprus oil through the ear on the side where the pain is; also by the left eye-tooth of a dog, the aching tooth being scraped round with it; by one of the vertebrae of the draco or of the

e spina, item enhydridis, est autem serpens masculus
et albus. huius maximo dente circumscarifant, aut
in superiorum dolore duos superiores adalligant, e
22 diverso inferiores. huius adipe perunguuntur qui
crocodilum captant. dentes scarifant et ossibus
lacertae ex fronte luna plena exemptis ita ne terram
adtingant. colluunt dentibus caninis decoctis in vino
ad dimidias partes. cinis eorum pueros tarde
dentientes adiuvat cum melle. fit eodem modo et
dentifricium. cavis dentibus cinis e murino fimo
23 inditur, vel iocur lacertarum aridum. anguinum cor
si mordeatur adalligeturve efficax habetur. sunt
inter eos qui murem bis in mense iubeant mandi
doloresque ita caveri. vermes terreni decocti in oleo
infusique auriculae cuius a parte doleat praestant
levamentum. eorundem cinis exesis dentibus coniectus [1] ex facili [2] cadere eos cogit, integros dolentes
inlitus iuvat. comburi autem oportet in testo. prosunt et cum mori radice in aceto scillite decocti ita ut
24 colluantur dentes. is quoque vermiculus qui in
herba Veneris labro appellata invenitur cavis dentium
inditus mire prodest. nam urucae brassicae eius
contactu cadunt, et a malva cimices infunduntur
auribus cum rosaceo. harenulae quae inveniuntur
in cornibus coclearum cavis dentium inditae statim

---

[1] coniectus r *Pl. Iun., Mayhoff*: coiectus E: collectus d,
Detlefsen: collectis *aliquot codd.*: colutis *Ianus.*

[2] ex facili *aliquot codd., Detlefsen, Mayhoff*: ex facile VGR.
Marcellus (XII 31) " insertus et cera opertus facile cadere eos
cogit." *Fortasse* coniectus et cera contectus facile. *Warmington* coniectus facile excidere *coni.*

---

[a] The true text is very hard to discover. The general
sense is plain, but the parallel passage in Marcellus XII. 31
seems to suggest that a phrase like " covered with wax " has

enhydris, the serpent being a white male. With this eye-tooth they scrape all round the painful one, or they make an amulet of two upper teeth, when the pain is in the upper jaw, using lower teeth for the lower jaw. With its fat they rub hunters of the crocodile. They also scrape teeth with bones extracted from the forehead of a lizard at a full moon, without their touching the earth. They rinse the mouth with a decoction of dogs' teeth in wine, boiled down to one-half. The ash of these teeth with honey helps children who are slow in teething. A dentifrice also is made with the same ingredients. Hollow teeth are stuffed with the ash of mouse dung or with dried lizards' liver. A snake's heart, eaten or worn as an amulet, is considered efficacious. There are among them some who recommend a mouse to be chewed up twice a month to prevent aches. Earthworms, boiled down in oil and poured into the ear on the side where there is pain, afford relief. These also, reduced to ash and plugged into decayed teeth, force them to fall out easily,[a] and applied to sound teeth relieve any pain in them. They should be burnt, however, in an earthen pot. They also benefit if boiled down in squill vinegar with the root of a mulberry tree, so as to make a wash for the teeth. The maggot also, which is found on the plant called Venus' Bath, plugged into hollow teeth, is wonderfully good. But they fall out at the touch of the cabbage caterpillar, and the bugs from the mallow are poured into the ears with rose oil. The little grains of sand, that are found in the horns of snails, if put into hollow teeth, free them at once

been lost. My own guess presupposes a loss of *cera contectus* after *coniectus*. Warmington's coniecture is attractive.

liberant dolore. coclearum inanium cinis cum murra gingivis prodest, serpentis cum sale in olla exustae cinis cum rosaceo in contrariam aurem infusus, anguinae vernationis membrana cum oleo taedaeque
25 resina calefacta et auri alterutri infusa—adiciunt aliqui tus et rosaceum—eadem cavis indita ut sine molestia cadant praestat. vanum arbitror esse circa canis ortum angues candidos membranam eam exuere, quoniam ante ortum [1] in Italia visum est, multoque minus credibile in tepidis regionibus tam sero exui. hanc autem vel inveteratam cum cera celerrime evellere tradunt, et dens anguium adalli-
26 gatus dolores mitigat. sunt qui et araneum animal ipsum sinistra manu captum tritumque in rosaceo et in aurem infusum cuius a parte doleat prodesse arbitrentur. ossiculi gallinarum in pariete servati fistula salva; [2] tacto dente vel gingiva scarifata proiectoque ossiculo statim dolorem abire tradunt, item fimo corvi lana adalligato vel passerum cum oleo calefacto et proximae auriculae infuso. pruri-tum quidem intolerabilem facit et ideo utilius est passeris pullorum sarmentis crematorum cinerem ex aceto infricare.
27 IX. Oris saporem commendari adfirmant, murino cinere cum melle si fricentur dentes. admiscent quidam marathi radices. pinna vulturis si scalpantur

---

[1] ante ortum *Mayhoff*: neutrum *codd.*, *Detlefsen*.
[2] in pariete servati fistula salva] *Nescioquo loco latet error nondum sanatus. Vide notam.*

---

[a] Both the structure and the sense are difficult. Mayhoff conjectures *panno* or *puxide* for *pariete*, but the last occurs in similar cures in § 51 and elsewhere. I translate as though

from pain. Empty snail shells, reduced to ash and myrrh added, are good for the gums, as is the ash of a serpent burnt with salt in an earthen pot, poured with rose oil into the opposite ear, or the slough of a snake with oil and pitch-pine resin warmed and poured into either ear—some add frankincense and rose oil—and if put into hollow teeth it also makes them fall out without trouble. I think it an idle tale that white snakes cast their slough about the rising of the Dog-star, since the casting has been seen in Italy before the rising, and in warm regions it is much less probable for sloughing to be so late. But they say that this slough, even when dry, combined with wax forces out teeth very quickly. A snake's tooth also, worn as an amulet, relieves toothache. There are some who think that a spider also is beneficial, the animal itself, caught with the left hand, beaten up in rose oil, and poured into the ear on the side of the pain. The little bones of hens have been kept hanging on the wall of a room with the gullet intact; [a] if a tooth is touched, or the gum scraped, and the bone thrown away, they assure us that the pain at once disappears, as it does if a raven's dung, wrapped in wool, is worn as an amulet, or if sparrows' dung is warmed with oil and poured into the ear nearer the pain. This however causes unbearable itching, and so it is better to rub the part with vinegar and the ash of a sparrow's nestlings burnt on twigs.

IX. They assert that the taste in the mouth is made agreeable if the teeth are rubbed with the ash of burnt mice mixed with honey; some add fennel root. If the teeth are picked with a vulture's

servati were a finite verb, and a new sentence began at *tacto*. This gives the general sense.

dentes, acidum halitum faciunt. hoc idem hystricis spina fecisse ad firmitatem pertinet. linguae ulcera et labrorum hirundines in mulso decoctae sanant, adeps anseris aut gallinae rimas, oesypum cum galla, araneorum telae candidae et quae in trabibus [1] parvae texuntur. si ferventia os intus exusserint, lacte canino statim sanabuntur.

28   X. Maculas in facie oesypum cum melle Corsico quod asperrimum habetur extenuat, item scobem cutis in facie cum rosaceo inpositum vellere—quidam et butyrum addunt—si vero vitiligines sint, fel caninum prius acu conpunctas, ad liventia et suggillata pulmones arietum pecudumque in tenues
29 consecti membranas calidi inpositi, vel columbinum fimum. cutem in facie custodit adeps anseris vel gallinae. lichenas et murino fimo ex aceto inlinunt et cinere irenacei ex oleo. in hac curatione prius nitro ex aceto faciem foveri praecipiunt. tollit ex facie vitia et coclearum quae latae et minutae passim inveniuntur cum melle cinis. omnium quidem coclearum cinis spissat,[a] calfacit, smectica vi, et ideo causticis miscetur, psorisque et lepris et lentigini inlinitur. invenio et formicas Herculaneas appellari quibus tritis adiecto sale exiguo talia vitia sanentur.
30 buprestis animal est rarum in Italia, simillimum scarabaeo longipedi. fallit inter herbas bovem maxime, unde et nomen invenit, devoratumque tacto

---

[1] in trabibus *Hermolaus Barbarus, Mayhoff*: intra bulbus *codd.*

---

[a] *Spissare*, a favourite word of Pliny, is often of uncertain meaning and difficult to translate. Here perhaps there is reference to the drying up of morbid humours.

feather, they make the breath sour. To pick them
with a porcupine's quill conduces to their firmness.
Sores on the tongue or lips are healed by a decoction
of swallows in honey wine; chaps on them by goose
grease or hen's grease, by oesypum with gall nut, by
white webs of spiders, or by the small webs spun on
rafters. If the mouth has been scalded by over-hot
things, bitch's milk will give an immediate cure.

X. Spots on the face are removed by oesypum *Facial*
with Corsican honey, which is considered the most *remedies.*
acrid; scurf on the skin of the face by the same
with rose oil on a piece of fleece; some add also
butter. If however there is psoriasis, dog's gall is
applied to the spots, which are first pricked with a
needle; to livid spots and bruises rams' or sheep's
lungs are applied hot and cut into thin slices, or else
pigeon's dung. The skin of the face is preserved by
goose grease or hen's. To lichen is also applied
mouse dung in vinegar, or ash of the hedgehog in
oil; for this treatment they prescribe that the face
should first be fomented with soda and vinegar.
Facial troubles are also removed by the ash with
honey of the broad but small snails that are found
everywhere. The ash indeed of all snails, such is its
detergent property, thickens [a] and warms; for that
reason it is an ingredient of caustic preparations and
used as a liniment for itch, leprous sores, and freckles.
I find also that there are ants called Herculanean,
which beaten up and with the addition of a little salt
cure facial troubles. The buprestis is a creature
rarely found in Italy, and very similar to a long-
legged beetle. Oxen at pasture are very apt not to
see it—hence too its name—and should it be
swallowed it causes such inflammation on reaching

felle ita inflammat ut rumpat. haec cum hircino sebo
inlita lichenas ex facie tollit septica vi, ut supra
dictum est. vulturinus sanguis cum chamaeleontos
albae, quam herbam esse diximus, radice et cedria
tritus contectusque brassica lepras sanat, item pedes
locustarum cum sebo hircino triti, varos adeps gallinaceus cum cepa subactus. utilissimum et in facie
mel in quo apes sint inmortuae, praecipue tamen
faciem purgat atque erugat cygni adeps. stigmata
delentur columbino fimo ex aceto.

31   XI. Gravedinem invenio finiri, si quis nares
mulinas osculetur. uva[1] et faucium dolor mitigatur
fimo agnorum priusquam herbam gustaverint in
umbra arefacto, uva suco cocleae acu transfossae
inlita, ut coclea ipsa in fumo suspendatur, hirundinum
cinere cum melle. sic et tonsillis succurritur. tonsillas et fauces lactis ovilli gargarizatio adiuvat,
32   multipeda trita, fimum columbinum cum passo gar-
garizatum, etiam cum fico arida ac nitro inpositum
extra. asperitatem faucium et destillationes leniunt
cocleae—coqui debent inlotae, demptoque tantum
terreno conteri et in passo dari potu. sunt qui
Astypalaeicas efficissimas putent[2]—et cinis earum,
gryllus infricatus aut si quis manibus quibus eum
contriverit tonsillas attingat.

33   XII. Anginis felle anserino cum elaterio et melle
citissime succurritur, cerebro noctuae, cinere hirun-

---

[1] *an* uvae? *sic coni. Mayhoff.*
[2] putent—et cinis earum, gryllus (cinis menarum *Detlefsen*) *Urlichs, Detlefsen*: putent et minimas earum—, gryllus
Mayhoff: varia codd.

---

[a] XXIX. § 59.   [b] XXII. § 45.
[c] These are often mentioned. Slaves after manumission
might find them an embarrassment.

the gall that it bursts the animal. This insect applied
with he-goat suet removes lichen from the face by
its corrosive property, as I have already [a] said.
Vulture's blood, beaten up with cedar resin and the
root of the white chamaeleon, a plant I have already [b]
mentioned, and covered with a cabbage leaf, heals
leprous sores, as do the legs of locusts beaten up with
he-goat suet. Pimples are cured by poultry fat
kneaded with onion. Very useful too for the face is
honey in which bees have died, but the best thing
for clearing the complexion and removing wrinkles is
swan's fat. Branded marks [c] are removed by
pigeon's dung in vinegar.

XI. I find that a heavy cold clears up if the *Colds, etc.*
sufferer kisses a mule's muzzle. Pain in the uvula
and in the throat is relieved by the dung, dried in
shade, of lambs that have not yet eaten grass, uvula
pain by applying the juice of a snail transfixed by a
needle, so that the snail itself may be hung up in the
smoke, and by the ash of swallows with honey. This
also gives relief to affections of the tonsils. Gargling
with ewe's milk is a help to tonsils and throat, as is a
multipede beaten up, gargling with pigeon's dung
and raisin wine, and also an external application of it
with dried fig and soda. Sore throat and a running
cold are relieved by snails—they should be boiled
unwashed, and with only the earth taken off crushed
and given to drink in raisin wine; some hold that the
snails of Astypalaea are the most efficacious—by their
ash, and also by rubbing with a cricket or if anybody
touches the tonsils with hands that have crushed a
cricket.

XII. In quinsy very speedy relief is afforded by *Quinsy.*
goose gall with elaterium and honey, by the brain of

299

dinis ex aqua calida poto. huius medicinae auctor
est Ovidius poeta. sed efficaciores ad omnia quae ex
hirundinibus monstrantur pulli silvestrium—figura
nidorum eas deprehendit—multo tamen efficacissimi
ripariarum pulli. ita vocant in riparum cavis nidi-
ficantes. multi cuiuscumque hirundinis pullum edendum censent, ut toto anno non metuatur id malum.
34 strangulatos cum sanguine [1] comburunt in vase et
cinerem cum pane aut potu dant. quidam et
mustelae cinerem [2] pari modo admiscent. sic ad
strumae remedia dant et comitialibus cotidie potui.
in sale quoque servatae hirundines ad anginam
drachma bibuntur, cui malo et nidus earum mederi
35 dicitur potus. milipedam inlini anginis efficacissimum putant. alii XX tritas in aquae mulsae hemina
dari per harundinem, quoniam dentibus tactis nihil
prosint. tradunt et murem cum verbenaca excoctum,
si bibatur is liquor, remedio esse, et corrigiam cani-
nam ter collo circumdatam, fimum columbinum vino
et oleo permixtum. cervicis nervis et opisthotono ex
milvi nido surculus viticis adalligatus auxiliari dicitur,
36 strumis exulceratis mustelae sanguis, ipsa decocta in
vino; non tamen sectis admovetur. aiunt et in cibo
sumptam idem efficere, vel cinerem eius sarmentis

---

[1] cum sanguine] *Mayhoff* anginae *coni.*
[2] cinerem *Mayhoff*: cineres dEr *Detlefsen.*

---

[a] Perhaps " dog's lead."

an owl, and by the ash of a swallow taken in hot water. The last prescription is on the authority of the poet Ovid. But more efficacious for all ailments for which swallows are prescribed are the young of wild swallows, which are recognised by the shape of their nests, but by far the most efficacious are the young of sand martins, for so are called the swallows that build their nests in holes on river banks. Many hold that a young swallow of any kind should be eaten to banish the fear of quinsy for a whole year. They wring their necks, burn them blood and all in a vessel, and give the ash with bread or in drink. Some add also to the prescription an equal quantity of weasel ash. These preparations are given daily in drink for scrofula and for epilepsy. Preserved in salt also swallows are taken for quinsy in drachma doses, for which complaint their nest also, taken in drink, is said to be a cure. It is thought that an application of millepedes is very efficacious for quinsy; some think that twenty, beaten up in a hemina of hydromel, should be given through a reed, because if the teeth are touched the draught is thought to be useless. They also tell us that a mouse, well boiled with vervain, makes a broth that is a remedy, as does a thong of dog leather [a] wrapped three times round the neck, or dove's dung thoroughly mixed with wine and oil. For neck-sinews and opisthotonus a twig of agnus castus taken from the nest of a kite and worn as an amulet, is said to help, for ulcerated scrofula a weasel's blood, or the weasel *Scrofula.* itself boiled down in wine, but it is not applied to sores that have been lanced. They say also that eating weasel in food has the same effect, or the animal burned over twigs and the ash mixed with

conbustae; miscetur axungia. lacertus viridis adalligatur, post dies XXX alium adalligatum oportet.
37 quidam cor eius in argenteo vasculo servant ad femineas [1] strumas et mares.[2] cocleae cum testa sua tusae inlinuntur, maxime quae frutectis adhaerent, item cinis aspidum cum sebo taurino inponitur, anguinus adeps mixtus oleo, item anguium cinis ex oleo inlitus vel cum cera. edisse quoque eos medios abscissis utrimque extremis partibus adversus strumas prodest, vel cinerem bibisse in novo fictili ita crematorum, efficacius multo inter duas orbitas
38 occisorum. et gryllum inlinere cum sua terra effossum suadent, item fimum columbarum per sese vel cum farina hordeacia aut avenacia ex aceto, talpae cinerem ex melle inlinere. alii iocur eiusdem contritum inter manus inlinunt et triduo non abluunt. dextrum quoque pedem eius remedio esse strumis adfirmant. alii praecidunt caput et cum terra a talpis excitata tusum digerunt in pastillos pyxide stagnea et utuntur ad omnia quae intumescant et quae apostemata vocant quaeque in cervice sint;
39 vesci suilla tunc vetant. tauri vocantur scarabaei terrestres ricino similes—nomen cornicula dedere, alii pediculos terrae vocant; ab his quoque terram egestam inlinunt strumis et similibus vitiis et podagris, triduo non abluunt. prodest haec medicina in annum, omniaque his adscribunt quae nos in gryllis

---

[1] femineas *Mayhoff*: feminas *codd.*: feminarum *Detlefsen*.
[2] mares *coni. Mayhoff e Marcello*: veteres *codd.*: strumas, et veteres cochlcae etc. *coni. Warmington*.

axle grease. A green lizard is attached as an
amulet; after thirty days the weasel should be
changed for another. Some keep a weasel's heart
in a small silver vessel for scrofula in woman or man.
An ointment is made of snails pounded with their
shells, especially those that cling to shrubs, or there
is applied the ash of asps with bull suet, snake's fat
mixed with oil, or an ointment of snake's ash in oil
or with wax. To eat also the middle part of a snake
after cutting off either end is good for scrofula, as is
to take in drink the ash of this middle burnt in new
earthenware, with much greater benefit if the snakes
have been killed between two wheel-ruts. They
recommend also the application of a cricket dug up
with its earth, also the application of dove's dung by
itself, or with barley meal or oatmeal in vinegar, or
of mole ash with honey. Some make an ointment of
a mole's liver crushed between the hands, and do not
wash it off for three days. They also assure us that
the right foot of the animal is a remedy for scrofula.
Others cut off the head, pound it with the earth of a
mole-hill, work into lozenges in a pewter box, and
use for all swellings, for what are called apostemata,
and for affections of the neck; during the treatment
the eating of pork is forbidden. There are earth
beetles like ticks that are called " bulls "—a name
given because of their little horns—and by some
" earth lice." These too throw up earth that is
applied to scrofulous and similar sores, and also to
gouty parts, not being washed off for three days.
The efficacy of this treatment lasts for a year. To
these creatures are assigned all the properties I have
mentioned when speaking of crickets. Some also use
for this purpose the earth thrown up by ants, others

rettulimus. quidam et a formicis terra egesta sic
utuntur, alii vermes terrenos totidem quot sint
strumae adalligant pariterque cum his arescunt.
40 alii viperam circa canis ortum circumcidunt ut diximus, dein mediam comburunt, cinerem eum dant
bibendum ter septenis diebus quantum prenditur
ternis digitis, sic strumis medentur, aliqui vero
circumligantes lino quo praeligata infra caput vipera
pependerit donec exanimaretur. et milipedis utuntur addita resinae terebinthinae parte quarta, quo
medicamento omnia apostemata curari iubent.
41   XIII. Umeri doloribus mustelae cinis cum cera
medetur. ne sint alae hirsutae formicarum ova
pueris infricata praestant, item mangonibus, ut
lanugo tardior sit pubescentium, sanguis e testiculis
agnorum, cum castrantur, qui evulsis pilis inlitus et
contra virus proficit.
42   XIV. Praecordia vocamus uno nomine exta in
homine, quorum in dolore cuiuscumque partis si
catulus lactens admoveatur adprimaturque his partibus, transire in eum dicitur morbus, idque exinterato
perfusoque vino deprehendi vitiato viscere illo quod
43 doluerit homini, sed obrui tales religio est. hi quoque quos Melitaeos vocamus stomachi dolorem sedant
adplicati saepius. transire morbos aegritudine
eorum intellegitur, plerumque et morte. pulmonum[1] vitiis medentur et[2] mures, maxime Africani,

---

[1] *Post* pulmonum *addunt* quoque *multi codd.,* Mayhoff: om.
d E r, *Detlefsen.*
[2] et E r, *Detlefsen* : id VGRd : iidem *Ianus* : item *Mayhoff.*

---

[a] XXIX. §§ 70 and 121.
[b] From the Dalmatian island of Melita.

tie as an amulet as many earth worms as there are
sores, which dry up as the worms shrivel. Others
about the time of the Dog-star cut off, as I have said,[a]
the ends of a viper, then burn the middle part and
give a three-finger pinch of the ash to be taken in
drink for thrice seven days, treating scrofulous sores
in this way; some however do so by tying round them
a linen thread by which a viper has been suspended
by the neck until it died. They also use millepedes
with a fourth part of terebinth resin, a medicament
which they recommend for the treatment of all
apostemata.

XIII. Good treatment for pains in the shoulder is
weasel ash and wax. Rubbing with ants' eggs prevents hair in the arm-pits of children, and dealers, to
delay growth of downy hair on adolescents, use blood
that comes from the testicles of lambs when they are
castrated. Applications of this blood after the hair
has been pulled out also do away with the rank
smell of the arm-pits.

XIV. *Praecordia* is a comprehensive name we use
for the vital organs of the human body. When any
one of them is in pain, the application of a sucking
puppy pressed close to that part is said to transfer
the malady to it; they add that, if the organs of the
puppy are taken out and washed with wine, by the
diseased aspect of those organs can be detected the
source of the patient's pain; but the burial of an
animal so used is an essential part of the ritual.
Those puppies too that we call Melitaean [b] relieve
stomach-ache if laid frequently across the abdomen.
That the disease is transferred to the puppy is seen
by its sickening, usually even by its death. Lung
complaints are also cured by mice, especially African;

Shoulders and depila- tories.

Cures for the internal organs.

detracta cute in oleo et sale decocti atque in cibo sumpti. eadem res et purulentis vel cruentis ex-
44 creationibus medetur, XV. praecipue vero coclearum cibus stomacho. in aqua eas subfervefieri intacto corpore earum oportet, mox in pruna torreri nihilo addito, atque ita e vino garoque sumi, praecipue Africanas. nuper hoc conpertum plurimis prodesse. id quoque observant ut numero inpari sumantur. virus tamen earum gravitatem halitus facit. prosunt et sanguinem excreantibus dempta testa tritae in
45 aqua[1] potu. laudatissimae autem sunt Africanae— ex his Iolitanae—Astypalaeicae,[2] Siculae modicae, quoniam magnitudo duras facit et sine suco, Baliari-cae, quas cavaticas vocant, quoniam in speluncis nascuntur. laudatae ex[3] insulis et[4] Caprearum, nullae[5] autem cibis gratae neque veteres neque recentes. fluviatiles et albae virus habent, nec silvestres stomacho utiles, alvum solvunt, item omnes minutae. contra marinae stomacho utiliores, effica-cissimae tamen in dolore stomachi e laudatis tra-
46 duntur quaecumque vivae cum aceto devoratae. praeterea sunt quae ἀκέρατοι vocantur, latae, multi-fariam nascentes, de quarum usu dicemus suis locis.

[1] aqua *Mayhoff*: aquae *codd. Cf.* XXVIII. § 202.
[2] *Ante* Siculae *addunt* et ne VGR: *om.* d E r, *Detlefsen*: Aetnaeae *Gronovius, Sillig*: item *Mayhoff*.
[3] et ex *codd.*: ex *Detlefsen, Mayhoff*.
[4] *Ante* Caprearum *addunt codd.* et *aut* ex: et *Detlefsen, Mayhoff*.
[5] nullae d r, *Mayhoff*: nullis VGR[1] *Detlefsen*.

---

[a] A sauce made of small fish.
[b] The phrase *in aquae potu* occurs in XXVIII. § 202, but not depending on *tritae*.

they are skinned, boiled down in oil and salt, and
taken in food. The same preparation is also a cure
for expectoration of pus or blood. XV. The best
medicine, however, for the stomach is a diet of snails. *Snails.*
They should be gently boiled in water, African snails
by preference, with their bodies whole, then with
nothing added grilled over a coal fire, and so taken in
wine and garum.[a] Recently this treatment has been
found to benefit very many sufferers, who are also
careful that the number of the snails taken is odd.
Their rank juice, however, makes the breath foul.
Pounded without their shells and taken [b] in water
they are also good for the spitting of blood. The
most prized snails are the African, especially those of
Iol, those of Astypalaea, moderate sized Sicilian (for
the large are hard, and without juice), and those of
the Baliaric islands, called *cavaticae* because they
breed in caverns. Those from the islands and of
Capreae are prized, but none whether preserved or
fresh make pleasant eating.[c] River snails and white
snails have a rank taste; wood snails are not good
for the stomach, relaxing the bowels, and so with all
small snails. On the other hand sea snails [d] are
rather beneficial for the stomach, but of the prized
snails the most efficacious for stomach-ache are said
to be all that are swallowed alive in vinegar. Moreover, there are some snails called ἀκέρατοι,[e] which
are broad, and breed in many places; of these I shall

[c] The text in this part of the chapter is uncertain as well
as the punctuation. Dioscorides (II. 9) does not help, except
once in showing that a full stop should be placed with Mayhoff
after *recentes*.

[d] Periwinkles.

[e] I.e. " hornless."

gallinaceorum ventris membrana inveterata et inspersa potioni destillationes pectoris, et umidam tussim vel recens tosta lenit. cocleae crudae tritae cum aquae tepidae cyathis tribus si sorbeantur, tussim sedant. destillationes sedat et canina cutis cuilibet digito circumdata. iure perdicum stomachus recreatur.

47 XVI. Iocinerum doloribus medetur mustela silvestris in cibo sumpta vel iocinera eius, item viverra porcelli modo inassata, suspiriosis multipedae ita ut ter septenae in Attico melle diluantur et per harundinem bibantur, omne enim vas nigrescit contactu. quidam torrent sextarium in patina donec candidae fiant, tunc melle miscent [alii centipedam vocant] [1]

48 et ex aqua calida dari iubent. cocleae in cibo [2] iis quos linquit animus aut quorum alienatur mens aut quibus vertigines fiunt, ex passi cyathis tribus singulae contritae cum sua testa et calefactae in potu datae diebus plurimum novem, aliqui singulas primo die dedere, sequenti binas, tertio ternas, quarto duas,

49 quinto unam. sic et suspiria emendant et vomicas. esse animal locustae simile sine pennis, quod trixallis Graece vocetur, Latinum nomen non habeat, aliqui arbitrantur, nec pauci auctores, hoc esse quod grylli vocentur. ex his XX torreri iubent ac bibi e mulso contra orthopnoeas. sanguinem expuentibus cocleae; [3] si qui inlotis protropum infundat, vel marina aqua ita decoquat et in cibo sumat, aut si

---

[1] alii centipedam vocant] *In uncis Mayhoff.*

[2] iubent in cibo. cocleae *Mayhoff.*

[3] si qui *Mayhoff* est qui *plerique codd., Detlefsen.*

---

[a] The part in brackets (clearly a gloss on *multipedae*) means: " some call it centipede."

speak in the appropriate places. The skin of the crop of poultry, sprinkled into the drink when dried, or roasted if fresh, relieves chest catarrhs and moist coughs. A cough is relieved by pounded raw snails swallowed in three cyathi of tepid water, running colds also by a piece of dog skin put round any finger. Partridge broth acts as a tonic on the stomach.

XVI. Pains in the liver are treated by the wild weasel, or its liver, taken in food, also by a ferret roasted as is a sucking pig; asthma by thrice seven multipedes, soaked in Attic honey and sucked through a reed, for every vessel they touch they turn black. Some roast a sextarius of them in a pan until they turn white, then they mix them with honey and recommend giving them in warm water.*a* Snails in food have been given to those subject to fainting, aberration of the mind, or vertigo, a dose being one snail in three cyathi of raisin wine, pounded with the shell, warmed, and taken in drink for nine days at most; some have given one on the first day, two on the next, three on the third, two on the fourth, and one on the fifth. This treatment is also good for asthma and abscesses. Some hold that there is a creature like a locust, but without wings, called *trixallis* in Greek but without a name in Latin; some, and not a few authorities, maintain that it is what is called in Latin *gryllus* (cricket); twenty of these they recommend to be roasted and taken in honey wine for orthopnoea. A cure for spitting of blood are snails, if the patient pours *protropum* *b* on them unwashed, or if he boils them down in sea-water, and takes them

b *Protropum* was the must that came from the grape clusters before they were pressed. The text here seems incapable of restoration, but the meaning of the passage is plain.

tritae cum testis suis sumantur cum protropo; sic et
tussi medentur. vomicas privatim sanat mel in quo
50 apes sint demortuae. sanguinem reicientibus pulmo
vulturinus vitigineis lignis conbustus adiecto flore
Punici mali ex parte dimidia, item cotoneorum lilior-
umque isdem portionibus potus mane atque vesperi
e vino, si febres absint, si minus, ex aqua in qua
cotonea decocta sint.

51 XVII. Pecudis lien recens magicis praeceptis
super dolentem lienem extenditur dicente eo qui
medeatur lieni se remedium facere. post hoc iubent
in pariete dormitorii eius tectorio includi et obsignari
anulo ter novies eademque [1] dici. caninus si viventi
eximatur et in cibo sumatur, liberat eo vitio. quidam
52 recentem superinligant. alii duum dierum catuli ex
aceto scillite dant ignoranti, vel irenacei lienem,
item coclearum cinerem cum semine lini et urticae
addito melle, donec persanet. liberat et lacerta
viridis viva in olla ante cubiculum dormitorium eius
cui medeatur suspensa, ut egrediens revertensque
attingat manu, cinis e capite bubonis cum unguento,
mel in quo apes sint mortuae, araneus, et maxime
qui lycos vocatur.

53 XVIII. Upupae cor lateris doloribus laudatur,
coclearum cinis in tisana decoctarum—et per se
inlinuntur—canis rabiosi calvariae cinis potioni
inspergitur. lumborum dolori stelio transmarinus

---

[1] eademque *Mayhoff*: carmenque *Detlefsen*: carmen d(?)
vulg.: earumque (—quae E) VRGE: anulo, terque novies
eadem dici. *coni. Warmington.*

in food, or if pounded with their shells they are taken with *protropum*; these preparations also cure a cough. Specific for abscesses is honey in which bees have died. For coughing up blood a vulture's lung burnt over vine wood, with half as much pomegranate blossom and the same quantity of quince blossom and of lilies, taken morning and evening in wine, if there is no fever, otherwise in water in which quinces have been boiled.

XVII. The fresh spleen of a sheep is placed, by a Magian prescription, over the painful spleen of a patient, the attendant saying that he is providing a remedy for the spleen. After this the Magi prescribe that it should be plastered into the wall of the patient's bedroom, sealed with a ring thrice nine times and the same words repeated. If a dog's spleen is cut out of the living animal and taken in food it cures splenic complaints; some bind it when fresh over the affected part. Others without the patient's knowledge give in squill vinegar the spleen of a two-days-old puppy, or that of a hedgehog, also the ash of snails with linseed, nettle seed, and honey, until there is a complete cure. Another remedy is a live green-lizard, hung up in a pot before the door of the bedroom of the patient, that he may touch it with his hand on going out and coming in, the ash of a horned owl's head with an unguent, honey in which bees have died, or a spider, especially that called " wolf."

XVIII. The heart of a hoopoe is a prized remedy *Lumbago, sciatica, etc.* for pains in the side, as is the ash of snails boiled down in barley water; these are also used by themselves as a liniment. The skull of a mad dog is reduced to ash and sprinkled in drink. For lumbago an overseas

capite ablato et intestinis decoctus in vino cum papaveris nigri denarii pondere dimidio eo suco bibitur. lacerti [1] virides decisis pedibus et capite in cibo sumuntur, cocleae tres contritae cum testis suis atque in vino decoctae cum piperis granis XV.
54 aquilae pedes evellunt in aversum a suffragine ita ut dexter dextrae partis doloribus adalligetur, sinister laevae. multipeda quoque, quam oniscon appellavimus, medetur denarii pondere ex vini cyathis duobus pota. vermem terrenum catillo ligneo ante fisso et ferro vincto inpositum aqua excepta [2] perfundere et defodere unde effoderis Magi iubent, mox aquam bibere catillo, mire id prodesse ischiadicis adfirmantes.

55   XIX. Dysintericos recreant femina pecudum decocta cum lini semine ea [3] aqua pota, caseus ovillus vetus, sebum ovium decoctum in vino austero. hoc et ileo medetur et tussi veteri dysintericis stelio transmarinus, ablatis intestinis et capite pedibusque ac cute, decoctus aeque et in [4] cibo sumptus, cocleae

---

[1] lacerti dE *Detlefsen* : lacertae R *vulg.*, *Mayhoff* : lacerte VG.

[2] inpositum aqua excepta] *coni.* aqua perfundere et exceptum *Mayhoff*.

[3] ea *Urlichs, Detlefsen, Mayhoff* : *om. codd.*

[4] in *vulg., Mayhoff* : *om. codd., Detlefsen.*

---

[a] See note on XXVI. § 67.

[b] It is not clear who " they " are, but most of this part of Pliny seems taken from the same source as that from which he took his account of the Magi.

[c] See XXIX. § 136.

[d] Mayhoff's reading would mean : " soaked in water, taken out, and buried, etc." The word *exceptum*, written as *exceptū*,

spotted lizard, with head and intestines removed, is
boiled down in wine with half an ounce by weight of
black *ᵃ* poppy, and this broth is drunk. Green
lizards, with feet and head cut off, are taken in food,
or three snails, beaten up with their shells and boiled
down in wine with fifteen peppercorns. They *ᵇ* break
off, in the opposite way to the joint, the feet of an
eagle, so that the right foot is attached as an amulet
for pains in the right side, the left foot for those in
the left side. The multipede too, that I have called
oniscos,*ᶜ* is another remedy, the dose being a denarius
by weight taken in two cyathi of wine. The Magi
prescribe that an earth-worm should be placed upon
a wooden plate that has been split beforehand and
mended with a piece of iron, soaked in water that
has been taken *ᵈ* up in the dish, and buried in the
place from which it was dug out. Then the water in
the plate is to be drunk, which they say is a wonderful
remedy for sciatica.

XIX. Dysentery is relieved by a leg of mutton *Dysentery.*
boiled down with linseed, the broth of which is drunk,
by old cheese made with ewe's milk, and by mutton
suet boiled down in a dry wine. By this are also
benefited ileos and chronic cough, and dysentery by
a spotted lizard from overseas, boiled down with its
intestines, head, feet, and skin removed *ᵉ*—it is
as efficacious in food also as decocted—by two snails

might easily be taken for *excepta*; the transposition would
naturally follow.

ᵉ In § 53 is practically the same remedy, but *in vino* comes
after *decoctus*. In such expressions *in* with a noun is usual,
so that perhaps *aeque* is a mistake for *in aqua*. I have not
adopted it because an easy reading like *in aqua* is unlikely
to have been changed to *aeque*. The meaning " steadily ",
which would make good sense, seems without a parallel.

duae cum ovo, utraque cum putamine contrita atque
in vase novo addito sale et passi cyathis duobus aut
palmarum suco et aquae cyathis tribus subfervefacta
56 et in potu data.[1] prosunt et combustae, ut cinis
earum bibatur in vino addito resinae momento.
cocleae nudae, de quibus diximus—in Africa maxime
inveniuntur—utilissimae dysintericis, quinae combustae cum denarii dimidii pondere acaciae; ex eo
cinere dantur coclearia bina in vino myrtite aut
57 quolibet austero cum pari modo caldae. quidam
omnibus Africanis ita utuntur, alii totidem Africanas
vel latas [2] infundunt potius et, si maior fluctio sit,
addunt acaciam fabae magnitudine. senectus anguium dysinteriae et tenesmis in stagneo vase decoquitur cum rosaceo, vel si in alio, cum stagno inlinitur.
ius ex gallinaceis isdem medetur, sed veteris galli-
58 nacei vehementius salsum ius alvum ciet. membrana
gallinarum tosta et data in oleo ac sale coeliacorum
dolores mulcet—abstinere autem frugibus ante et
gallinam et hominem oportet—fimum columbinum
tostum potumque. caro palumbis in aceto decocta
dysintericis et coeliacis medetur, turdus inassatus
cum myrti bacis dysintericis, item merulae, mel in
quo apes sint inmortuae decoctum.

---

[1] subfervefacta . . . data *Mayhoff cum vet. Dal.*: -tis . . .
-tis *codd., Detlefsen.*
[2] vel latas *codd., Detlefsen*: velatas (*opp.* nudas) *Mayhoff,
qui et* latas *sine* vel *coni.*

---

[a] See XXIX. § 112.

with egg, each beaten up with its shell, allowed to
simmer in a new vessel with salt, two cyathi of raisin
wine or date juice, and three cyathi of water; this
preparation is taken in drink. Snails are also
beneficial when burnt, and their ash taken in wine
with a small piece of resin. Snails without shells,
about which I have spoken [a]—they are found chiefly
in Africa—are very useful in dysentery; five are
burnt and taken with half a denarius by weight of
gum acacia; of this ash two spoonfuls are given in
myrtle wine or any dry wine with an equal quantity
of hot water. Some, using all African snails, administer according to this recipe; others prefer to
inject the same number of African snails or broad
snails,[b] adding if the flux is severe gum acacia of the
size of a bean. The cast slough of snakes is boiled
down with rose oil for dysentery and tenesmus in a
pewter vessel; if in any other kind of vessel, the
application must be made with the help of pewter.
Chicken broth is good for these two complaints, but
broth made with an old cock, thoroughly salted, is
purgative. A hen's crop, roasted and given in oil
and salt, soothes the pains of coeliac troubles—but
previously hen and patient must both abstain from
cereals [c]—as does dove's dung roasted and taken in
drink. The flesh of a wood-pigeon boiled in vinegar
is good for dysentery and for coeliac troubles; for
dysentery too a thrush roasted with myrtle berries,
so are blackbirds and honey in which bees have died.

---

[b] Mayhoff's *velatas* would mean : " with shells," but I can
find no exact parallel.

[c] I think that the sense is that both hen and patient must
fast, and that *frugibus* is used as being peculiarly applicable
to *gallinam*, which is nearer to it than *hominem*.

59 XX. Gravissimum vitium[1] alvi ileos[2] appellatur.
huic resisti aiunt discerpti vespertilionis sanguine,
etiam inlito ventre subveniri. sistit alvum coclea
sicut diximus in suspiriosis temperata, item cinis
earum quae vivae crematae sint potus ex vino austero,
gallinaceorum iocur assum aut ventriculi membrana
60 quae abici solet inveterata admixto papaveris suco—
alii recentem torrent ex vino bibendam—ius perdicium et per se ventriculus contritus ex vino nigro,
item palumbis ferus ex posca decoctus, lien pecudis
tostus et in vino tritus, fimum columbinum cum
melle inlitum, ossifragi venter arefactus et potus, iis
qui cibos non conficiant utilissimus, vel si manu tantum teneant capientes cibum. quidam adalligant
ex hac causa, sed continuare non debent, maciem
enim facit. sistit et anatum mascularum sanguis.
61 inflationes discutit coclearum cibus, tormina lien
ovium tostus atque e vino potus, palumbus ferus ex
posca decoctus, adips otidis ex vino, cinis ibide sine
pennis cremata potus. quod praeterea traditur in
torminibus mirum est, anate adposita ventri transire
62 morbum anatemque emori. tormina et melle curantur in quo sunt apes inmortuae decocto. coli vitium
efficacissime sanatur ave galerita assa in cibo sumpta.
quidam in vase novo cum plumis exuri iubent con-
terique in cinerem, bibi ex aqua coclearibus ternis

[1] vitium d E, *Detlefsen, Mayhoff*: vulnus vitium VGR:
ventris vitium *Urlichs*.
[2] alvi ileos *Ianus, Detlefsen, Mayhoff*: apu (apii VG) illi
eos VGE: apuleius R.

<hr>

[a] See § 48.

XX. The most serious disease of the abdomen is *Ileos and* *troubles of* *the abdomen.* ileos. It may be combated, they say, by tearing a bat apart and drinking its blood; it is also a help to rub the belly with it. Looseness of the bowels is checked by a snail prepared according to my prescription [a] for asthma, and also by the ash, taken in a dry wine, of snails that have been burnt alive. Other remedies are: the roasted liver of cocks or the skin of their crop, usually thrown away, mixed with poppy juice if dried, while some roast it fresh to be given in wine, partridge broth and its crop pounded by itself in dark wine, also wild wood-pigeon boiled down in vinegar and water, spleen of a sheep roasted and beaten up in wine, pigeon's dung applied with honey, the gizzard of an osprey dried and taken in drink, very beneficial to those who cannot digest their food, even if they only hold it in their hand while eating. Some use it as an amulet for this purpose, but it must not be so used continuously, for it makes the body thin. Looseness is also checked by the blood of drakes. Flatulence is dispersed by a diet of snails, griping by the spleen of sheep, roasted and taken in wine, wild wood-pigeon boiled down in vinegar and water, the fat of a bustard in wine, the ash of an ibis burnt without the feathers and taken in drink. Another prescription for griping is of a marvellous character: it is said that if a duck is laid on the belly, the disease is transferred to the duck, which dies. Good for griping is also boiled honey in which bees have died. Colic is effectively cured by a crested lark, roasted and taken in food. Some recommend that it should be burnt with the feathers in a new vessel, ground to dust and taken in water, three spoonfuls daily for four days,

per quadriduum, quidam cor eius adalligari femini,
alii recens tepensque adhuc devorari.[1] consularis
63 Asprenatum domus est in qua alter e fratribus colo
liberatus est ave hac in cibo sumpta et corde eius
armilla aurea incluso, alter sacrificio quodam facto
crudis laterculis ad formam camini atque, ut sacrum
peractum erat, obstructo sacello. unum est ossifrago
intestinum mirabili natura omnia devorata con-
ficienti. huius partem extremam adalligatam prodesse contra colum constat. sunt occulti inter-
64 aneorum morbi de quibus mirum proditur. si catuli
priusquam videant adplicentur triduo stomacho
maxime ac pectori et ex ore aegri suctum lactis acci-
piant, transire vim morbi, postremo exanimari dis-
sectisque palam fieri aegri[2] causas, † mori et †[3]
humari debere eos obrutos terra. Magi quidem
vespertilionis sanguine contacto ventre in totum
annum caveri tradunt, aut in dolore[4] si quis aquam[5]
pedes eluens[6] haurire sustineat.

65   XXI. Murino fimo contra calculos inlinere ventrem prodest. irenacei carnem iucundam esse aiunt,
si capite percusso uno ictu interficiatur priusquam in

    [1] devorari d(?) Detlefsen : devoratur reliqui codd. et Mayhoff,
qui aliis pro alii scribit.
    [2] aegri om. Urlichs et Detlefsen: aegritudinis Warmington.
    [3] mori et codd. : morbi et Ianus, Detlefsen : monent May-
hoff: mox et coni. Warmington.
    [4] in dolore fere omnes codd., Mayhoff : per dolorem E,
Gelenius, Detlefsen.
    [5] per post aquam codd. : del. Detlefsen : ter Mayhoff.
    [6] eluens Mayhoff, qui eluentis coni. : eluentem Detlefsen :
fluentes aut fluentis codd.: aquam per pedes fluentem Warmington.

others that a lark's heart should be tied as an amulet
to the patient's thigh, and others that it should be
swallowed while fresh and still warm. The Aspren-
ates are a consular family in which one of two
brothers was cured of colic by this bird taken in food
and its heart worn in a golden bracelet, the other by
performing a certain sacrifice in a shrine of unbaked
bricks built in the shape of an oven, and when a certain rite was over blocking it up. The osprey has
only one gut, which through its wonderful character
digests everything that the bird eats; the end of it
attached as an amulet is well known to be excellent
for colic. There are some obscure diseases of the
intestines, for which is prescribed a wonderful cure.
If, before they can see, puppies are applied for three
days especially to the stomach and chest of a
patient, and suck milk from his mouth, the power of
the disease is transferred to them; finally they die
and dissection makes clear the patient's trouble [a];
the puppies must be buried in the earth. The
Magi indeed tell us that if the belly is touched with
a bat's blood there is protection from colic for a
whole year; should there be pain, it is sufficient if
the patient can bring himself to drink [b] the water in
which he washes his feet.

XXI. Mouse dung rubbed on the belly is good for **Bladder troubles.**
stone in the bladder. The flesh of a hedgehog is
said to be pleasant to eat if it is killed by one blow

---

[a] *Causas* seems to be here the equivalent of *morbos*. The
emendation *morbi* of Jan was due to his taking *causas* in its
usual sense, but see XXVIII. § 218.

[b] Mayhoff's *ter* would give : " to drink three times of the
water, etc." The text at the end of this chapter is very
uncertain, but the general sense is clear. I think that *per*
before *pedes* is dittography. See Note G, p. 566.

se urinam reddat. haec caro ad hunc modum occisi
stillicidium [1] vesicae [2] emendat, item suffitus ex
eodem. quod si urinam in se reddiderit, eos qui
carnem comederint stranguriae morbum contrahere
66 traditur. iubent et vermes terrenos bibi ex vino aut
passo ad comminuendos calculos vel cocleas decoctas
ut in suspiriosis, easdem exemptas testis tres tritasque
in vini cyatho bibi, sequenti die duas, tertio die unam,
ut stillicidia urinae emendent, testarum vero inanium cinerem ad calculos pellendos, item hydri iocur
bibi vel scorpionum cinerem aut in pane sumi, [vel
67 si quis ut locusta edit,] [3] lapillos qui in gallinaceorum
vesica aut in palumbium ventriculo inveniantur conteri et potioni inspergi, item membranam e ventriculo gallinacei aridam vel, si recens sit, tostam,
fimum quoque palumbinum in faba sumi contra
calculos et alias difficultates vesicae, similiter plumarum cinerem palumbium ferorum ex aceto mulso et
intestinorum ex his cinerem coclearibus tribus, e nido
68 hirundinum glaebulam dilutam [4] aqua calida, ossifragi
ventrem arefactum, turturis fimum in mulso decoctum
vel ipsius discoctae ius. turdos quoque edisse cum
bacis myrti prodest urinae, cicadas tostas in patellis,
milipedam oniscon bibisse et in vesicae doloribus
decoctum agninorum pedum. alvum ciet et gallin-

---

[1] stillicidium *Mayhoff*: stillicidia d, *Detlefsen*: stillicidi in
reliqui codd.

[2] vesicae *Mayhoff*: vessicam *multi codd.*

[3] vel si quis ut locusta edit *in uncis Mayhoff*: *pro* ut *habet*
cum *vulg.*: vel siquis VI locustas edit *Detlefsen.*

[4] glaebulam dilutam *ex Pl. iun. et Marcello Hard.*: fimum
dilutum *Detlefsen*: grillum dirutum *multi codd.*

on the head before it can void its urine on itself. The
flesh of hedgehogs killed in this manner is a remedy
for obstruction to the urine; another is fumigation
with the same animal. Should however it have
voided its urine on itself those who have eaten the
flesh are said to be attacked by strangury. It is
also recommended, in order to break up stone, to
take earthworms in wine or raisin wine, or snails
boiled down as for asthma *a*; three snails taken from
their shells, pounded, and given in a cyathus of wine,
on the next day two, and on the third day one, for
removing difficulty of urination; but the ash of the
empty shells for expelling stone; the liver of a water
snake or the ash of scorpions to be taken in drink or
in bread,*b* the grits to be found in the gizzard of
poultry or in the crop of wood-pigeons to be crushed
and sprinkled on drink, also the skin of the crop of
poultry. When dried, or roasted when fresh, the
dung too of wood-pigeons to be taken in beans for
stone and other bladder trouble; the ash too of wild
wood-pigeon's feathers in oxymel, three spoonful-
doses of their intestines reduced to ash, a bit of earth *c*
from a swallow's nest diluted with warm water, the
crop of an osprey dried, dung of a turtle-dove boiled
down in honey wine, or the broth of the bird itself.
To eat thrushes also with myrtle berries is good for
the urine, cicadas roasted in a shallow pan, to take in
drink the millepede *oniscos*, and for pains in the
bladder the broth of lambs' trotters. Chicken broth

---

a See § 48 of this Book.

b The part in brackets would mean: " or if taken with a
locust (*cum locusta*)," " or if six locusts are eaten " (Detlefsen).

c Detlefsen's reading: " diluted dung ": that of the
MSS.: " a cricket taken."

321

aceorum discoctorum ius et acria mollit, ciet et
hirundinum fimum adiecto melle subditum.

69   XXII. Sedis vitiis efficacissima sunt oesypum—
quidam adiciunt pompholygem [1] et rosaceum—canini
capitis cinis, senecta serpentis ex aceto, si rhagades
sint, cinis fimi canini candidi cum rosaceo—aiunt inventum Aesculapii esse eodemque et verrucas
efficacissime tolli—murini fimi cinis, adeps cycni,
adeps bovae. procidentia ibi sucus coclearum
70 punctis evocatus inlitu repellit. adtritis medetur
cinis muris silvatici cum melle, fel irenacei cum
vespertilionis cerebro et canino lacte, adeps anserinus
cum cerebro et alumine et oesypo, fimum columbinum cum melle, condylomatis privatim araneus
dempto capite pedibusque infricatus; ne acria
perurant, adeps anserinus cum cera Punica, cerussa,
rosaceo, adeps cycni. hic et haemorroidas sanare
71 dicitur. ischiadicis cocleas crudas tritas cum vino
Aminneo et pipere potu prodesse dicunt, lacertam
viridem in cibo ablatis pedibus, interaneis, capite, sic
et stelionem adiectis huic papaveris nigri obolis tribus, ruptis, convulsis fel ovium cum lacte mulierum.
72 verendorum formicationibus verrucisque medetur
arietini pulmonis inassati sanies, ceteris vitiis vellerum
eius vel sordidorum cinis ex aqua, sebum ex omento

---

[1] pompholygem *Hermolaus Barbarus* : *varia codd.* : *cf.*
§ 106.

---

[a] A deposit from the smoke of smelting furnaces.

too is laxative and softens acridities, laxative too is the dung of swallows with honey used as a suppository.

XXII. For complaints of the anus very efficacious are wool grease—some add pompholyx [a] and rose oil—dog's head reduced to ash, a serpent's slough in vinegar, if there are chaps, the ash of white dog's-dung with rose oil—it is said to have been a discovery of Aesculapius, removing warts also very efficaciously—ash of mouse dung, fat of a swan, fat of a boa. Prolapsus there is reduced by an application of snail juice extracted by pricks. Chafings are relieved by the ash of a field mouse with honey, the gall of a hedgehog with the brain of a bat and bitch's milk, by goose grease with goose brain, alum and wool grease, and by pigeon dung with honey; specific for condylomata is a spider rubbed on the place when the head and feet have been removed; to prevent the smart from acrid juices, apply goose grease with Punic wax, white lead, rose oil, and swan fat. This fat is said also to cure haemorrhoids. They say that beneficial for sciatica are raw snails, pounded with Amminean [b] wine and pepper and taken in drink, a green lizard taken in food, but with feet, bowels and head removed, also so treated a spotted lizard with the addition of three oboli of black poppy [c]; for ruptures and sprains, sheep's gall with woman's milk. Itching eruptions and warts on the privates are treated with the gravy from the roasted lung of a ram, other genital affections by the ash, applied with water, of raw, even unwashed, ram's wool, by

Anus complaints

Complaints of the genitals, etc.

[b] Mayhoff has a note (XXXIV. § 103) on this word. He prefers the spelling " Amminean."

[c] For " black poppy " see note on XXVI. § 67 (vol. VII. p. 313).

pecudis, praecipue a renibus, admixto pumicis cinere
et sale, lana sucida ex aqua frigida, carnes pecudis
combustae ex aqua, mulae ungularum cinis, dentis
caballini contusi farina inspersa, testibus vero farina
ex ossibus capitis sine carne tusis. si decidat testium
alter, spumam coclearum inlitam in remedio esse tra-
73 dunt. taetris ibi ulceribus et manantibus auxiliantur
canini capitis recentes cineres, cocleae parvae latae
contritae ex aceto, senectus anguium ex aceto vel
cinis eius, mel in quo apes sint inmortuae cum resina,
cocleae nudae, quas in Africa gigni diximus, tritae
cum turis polline et ovorum albo. XXX die resol-
74 vunt; aliqui pro ture bulbum admiscent. hydro-
celicis stelionis mire prodesse tradunt capite, pedibus, interaneis ademptis relicum corpus inassatum—
in cibo id saepius datur—sicut ad urinae incontinentiam caninum adipem cum alumine schisto
fabae magnitudine, cocleas Africanas cum sua carne
et testa crematas poto cinere, anserum trium linguas
inassatas in cibo. huius rei auctor est Anaxilaus.
75 at panos aperit sebum pecudum cum sale tosto, murinum fimum admixto turis polline et sandaraca discutit, lacertae cinis et ipsa divisa inposita, item multipeda contrita admixta resina terebinthina ex parte
tertia—quidam et sinopidem admiscent—cocleae
contusae per se, cinis inanium coclearum cerae

---

[a] See § 56.
[b] A Pythagorean banished by Augustus for magic practices.

the suet from the caul of a sheep, especially that of the kidneys, mixed with salt and the ash of pumice, by greasy wool in cold water, by the burnt flesh of sheep in water, by the ash of a she-mule's hoofs, by the tooth of a horse, ground to powder and dusted on the parts, and complaints of the testicles by the bones of a horse's head ground to powder without the flesh. If either testicle hangs down, we are told that a remedy is found in applying the slime of snails. Foul and running ulcers on these parts are relieved by the fresh ashes of a dog's head, by the small broad kind of snail beaten up in vinegar, by the slough of a snake or its ash in vinegar, by honey in which bees have died mixed with resin, by the shell-less kind of snail, which I have said [a] breeds in Africa, beaten up with powdered frankincense and white of eggs; the application is removed on the thirtieth day, and some add a bulb instead of frankincense. Hydrocele, they tell us, is wonderfully benefited by the spotted lizard: head, feet, and bowels are removed, and the rest of the body is roasted—frequent doses are given in food—in food too for incontinence of urine they prescribe dog fat with split alum in doses the size of a bean, African snails burnt with their flesh and shell, the ash being taken in drink, three roasted geese tongues taken in food. Sponsor for this treatment is Anaxilaus.[b] But superficial abscesses are opened by mutton suet and roasted salt; they are dispersed by mouse dung mixed with powdered frankincense and sandarach, by ash of a lizard or the lizard itself, split and applied, also by multipedes pounded and mixed with one third part of terebinth resin—some add also red ochre of Sinope—by crushed snails by themselves, or by the ash of empty snail-shells mixed

mixtus. discussoriam vim habet fimum columbarum per sese vel cum farina hordeacia aut avenacia inlitum. cantharides mixtae calce panos scalpelli vice auferunt, inguinum tumorem cocleae minutae cum melle inlitae leniunt.

76 XXIII. Varices ne nascantur, lacertae sanguine pueris crura ieiunis a ieiuno inlinuntur. podagras lenit oesypum cum lacte mulieris et cerussa, fimum pecudum quod liquidum reddunt, pulmones pecudum, fel arietis cum sebo, mures dissecti inpositi, sanguis mustelae cum plantagine inlitus et vivae combustae cinis ex aceto et rosaceo [1]—penna inlinatur vel si cera et rosaceum admisceatur—fel caninum ita ne manu attingatur, sed penna inlinatur, fimum gallinarum, vermium terrenorum cinis cum melle ita ut tertio

77 die solvantur. aliqui [2] ex aqua inlinere malunt, alii ipsos—acetabuli [3] mensura [4] cum mellis cyathis tribus, pedibus ante rosaceo perunctis. cocleae latae potae tollere dicuntur pedum et articulorum dolores. bibuntur autem binae in vino tritae. eaedem inlinuntur cum helxines herbae suco. quidam ex aceto intrivisse contenti sunt. sale † quidam cum vipera crematus † [5] in olla nova saepius sumpto aiunt

---

[1] *Hic add.* si E r : *om. ceteri codd.*
[2] aliqui VGRdT : alium E r : ali eum *Detlefsen.*
[3] acetabuli *vet. Dal.* : aceto *codd.*
[4] mensura *aut* mensuram *codd.* : macerant *Detlefsen.*
[5] quidam . . . crematus *codd.* : quidam . . . cremata *Urlichs, Detlefsen:* qui una . . . crematus sit *Mayhoff:* cremato *Warmington: ego obelos addo.*

---

[a] Or : " or it may be made into ointment with wax and oil " ; a puzzling sentence with a parenthesis of uncertain length, Detlefsen ending it at *inlinatur.*

with wax. Power to disperse is possessed by pigeon's dung, applied by itself or with barley meal or with oatmeal. Cantharides mixed with lime remove superficial abscesses as well as the lancet; swelling of the groin is relieved by an application of small snails with honey.

XXIII. To prevent varicose veins the legs of children are rubbed with lizard's blood, but both patient and rubber must be fasting. Gouty pains are soothed by oesypum with woman's milk and white lead, by the dung of sheep that they pass liquid, by lungs of sheep, by ram's gall with ram's suet, by mice split and laid on the parts, by blood of a weasel applied with plantain and the ash of a weasel burnt alive with vinegar and rose oil—the remedy should be applied with a feather even *a* if wax and oil are made ingredients—by dog's gall, which must not be touched by hand but applied with a feather, by dung of hens, by ash of earth-worms with honey, taken off on the third day. Some prefer to apply the worms in water, others prefer to rub the feet first with rose oil and then to apply without water an acetabulum *b* of worms with three cyathi of honey. Snails of the broad kind taken in drink are said to banish pains of the feet and joints; the dose is two pounded in wine. They are also applied with juice of the plant helxine; some are content to beat them up in vinegar. Salt, burnt *c* with a viper in a new jar and taken fre-

Varicose veins, gout, etc.

----

b With Detlefsen's reading : " they macerate the worms themselves in vinegar."

c I have added daggers because, although the sense is plain, the actual words of Pliny are more than uncertain. The origin of the trouble seems to be the intrusion *of quidam* repeated from the preceding sentence. Pliny may be referring to salt in which a viper has been preserved; cf. § 117.

podagra liberari, utile esse et adipe viperino pedes
78 perungui. et de milvo adfirmant, si inveterato trito-
que quantum tres digiti capiant bibatur ex aqua, aut
si pedes sanguine eius perunguantur. inlinuntur et
columbarum sanguine[1] cum urtica, vel pennis earum
cum primum nascentur tritis cum urtica. quin et
fimus earum articulorum doloribus inlinitur, item
cinis mustelae aut coclearum, et cum amylo vel
tragacantha. incussos articulos aranei telae com-
modissime curant. sunt qui cinere earum uti malint
sicut fimi columbini cinere cum polenta et vino albo.
79 articulis luxatis praesentaneum est sebum pecudis
cum cinere e capillo mulierum. pernionibus quoque
inponitur sebum pecudum cum alumine, canini
capitis cinis aut fimi murini. quod si pura sint,
ulcera cera addita ad cicatricem perducunt . . .[2] vel
glirium crematorum favilla ex oleo, item muris silvatici cum melle, vermium quoque terrenorum cum
oleo vetere et cocleae quae nudae inveniuntur.
80 ulcera omnia pedum sanat cinis earum quae vivae
combustae sint, fimi gallinarum cinis, exulcerationes
columbini fimi ex oleo. adtritus calciamentorum
veteris soleae[3] cinis, agninus pulmo et arietis sanant,
dentis caballini contusi farina privatim subluviem,
lacertae viridis sanguis subtritus et hominum et

---

[1] eius perunguantur . . . sanguine *add. Mayhoff*: milvi vel columbarum unguantur *Urlichs, Detlefsen*: *lacunam indicat Sillig.*

[2] *Ego lacunam indico*: soricum *add. Mayhoff.*

[3] soleae *vulg. e Pl. iun. et Marcello*: soli RdE, *Detlefsen.*

---

[a] I have translated the words added by Mayhoff, because they are rather more likely than the addition of Urlichs adopted by Detlefsen.

quently, frees they say from gout, adding that it is
also beneficial to rub the feet with viper fat. They
assure us also that the kite is a remedy; it is dried,
pounded, and a three-finger pinch taken in water,
or the feet are rubbed with its blood. To the feet is
also applied the blood of pigeons [a] with nettles, or
their feathers may be used when they are just
sprouting, beaten up with nettles. Moreover their
dung is applied to painful joints, also the ash of a
weasel or of snails, and with starch or tragacanth.
Bruised joints are treated very effectively with
spider's web; some prefer to use the ash of it, or
else that of pigeon's dung with pearl barley and
white wine. For dislocations a sovereign remedy is
mutton suet with ash of woman's hair. For chilblains too is applied mutton suet with alum, or the
ash of a dog's head or of mouse dung. But if they
are clean, ulcers are brought to cicatrize ⟨by these⟩ [b]
with the addition of wax, or by the warm ash in oil
of burnt dormice, also by that of field mice with
honey, and by that of earth-worms also with old oil
and [c] the snails that are found without shells. All
sores of the feet are healed by the ash of those snails
that have been burnt alive, by the ash of hens' dung,
and ulcerations by the ash of pigeon's dung in oil.
Chafings caused by foot-wear are healed by the ash
of an old shoe, by the lung of a lamb and of a ram;
for whitlows is specific a horse's tooth ground to
powder; chafings under the feet of man or beast are
healed by applying a green lizard's blood, corns on

[b] Some plural subject is required to go with *perducunt*;
perhaps *haec*.
[c] The *et* would be strange unless it joins the two ingredients,
favilla and *cocleae*.

329

iumentorum pedes sublitus, clavos pedum urina muli
mulaeve cum luto suo inlita, fimum ovium, iocur
lacertae viridis vel sanguis flocco inpositus, vermes
terreni ex oleo, stelionis caput cum viticis pari modo
tritum ex oleo, fimum columbinum decoctum ex
81 aceto, verrucas omnium generum urina canis recens
cum suo luto inlita, fimi canini cinis cum cera, fimum
ovium, sanguis recens murinus inlitus vel ipse mus
divolsus, irenacei fel, caput lacertae vel sanguis vel
cinis totius, membrana senectutis anguium, fimum
gallinae cum [1] oleo ac nitro. cantharides cum uva
taminia intritae exedunt, sed ita erosas aliis quae ad
persananda ulcera demonstravimus curari oportet.

82 XXIV. Nunc praevertemur ad ea quae totis corporibus metuenda sunt. fel canis nigri masculi
amuletum esse dicunt Magi domus totius suffitae eo
purificataeve contra omnia mala medicamenta, item
sanguinem [2] canis respersis parietibus genitaleque [3]
eius sub limine ianuae defossum.[4] minus mirentur
hoc qui sciunt foedissimum animalium in quantum
magnificent ricinum, quoniam uni nullus sit exitus
saginae nec finis alia quam morte, diutius in fame
viventi. septenis ita diebus durasse tradunt, at in
83 satietate paucioribus dehiscere; hunc ex aure sinistra
canis omnes dolores sedare adalligatum. indicium

---

[1] gallinae cum *Mayhoff*: gallinaceum (*sine* cum) *Detlefsen*
et VE: gallinaceum cum R d.

[2] sanguinem V, *Detlefsen, Mayhoff*: sanguine *plerique codd.*

[3] genitaleque *Sillig, Detlefsen, Mayhoff*: genitalique *codd.*

[4] defossum *Detlefsen, Mayhoff, multi codd.*: defosso
d (?) E.

---

[a] I have kept with misgiving the readings of both Detlefsen
and Mayhoff: ablatives absolute are perhaps more likely, for

the feet by applying the urine of a mule, male or female, with the mud made by it, by the dung of sheep, by the liver or blood of a green lizard laid on a piece of wool, by earth-worms in oil, by the head of a spotted lizard with an equal quantity of *agnus castus* beaten up in oil, by pigeon's dung boiled down in vinegar; all kinds of warts are cured by fresh dog's urine applied with its mud, by the ash of dog's dung with wax, by the dung of sheep, by the application of fresh mouse-blood, or of a mouse itself torn asunder, by the gall of a hedgehog, by the head or blood of a lizard or the ash of the whole creature, by the slough of snakes, or by the dung of a hen with oil and soda. Cantharides beaten up with Taminian grapes eat away warts, but when corroded in this way they must be treated by the other remedies I have prescribed for the complete healing of ulcers.

XXIV. Now I will turn to those ills that threaten the whole body. The Magi say that the gall of a black male dog, if a house is fumigated or purified with it, acts as a talisman protecting all of it from sorcerers' potions; it is the same if the inner walls are sprinkled with the dog's blood or his genital[a] organ is buried under the threshold of the front door. Those would wonder less at this who know how highly the Magi extol that very loathsome animal the tick, on the ground that it is the only creature that has no vent for its gorging, nor yet any end save at death, living longer if it starves; they tell us that so it lasts for seven days, but if they eat to satiety they burst in a shorter time. They add that a tick from the left ear of a dog, worn as an amulet, relieves all

Diseases of the whole body.

que after a short *e* is most unusual. See Önnerfors, *Pliniana* p. 164.

in augurio vitalium habent, nam si aeger ei respon-
deat qui intulerit a pedibus stanti interrogantique
de morbo, spem vitae certam esse, moriturum nihil
respondere. adiciunt ut evellatur ex aure laeva
84 canis cui non sit alius quam niger color. Nigidius
fugere toto die canes conspectum eius qui e sue id
animal evellerit scriptum reliquit. rursus Magi tradunt lymphatos sanguinis talpae adspersu resipiscere,
eos vero qui a nocturnis diis Faunisque agitentur
draconis lingua et oculis et felle intestinisque in vino
et oleo decoctis ac sub diu nocte refrigeratis perunc-
tionibus matutinis vespertinisque liberari.
85   XXV. Perfrictionibus remedio esse tradit Nicander
amphisbaenam serpentem mortuam adalligatam vel
pellem tantum eius, quin immo arbori quae caedatur
adalligata non algere caedentes faciliusque sic
caedere. ita [1] sola serpentium frigori se committit,
prima omnium procedens et ante cuculi cantum.
aliud est cuculo miraculum: quo quis loco primum
audiat alitem illam si dexter pes circumscribatur ac
vestigium id effodiatur, non gigni pulices ubicumque
spargatur.
86   XXVI. Paralysim caventibus pinguia glirium de-
coctorum et soricum utilissima tradunt esse, mili-
pedas ut in angina diximus potas; phthisim sentientibus [2] lacertam viridem decoctam in vini sextariis

---

[1] ita E: itaque VRd *vulg.*: ista *Detlefsen*: ita. quae
Mayhoff.
[2] *Post* sentientibus *dist. plerique editores*; *post* potas *cum
Pl. iun. et Marcello Mayhoff.*

---

[a] *Theriaca* 372 foll. So named because it could move backwards or forwards.
[b] § 35.

pains. They also consider the tick a prognostication of life or death, for if the patient at the beginning of his illness makes reply when he who has brought in with him a tick, standing at his feet inquires about the illness, there is sure hope of recovery; should no reply be made the patient will die. They add that the tick must be taken from the left ear of a dog that is completely black all over. Nigidius has left it in writing that dogs run away for a whole day from the sight of one who has caught a tick on a pig. Again, the Magi tell us that sprinkling with mole's blood restores to their senses the delirious, while those who are haunted by night ghosts and goblins are freed from their terrors if tongue, eyes, gall, and intestines of a python are boiled down in wine and oil, cooled by night in the open air, and used as embrocation night and morning.

XXV. For feverish chills Nicander gives as a *Chills.* remedy a dead serpent, the amphisbaena,[a] worn as an amulet, or even its skin; nay, he says that, if it is fastened to a tree that is being felled, the fellers feel no cold and do their business more easily. So much does this, alone of serpents, stand up to the cold, being the first of all serpents to make its appearance, even before the cry of the cuckoo. One wonderful thing about the cuckoo is, that if, on the spot where that bird is heard for the first time, the print of the right foot is marked round and the earth dug out, no fleas breed wherever it is sprinkled.

XXVI. For those warding off paralysis the fats of *Paralysis.* decocted dormice and shrew mice are said to be very beneficial, as also millepedes taken in drink as I have prescribed[b] for quinsy; for consumptives a green lizard boiled down in three sextarii of wine to one

333

tribus ad cyathum unum, singulis coclearibus sumptis
per dies donec convalescant, coclearum cinerem
87 potum in vino, XXVII. comitialibus morbis oesypum cum murrae momento et vini cyathis duobus
dilutum magnitudine nucis abellanae, a balneo
potum, testiculos arietinos inveteratos tritosque
dimidio denarii pondere in aquae vel lactis asinini
hemina. interdicitur vini potus quinis diebus ante
88 et postea. magnifice laudatur et sanguis pecudum
potus, item fel cum melle, praecipue agninum, catulus
lactens sumptus absciso capite pedibusque ex vino et
murra, lichen mulae potus in oxymelite cyathis tribus,
stelionis transmarini cinis potus in aceto, tunicula
stelionis, quam eodem modo ut anguis exuit, in potu.
quidam et ipsum harundine exinteratum inveteratumque bibendum dederunt, alii, in cibo ligneis veri-
89 bus inassatum. operae pretium est scire quomodo
praeripiatur, cum exuatur, membrana hiberna alias
devoranti eam, quoniam nullum animal fraudulentius
invidere homini tradunt, inde stelionum nomine in
maledictum translato. observant cubile eius aestati-
bus—est autem in loricis ostiorum fenestrarumque
aut camaris sepulchrisve—ibi vere incipiente fissis
harundinibus textas opponunt ceu nassas [1] quarum
angustiis etiam gaudet, eo facilius exuens circumdatum torporem. sed relicto non potest remeare.
90 nihil ei remedio in comitialibus morbis praefertur.
prodest et cerebrum mustelae inveteratum potum-

---

[1] ceu nassas *Mayhoff*: casas *vulg.*, *Detlefsen*: quassas *codd.*

---

[a] A metaphorical meaning of *stelio* is " crafty person," or
" knave."

[b] Mayhoff makes a good emendation, for *nassa* was a
funnel-shaped trap into which fish could enter but from
which they could not escape.

cyathus, the daily dose being one spoonful until con-
valescence, or the ash of snails taken in wine;
XXVII. for epilepsy wool-grease with a morsel of
myrrh, diluted with two cyathi of wine, a piece the
size of a hazel nut being taken in drink, after the
bath, or the testicles of a ram dried and pounded,
half a denarius by weight being taken in a hemina
of water or of ass's milk; to drink wine is forbidden
for five days before and after. Very highly praised
also is the blood of sheep, taken by the mouth, the
gall of sheep, especially of a lamb, with honey, a
sucking puppy taken in wine and myrrh after the head
and feet have been cut off, the excrescence on the leg
of a she-mule taken in three cyathi of oxymel, the ash
of a spotted lizard from overseas taken in vinegar,
the coat of a spotted lizard, which it casts in the same
way as a snake, taken in drink. Some have also
given in drink the lizard itself, gutted with a reed
and dried, others in food the lizard roasted on wooden
spits. It is worth while knowing how, when cast,
the winter skin is hastily taken from the lizard,
which otherwise devours it, for no living creature,
they say, shows greater spite in cheating man, for
which reason its name [a] has been turned into a term
of abuse. They note in the summer time its nest,
which is in the cornices over doors and windows, or
in vaults or tombs. Over against the nest in the
beginning of spring they place cages like weels [b]
woven with split reeds, the narrow neck of which
gives the creature actual delight, as thereby it casts
off more easily the encumbrance of its covering, but
when this has been left no return is possible. No
remedy for epilepsy is preferred to this. A good one
too is a weasel's brain dried and taken in drink, or a

que et iocur eius, testiculi volvaeque aut ventriculus
inveteratus cum coriandro, ut diximus, item cinis,
silvestris vero tota in cibo sumpta. eadem omnia
praedicantur ex viverra. lacerta viridis cum condi-
mentis quae fastidium abstergeant, ablatis pedibus
et capite, coclearum cinis addito semine lini et
91 urticae cum melle unctu sanant. Magis placet
draconis cauda in pelle dorcadis adalligata cervinis
nervis vel lapilli e ventre hirundinum pullorum
sinistro lacerto adnexi. dicuntur enim excluso pullo
lapillum dare. quod si pullus is detur in cibo, quem
primum pepererit, cum quis primum temptatus sit,
liberatur eo malo. postea medetur hirundinum
sanguis cum ture vel cor recens devoratum. quin
et e nido earum lapillus inpositus recreare dicitur
92 confestim et adalligatus in perpetuum tueri. praedicatur et iocur milvi devoratum et senectus serpentium, iocur vulturis tritum cum suo sanguine ter
septenis diebus potum, cor pulli vulturini adalligatum. sed et ipsum vulturem in cibo dari iubent
et quidem satiatum humano cadavere. quidam
pectus eius bibendum censent in cerrino calice, aut
testes gallinacei ex aqua et lacte, antecedente
quinque dierum abstinentia vini, ob id inveteratos.[1]
fuere et qui viginti unam muscas rufas, et quidem a
mortuo,[2] in potu darent, infirmioribus pauciores.

---

[1] inveteratos *vulg.*: inveterant *Mayhoff*: inveterate *aut
inveteratae codd.*

[2] a mortuo Er *Detlefsen, Mayhoff*: mortuas *Sillig.*

---

[a] Pliny XXIX. § 60.
[b] The verb *devorare*, literally to swallow or devour, seems
sometimes, at least in Pliny, to be a synonym of *edere*.
[c] With the reading *mortuas*: " dead flies."

weasel's liver, testicles, uterus, or paunch, dried with coriander, as I have said [a]; likewise its ash, or a wild weasel taken whole in food. All the same good qualities are praised in the ferret. A green lizard, with seasonings to banish any nausea, the feet and head being taken off, and an application of snails, reduced to ash, with linseed, nettle seed, and honey, are also cures. The Magi recommend the tail of a python attached as an amulet in gazelle skin by deer sinews, or the bits of stone from the crops of baby swallows fastened to the left upper arm; for swallows are said to administer a bit of stone to each chick when hatched. But if, at the first attack of epilepsy, the chick from the first egg laid is given to the patient in food, he is freed from that complaint; afterwards the treatment is swallows' blood with frankincense, or eating [b] a fresh swallow's heart. Moreover, a little stone, taken from a swallow's nest and laid on the patient, is said to give immediate relief, and worn as an amulet permanent protection. Highly praised also is eating a kite's liver or a snake's slough, a vulture's liver pounded with its blood and taken in drink for thrice seven days, or the heart of a vulture's chick worn as an amulet. But they recommend also the vulture itself to be given in food, and that too when it has eaten its fill from a human corpse. Some are of opinion that a vulture's breast should be taken in drink in a cup made of Turkey-oak wood, or the testicles of a cock in water and milk, after abstinence from wine for five days; for this purpose the testicles are preserved. There have also been some who gave in drink twenty-one red flies, and that too from a corpse,[c] but fewer to weak patients.

93   XXVIII. Morbo regio resistunt sordes aurium aut
mammarum pecudis denarii pondere cum murrae
momento et vini cyathis duobus, canini capitis cinis
in mulso, multipeda in vini hemina, vermes terreni
in aceto mulso cum murra, gallina, si sit luteis pedibus, prius aqua purificatis, dein collutis vino quod
94 bibatur, cerebrum perdicis aut aquilae in vini cyathis
tribus, cinis plumarum aut interaneorum palumbis in
mulso ad coclearia tria, passerum cinis sarmentis
crematorum coclearibus duobus in aqua mulsa.
avis icterus vocatur a colore, quae si spectetur, sanari
id malum tradunt et avem mori. hanc puto Latine
vocari galgulum.

95   XXIX. Phreneticis prodesse videtur pulmo pecudum calidus circa caput alligatus. nam muris cerebrum dare potui ex aqua aut cinerem mustelae vel
etiam inveteratas carnes irenacei quis possit furenti,
etiamsi certa sit medicina? bubonis quidem oculorum cinerem inter ea quibus prodigiose vitam ludi-
ficantur acceperim, praecipueque febrium medicina
96 placitis eorum renuntiat. namque et in duodecim
signa digessere eam sole transmeante iterumque luna,
quod totum abdicandum paucis exemplis docebo,
siquidem crematis tritisque cum oleo perungui
iubent aegros, cum geminos transit sol, cristis et

---

^a The golden oriole.

XXVIII. Jaundice is combated by dirt from the *Cures for jaundice.*
ears or teats of a sheep, the dose being a denarius
by weight with a morsel of myrrh and two cyathi of
wine, by the ash of a dog's head in honey wine, by a
millepede in a hemina of wine, by earthworms in
oxymel with myrrh, by drinking wine that has
rinsed a hen's feet—they must be yellow—after they
have been cleansed with water, by the brain of a
partridge or eagle taken in three cyathi of wine, by
the ash of the feathers or intestines of a wood-
pigeon taken in honey wine up to three spoonfuls, or
by the ash of sparrows burnt over twigs taken in two
spoonfuls of hydromel. There is a bird called
" jaundice " from its colour. If one with jaundice
looks at it, he is cured, we are told, of that complaint
and the bird dies. I think that this bird is the one
called in Latin " galgulus." [a]

XXIX. For brain-fever appears to be beneficial a *Phrenitis, etc., Magical cures.*
sheep's lung wrapped warm round the patient's head.
But who could give to one delirious the brain of a
mouse to be taken in water, or the ash of a weasel,
or even the dried flesh of a hedgehog, even if the
treatment were bound to be successful? As for the
eyes of the horned owl reduced to ash, I should be
inclined to count this remedy as one of the frauds
with which magicians mock mankind, and it is
especially in fevers that true medicine is opposed to
the doctrines of these quacks. For they have
actually divided the art according to the passing of
the sun, and also that of the moon, through the
twelve signs of the Zodiac. That the whole theory
should be rejected I will show by a few examples. If
the sun is passing through Gemini, they recommend
the sick to be rubbed with the combs, ears, and

auribus et unguibus gallinaceorum, si luna, radiis
97 barbisque eorum; si virginem alteruter, hordei
granis, si sagittarium, vespertilionis alis, si leonem
luna, tamaricis fronde, et adiciunt sativae, si aquar-
ium, e buxo carbonibus tritis. ex istis confessa aut
certe verisimilia ponemus, sicuti lethargum olfac-
toriis excitari et inter ea fortassis mustelae testiculis
inveteratis[1] aut iocinere usto. his quoque pulmonem
pecudis calidum circa caput adalligari putant utile.
98 XXX. In quartanis medicina clinice propemodum
nihil pollet. quamobrem· plura eorum[2] remedia
ponemus primumque ea quae adalligari iubent:
pulverem in quo se accipiter volutaverit lino rutilo in
linteolo, canis nigri dentem longissimum. pseudo-
sphecem vocant vespam quae singularis volitat, hanc
sinistra manu adprehensam subnectunt, alii vero
quam quis eo anno viderit primam, viperae caput
abscissum in linteolo vel cor viventi exemptum,
99 muris rostellum auriculasque summas russeo panno
ipsumque dimittunt, lacertae vivae dextrum oculum
effossum, muscam capite suo deciso in pellicula
caprina, scarabaeum qui pilulas volvit. propter
hunc Aegypti magna pars scarabaeos inter numina
colit, curiosa Apionis interpretatione, qua colligat
Solis operum similitudinem huic animali esse, ad

---

[1] inveteratis *vulg.*, *Mayhoff*: inveteratum *codd.*, *Detlefsen.*
[2] eorum] Magorum *coni. Warmington.*

---

[a] See List of Diseases.
[b] Literally : " bed-side medicine."
[c] " Bastard wasp."

claws of cocks, burnt and pounded with oil; if it is
the moon, the cocks' spurs and wattles must be used.
If either sun or moon is passing through Virgo,
grains of barley must be used; if through Sagit-
tarius, a bat's wings; if the moon is passing through
Leo, leaves of tamarisk, and they add that it must
be the cultivated shrub; if through Aquarius, box-
wood charcoal, pounded. Of these remedies I shall
include only those recognised, or at least thought
probable: for example, to rouse the victims of
lethargus *a* by pungent smells, among which perhaps
I would put the dried testicles of a weasel or the
fumes of his burnt liver. For these patients also
they consider it useful to wrap round the head the
warm lung of a sheep.

XXX. In quartans ordinary medicines *b* are *Quartans.*
practically useless; for which reason I shall include
several of the magicians' remedies, and in the first
place the amulets they recommend: the dust in
which a hawk has rolled himself tied in a linen cloth
by a red thread, and the longest tooth of a black dog.
The wasp they call pseudosphex,*c* that flies about by
itself, they catch with the left hand and hang under
the chin, and others use the first wasp seen in that
year; a severed viper's head attached in a linen
cloth, or the heart taken from the creature while
still alive; the snout and ear tips of a mouse, wrapped
in red cloth, the mouse itself being allowed to go
free; the right eye gouged out of a living lizard; a
fly in a bit of goat skin, with its head cut off; or the
beetle that rolls little pellets. Because of this beetle
the greater part of Egypt worships the beetle as one
of its deities. Apion gives an erudite explanation:
he infers that this creature resembles the sun and

341

100 excusandos gentis suae ritus. sed et alios adalligant Magi: cui sunt cornicula reflexa, sinistra manu
collectum; tertium, qui vocatur fullo, albis guttis,
dissectum utrique lacerto adalligant, cetera sinistro;
cor anguium sinistra manu exemptum viventibus,
scorpionis caudae quattuor articulos cum aculeo,
panno nigro, ita ut nec scorpionem dimissum nec eum
qui adalligaverit videat aeger triduo, post tertium
101 circuitum id condat. erucam in linteolo ter lino cir-
cumdant totidem nodis ad singulos dicente quare
faciat qui medebitur, limacem in pellicula vel quattuor limacum capita praecisa harundine, multi-
pedam lana involutam, vermiculos ex quibus tabani
fiunt, antequam pennas germinent, alios e spinosis
frutectis lanuginosos. quidam ex illis quaternos
102 inclusos iuglandis nucis putamine adalligant. cocleas
quae nudae inveniuntur, stelionem inclusum[1] capsulis subiciunt capiti et sub decessu febris emittunt.
devorari autem iubent cor mergi marini sine ferro
exemptum inveteratumque conteri et in calida aqua
bibi, hirundinum corda cum melle, alii fimum
drachma una in lactis caprini vel ovilli vel passo
cyathis tribus ante accessiones, sunt qui totas cen-
103 seant devorandas. aspidis cutem pondere sexta

---

[1] inclusum d(?) *vulg.*: inclusos *Detlefsen*: cum incluserunt
Mayhoff: incluserant VRE.

---

[a] " The fuller."
[b] The plural *capsulis* because two kinds of amulet are
referred to.

its revolutions, seeking to find an excuse for the
religious customs of his race. But the Magi also
make amulets of other beetles. There is one with
bent-back little horns, which they take up in the left
hand; a third kind, called *fullo*,[a] with white spots,
they cut in two and wear as an amulet on either
upper arm; all the rest are worn on the left arm;
the heart, taken out with the left hand from a living
snake; four joints of a scorpion's tail, with the sting,
wrapped in black cloth, care being taken that the
sick man does not see, for three days, either the
scorpion when set free or him who attaches the
amulet; after the third paroxysm he must hide it
away. They tie a thread three times round a caterpillar in a linen cloth, and with three knots, the
ministering attendant saying at each knot the reason
for so doing. Other amulets are: a slug in a piece
of skin, or four slugs' heads cut off with a reed, a
multipede wrapped up in wool, the grubs from
which gad-flies are born, before they develop wings,
or other hairy grubs found on thorny bushes. Some
shut up four of these grubs in a walnut shell and
attach as an amulet. Snails that are found without
shells, or a spotted lizard shut up in a little box,[b]
they place under the patient's head and let out when
the fever goes down. They also recommend the
heart of a sea-diver, cut out without iron, dried and
pounded, to be taken in warm water, or the hearts of
swallows with honey; others swallows' dung in doses
of one drachma in three cyathi of goat's or sheep's
milk or in raisin wine, to be taken before the
paroxysms. Some hold that the entire swallow
should be taken. An asp's skin, in doses of one
sixth of a denarius by weight with an equal quantity

parte denarii cum piperis pari modo Parthorum gentes in remedium quartanae bibunt. Chrysippus philosophus tradidit phryganion adalligatum remedio esse quartanis. quod esset animal neque ille descripsit nec nos invenimus qui novisset. demonstrandum tamen fuit a tam gravi auctore dictum, si cuius cura

104 efficacior esset inquirendi. cornicis carnes esse et nidum [1] inlinere in longis morbis utilissimum putant. et in tertianis fiat potestas experiendi, quoniam miserias copia spei delectat, anne aranei, quem lycon vocant, tela cum ipso in spleniolo resinae ceraeque inposita utrisque temporibus et fronti prosit, aut ipse calamo adalligatus, qualiter et aliis febribus prodesse traditur, item lacerta viridis adalligata viva in eo vase quod capiat, quo genere et recidivas frequenter abigi adfirmant.

105 XXXI. Hydropicis oesypum ex vino addita murra modice potui datur, nucis abellanae magnitudine. aliqui addunt et anserinum adipem ex vino myrteo. sordes ab uberibus ovium eundem effectum habent, item carnes inveteratae irenacei sumptae. vomitus quoque canum inlitus ventri aquam trahere promittitur.

106 XXXII. Igni sacro medetur oesypum cum pompholyge et rosaceo, ricini sanguis, vermes terreni ex aceto inliti, grillus contritus in manibus—quo genere praestat ut qui id fecerit, antequam incipiat

---

[1] nidum] *coni.* fimum *Warmington.*

---

[a] Chrysippus of Soli was the third head of the Stoic school.
[b] With Warmington's emendation: " dung."

of pepper, is taken by Parthian tribes as a cure for
a quartan. Chrysippus [a] the philosopher has told us
that wearing a phryganion as an amulet is a cure for
quartans: but what the animal is Chrysippus has
left no account, and I have met nobody who knew.
Yet a statement made by so great an authority it
was necessary to mention, in case somebody's
research should meet with better success. To eat
the flesh of a crow or to apply its nest [b] as a friction
they think very beneficial in chronic diseases. In
tertians too it may be worth while to try whether
there is any benefit (so much does suffering delight
in hoping against hope) in the spider called *lycos*
(wolf) applied with its web in a small plaster of resin
and the wax to both temples and to the forehead, or in
the spider itself attached as an amulet in a reed, in
which form it is also said to be beneficial for other
fevers. A green lizard too may be tried, attached
alive, in a vessel just large enough to contain it; by
which method we are assured that recurrent fevers
also are often banished.

XXXI. For dropsy is given in drink wool grease in *Dropsy.*
wine mixed with a little myrrh, in doses the size of a
hazel nut. Some also add goose grease in myrtle
wine. The dirt from the udders of sheep has the
same effect, as has the dried flesh of a hedgehog
taken by the mouth. An application too of dogs'
vomit to the abdomen brings away, we are assured,
the dropsical fluid.

XXXII. Erysipelas is benefited by wool grease *Erysipelas.*
with pompholyx and rose oil, by the blood of a tick,
by earth-worms applied in vinegar, by a cricket
crushed between the hands—he who succeeds in
doing this before the complaint shows itself is pro-

vitium, † toto eo anno accidat; † [1] oportet autem eum ferro cum terra cavernae suae tolli—anseris adeps, viperae caput aridum adservatum et combustum, dein ex aceto inpositum, senectus serpentium ex aqua inlita a balneo cum bitumine et sebo agnino.

107 XXXIII. Carbunculus fimo columbino aboletur per se inlito vel cum lini semine ex aceto mulso, item apibus quae in melle sint mortuae inpositis polentaque inspersa.[2] si in verendis sit ceterisque ibi ulceribus occurrit ex melle oesypum cum plumbi squamis, item fimum pecudum incipientibus carbunculis. tubera et quaecumque molliri opus sit efficacissime anserino adipe curantur, idem praestat et gruum adeps.

108 XXXIV. Furunculis mederi dicitur araneus priusquam nominetur [3] inpositus et tertio die solutus, mus araneus pendens enecatus sic ut terram ne postea attingat, ter circumlatus furunculo, totiens expuentibus medente et cui is medebitur, ex gallinaceo fimo quod est rufum maxime recens inlitum ex aceto, ventriculus ciconiae ex vino decoctus, muscae inpari numero infricatae digito medico, sordes ex pecudum auriculis, sebum ovium vetus cum cinere

---

[1] toto eo anno accidat] *obelos ego addo* : toto eo anno non accipiat *Detlefsen* : toto ei anno non accidat *Mayhoff, qui ne pro* ut *ante qui coni.*

[2] inspersa *Detlefsen* : inposita insuper *Mayhoff* : inposita inspersa *codd.* (si *add.* E).

[3] nominetur *codd., Mayhoff* : stamen netur *Detlefsen.*

---

[a] With the MSS. reading *accidat* there is required a dative, but Mayhoff's *ei* is strangely placed, while Detlefsen's *accipiat* is not very attractive. Mayhoff's *ne* for *ut* would obviate the addition of *non.* Warmington translates: " in this connection it guarantees that he who succeeds in doing this. . . ."

tected from an attack for the whole of that year,[a] but the cricket must be lifted with iron along with the earth of its hole—by goose grease, by the head of a viper, kept till dry, burnt, and then applied in vinegar, by a serpent's slough applied in water with bitumen and lamb suet after a bath.

XXXIII. A carbuncle is removed by pigeon's *Carbuncles.* dung, applied by itself or with linseed in oxymel, also by bees that have died in honey, applied and sprinkled with pearl barley. If a carbuncle or other sore is on the privates, the remedy is wool grease with lead scales [b] in honey, and sheep dung for incipient carbuncles. Hard swellings and whatever needs to be softened are treated very efficaciously with goose grease, and equally good results are also given by the grease of cranes.

XXXIV. Boils are said to be cured by a spider, *Boils.* applied before its name has been mentioned [c] and taken off on the third day, by a shrew mouse, killed and hung up so that it does not touch earth after death, and passed three times round the boil, both the attendant and the patient spitting the same number of times, by the red part of poultry dung, best applied fresh in vinegar, by a stork's crop boiled down in wine, by an odd number of flies rubbed on with the medical finger [d] by dirt from the ears of sheep, by stale mutton [e] suet with the ash of woman's

---

[b] Some oxide of lead.

[c] With Detlefsen's emendation : " before its web is spun." This is a clever conjecture, but we should expect the subjunctive, while " naming " is not unusual in magical remedies.

[d] The finger next the little finger.

[e] Perhaps here " suet of ewes," because of *pecudum* preceding. See § 123.

capilli mulierum, sebum arietis cum cinere pumicis et salis pari pondere.

109 XXXV. Ambustis canini capitis cinis medetur, item glirium cum oleo, fimum ovium cum cera, murium cinis, coclearum quoque sic ut ne cicatrix quidem appareat, adips viperinus, fimi columbini

110 cinis ex oleo inlitus, XXXVI. nervorum nodis capitis viperini cinis in oleo cyprino, terreni vermes cum melle inliti. dolores eorum ⟨sedat . . .⟩[1] adips, amphisbaena mortua adalligata, adips vulturinus cum ventre arefactus tritusque cum adipe suillo inveterato, cinis e capite bubonis in mulso potus cum lilii radice, si Magis credimus. in contractione nervorum caro palumbina in cibis prodest [et][2] inveterata, irenacei spasticis, item mustelae cinis—serpentium senectus in pelle taurina adalligata spasmos fieri prohibet—opisthotonicis milvi iocur aridum tribus obolis in aquae mulsae cyathis tribus potum.

111 XXXVII. Reduvias et quae in digitis nascuntur pterygia tollunt canini capitis cinis aut vulva decocta in oleo, superinlito butyro ovillo cum melle, item folliculus cuiuslibet animalium fellis, unguium scabritiam cantharides cum pice tertio die solutae aut locustae frictae cum sebo hircino, pecudum sebum.

[1] sedat . . . *add. Mayhoff.*
[2] et *delere velim.*

---

[a] Here the name of an animal must be supplied.
[b] If *et* is kept it must, I think, mean " even." But it seems to be a duplication from -*est*.

hair, and by ram's suet with ash of burnt pumice and an equal quantity of salt.

XXXV. Burns are treated with ash of a dog's head, *Burns.* the ash of dormice and oil, sheep dung and wax, the the ash of mice; with the ash of snails so well that not even a scar is to be seen, with viper fat, and with the ash of pigeon's dung applied in oil. XXXVI. Hard lumps in the sinews are treated with the ash of a viper's head in cyprus oil, and by an application of earth-worms and honey. Pains in the sinews ⟨are soothed by . . .⟩ *a* fat, by a dead amphisbaena attached as an amulet, by vulture's fat with its crop, dried and pounded with stale pig's fat, by the ash of a horned-owl's head taken in honey wine with the root of a lily, if we believe the Magi. For cramp in the sinews wood-pigeon's flesh dried and *b* taken in the food, for cramping spasms hedgehog's flesh, also the ash of a weasel—a serpent's slough attached as an amulet in a piece of bull's leather prevents such spasms *c*—for opisthotonic tetanus the dried liver of a kite, the dose being three oboli taken in three cyathi of hydromel.

XXXVII. Hangnails and whitlows that form on *Hangnails.* the fingers *d* are removed by the ash of a dog's head, or by the uterus boiled down in oil, with a layer on top of butter from ewe's milk with honey, as also by the gall bladder of any animal; roughness of the nails by cantharides and pitch, taken off on the third day, or by locusts fried with he-goat suet, and by mutton suet. Some mix with the ingredients

c Detlefsen's parenthesis seems the best way of treating this clumsy sentence.

d This clause is added because *pterygium* may mean an eye affection. See List of Diseases.

aliqui miscent viscum et porcillacam, alii aeris florem et viscum ita ut tertio die solvant.

112 XXXVIII. Sanguinem sistit in naribus sebum ex omento pecudum inditum, item coagulum ex aqua, maxime agninum subductum vel infusum, etiam si alia non prosint, adips anserinus cum butyro pari pondere pastillis ingestus, coclearum terrena, sed et ipsae extractae testis; e naribus fluentem cocleae contritae fronti inlitae, aranei telae, gallinacei cerebellum vel sanguis profluvia ex cerebro, item columbinus ob id servatus concretusque. si vero ex vulnere inmodice fluat, fimi caballini cum putaminibus ovorum cremati cinis inpositus mire sistit.

113 XXXIX. Ulceribus medetur oesypum cum hordei cinere et aerugine aequis partibus, ad carcinomata quoque ac serpentia valet. erodit et ulcerum margines, carnesque exscrescentes ad aequalitatem redigit. explet quoque et ad cicatricem perducit. magna vis et in cinere pecudum fimi ad carcinomata, addito nitro, aut in cinere ex ossibus feminum agninorum, praecipue in his ulceribus quae cicatricem non trahunt, magna et pulmonibus, praecipue arietum: carnes excrescentes in ulceribus ad aequalitatem

114 efficacissime reducunt; fimo quoque ipso ovium sub testo calefacto et subacto tumor vulnerum sedatur, fistulae purgantur sananturque, item epinyctides. summa vero in canini capitis cinere: excrescentia

---

[a] Red oxide of copper.

[b] If there is any difference between *in naribus* here and *ex naribus* a few lines further on (this repetition may be carelessness), the second will denote a more violent flow of blood.

[c] I.e. from the skull.

[d] Night rashes. See List of Diseases.

mistletoe and purslane, others flowers of copper [a] and mistletoe, but remove the application on the third day.

XXXVIII. Bleeding in the nostrils [b] is arrested by *Epistaxis.* inserting suet from the cawl of a sheep, also by its rennet in water, especially by lamb's rennet, snuffed up or injected, even if other remedies do no good, by goose grease with an equal quantity of butter worked up into lozenges, by the earth off snails, but also by the actual snails themselves, taken from their shells; but when there is severe epistaxis it is stayed by snails beaten up and applied to the forehead, and also by spider's web; by the brain or blood of a cock are arrested fluxes from the brain,[c] also by pigeon's blood; it is stored and congealed for this purpose. If however there is violent haemorrhage from a wound, it is wonderfully arrested by an application of the ash of horse-dung burnt with egg shells.

XXXIX. Ulcers are healed by wool grease, barley *Ulcers.* ash, and copper rust, in equal parts; this is also equally efficacious for carcinomata and spreading sores. It cauterizes too the edges of ulcers, and levels out excrescences in the flesh; it also fills up hollows and forms scars. There is also great power to heal carcinomata in the ash of sheep's dung with soda added, or in the ash of a lamb's thigh bones, especially when ulcers refuse to cicatrize. There is great power too in the lungs, especially those of rams, which flatten out very efficaciously excrescences of flesh on ulcers; ewe dung too by itself, warmed under an earthen jar and kneaded, reduces swollen wounds, and cleans and heals fistulas and epinyctides.[d] The greatest power, however, is in the ash of a dog's head, which cauterizes and

omnia spodii vice erodit ac persanat. et murino
fimo eroduntur, item mustelae fimi cinere. duritias
in alto ulcerum et carcinomata persequitur multipeda
trita admixta resina terebinthina et sinopide. eadem
utilissima sunt in his ulceribus quae vermibus peri-
115 clitentur. quin et vermium ipsorum genera miran-
dos usus habent. cosses qui in ligno nascuntur sanant
ulcera omnia, nomas vero combusti cum pari pondere
anesi ex oleo inliti. vulnera recentia conglutinant
terreni adeo ut nervos quoque abscisos inlitis solidari
intra septimum diem persuasum sit; itaque in melle
servandos censent. cinis eorum margines ulcerum
duriores absumit cum pice liquida vel symphyto et
116 melle. quidam arefactis in sole ad vulnera ex aceto
utuntur nec solvunt nisi biduo intermisso. eadem
ratione et coclearum terrena prosunt, totaeque
exemptae recentia vulnera tusae inpositae conglutinant et nomas sistunt. herpes quoque animal a
Graecis vocatur quo praecipue sanantur quaecumque
serpunt. cocleae quoque prosunt eis cum testis suis
tusae, cum murra quidem et ture etiam praecisos
117 nervos sanare dicuntur. draconum quoque adeps siccatus in sole magnopere prodest, item gallinacei cerebrum recentibus plagis. sale viperino in cibo sumpto
tradunt et ulcera tractabiliora fieri ac celerius sanari.
Antonius quidem medicus cum incidisset insanabilia
ulcera, viperas edendas dabat miraque celeritate per-

---

^a See List of Diseases.
^b Perhaps : " on the same principle."
^c See List of Diseases.
^d It means " the creeper." Unidentified.
^e The salt in which vipers were preserved. Has *sale* arisen
from *sole* above ?

thoroughly heals all excrescences as well as does spodium. These are cauterized too by mouse dung, and also by the ash of weasel's dung. Indurations in deep-seated ulcers and carcinomata are penetrated by multipedes pounded and mixed with terebinth resin and earth of Sinope. The same remedies are very useful for those ulcers that are threatened by worms. Moreover, the various kinds of worms themselves have wonderful uses. The larvae that breed in wood heal all ulcers; and *nomae* [a] too if burnt with an equal weight of anise and applied in oil. Fresh wounds are united so well by earth worms that there is a general conviction that even severed sinews are by applying them made whole by the seventh day; accordingly it is thought that they should be preserved in honey. Their ash with liquid pitch or symphytum and honey removes too-hard edges of ulcers. Some dry them in the sun, use in vinegar to treat wounds, and do not take them off without an interval of two days. Used in the same way [b] the earth too off snails is beneficial, and snails taken out whole, beaten up, and applied, unite fresh wounds and arrest *nomae*.[c] There is also an insect called by the Greeks *herpes*,[d] which is specific for all creeping ulcers. Snails also are good for them, beaten up with their shells; with myrrh indeed and frankincense they are said to heal even severed sinews. The fat of a python also, dried in the sun, is of great benefit, as is a cock's brain for fresh wounds. By viper's salt [e] taken in food we are told that ulcers become more amenable to treatment and heal more rapidly. Indeed the physician Antonius after operating on ulcers without success gave vipers as food to bring about complete cures

353

sanabat. trixallidum cinis margines ulcerum duros
aufert cum melle, item fimi columbini cinis cum
arrhenico et melle; eadem [1] quae erodenda sunt.
118 bubonis cerebrum cum adipe anserino mire vulnera [2]
dicitur glutinare, quae vero vocantur cacoethe cinis
feminum arietis cum lacte muliebri, diligenter prius
elutis linteolis, ulula avis cocta in oleo, cui liquato
miscetur butyrum ovillum et mel. ulcerum labra
duriora apes in melle mortuae emolliunt, et elephantiasin sanguis et cinis mustelae. verberum vulnera
atque vibices pellibus ovium recentibus inpositis
obliterantur.
119 XL. Articulorum fracturis cinis feminum pecudis
peculiariter medetur cum cera—efficacius idem medicamentum fit maxillis simul ustis cornuque cervino
et cera mollita rosaceo—ossibus fractis caninum
cerebrum linteolo inlito, superpositis lanis quae
subinde [3] subfundantur, fere XIIII diebus solidat,
nec tardius cinis silvestris muris cum melle aut
vermium terrenorum, qui et ossa extrahit.
120 XLI. Cicatrices ad colorem reducit pecudum
pulmo, praecipue ex ariete, sebum ex nitro, lacertae
viridis cinis, vernatio anguium ex vino decocta,

---

[1] eadem quae erodenda sunt *codd.*: ea quae erodenda
sunt *vulg.*, *Detlefsen* : eademque erodentia sunt *Mayhoff.*
[2] vulnera *codd. et edd.*: ulcera *coni. Mayhoff.*
[3] *Inter* subinde *et* subfundantur *add.* oleo *Mayhoff*: subinde
oleo fundantur *coni. Warmington.*

with wonderful rapidity. The ash of the trixallis [a] with honey removes hard edges on ulcers, as does ash of pigeon's dung with arsenic and honey; these also remove all that needs a cautery. [b] The brain of a horned owl with goose grease is said to unite wounds wonderfully, as, with woman's milk, does the ash of a ram's thighs the ulcers called malignant, but the cloths must be first carefully washed, or the screech owl boiled in oil, with which when melted down are mixed ewe butter and honey. The lips of ulcers that are too hard are softened by bees that have died in honey, and elephantiasis by the blood and ash of a weasel. Wounds and weals made by the scourge are removed by an application of fresh sheep-skin.

XL. For fractures of the joints a specific is the ash of a sheep's thighs with wax—this medicament is more efficacious if there are burnt with the thighs the sheep's jawbones and a deer's horn, and the wax is softened with rose oil—specific for broken bones is a dog's brain, spread on a linen cloth, over which is placed wool, occasionally moistened underneath (with oil). In about fourteen days it unites the broken parts, as does quite as quickly the ash of a field-mouse with honey, or that of earth-worms, which also extracts fragments of bone. *Fractures.*

XLI. Scars are restored to the natural colour by the lungs of sheep, particularly of rams, by their suet in soda, by the ash of a green lizard, by a snake's slough boiled down in wine, and by pigeon's dung *Scars and skin diseases.*

---

[a] See § 49. Antonius is perhaps Antonius Castor (XXV. § 9).

[b] The reading of the MSS. can be just construed, with *eadem* subject, and *ea auferunt* understood.

fimum columbinum cum melle, item [1] vitiligines albas
ex vino, vitiliginem et cantharides cum rutae folio-
rum duabus partibus. in sole, donec formicet cutis,
tolerandae sunt, postea fovere oleoque perunguere
necessarium iterumque inlinire, idque pluribus diebus
121 facere, caventes exulcerationem altam. ad easdem
vitiligines et muscas inlini iubent cum radice eupatoriae,[2] gallinarum fimi candidum servatum in oleo
vetere cornea pyxide, vespertilionis sanguinem, fel
irenacei ex aqua. scabiem vero bubonis cerebrum
cum aphronitro, sed ante omnia sanguis caninus
sedant, pruritum cocleae minutae latae contritae
inlitae.

122   XLII. Harundines et tela quaeque alia extra-
henda sunt corpori evocat mus dissectus inpositus,
praecipue vero lacerta dissecta, et vel caput tantum
eius contusum cum sale inpositum, cocleae ex his
quae gregatim folia sectantur contusae inpositaeque
cum testis et eae quae manduntur exemptae testis,
sed cum leporis coagulo efficacissime ossa anguium.
eadem cum coagulo cuiuscumque quadripedis intra
tertium diem adprobant effectum. laudantur et
cantharides tritae cum farina hordei.

123   XLIII. In muliebribus malis membranae a partu
ovium proficiunt, sicut in capris rettulimus. fimum
quoque pecudum eosdem usus habet. locustarum

---

[1] item *codd. et edd.* : idem *coni. Mayhoff.*
[2] eupatoriae *Sillig coll.* XXV. § 65 : lupatoria *codd.*

---

[a] There were distinguished by the Romans three kinds of
vitiligo (psoriasis) : the dull white, the dark, and the bright
white.
[b] The word *easdem* seems to include both the *vitiligines
albas* and the *vitiliginem* of § 120.
[c] Perhaps : " bits of reed."

with honey; the last in wine does the same for both kinds [a] of white vitiligo; for vitiligo cantharides also with two parts of rue leaves. These must be kept on in the sun until the skin is violently irritated; then there must be fomentation and rubbing with oil, followed by another application. This treatment should be repeated for several days, but deep ulceration must be guarded against. For vitiligo of all kinds [b] they also recommend the application of flies with root of eupatoria, or the white part of hens' dung kept in old oil in a horn box, or bat's blood, or hedgehog's gall in water. Itch scab however is relieved by the brain of a horned owl with saltpetre, but best of all by dog's blood, and pruritus by the small, broad, kind of snail, crushed and applied.

XLII. Arrows,[c] weapons, and everything that *Things embedded in flesh.* must be extracted from the flesh, are withdrawn by a mouse split and laid on the wound, but especially by a split lizard, or even its head only, crushed and laid on the wound with salt, by the snails that attack leaves in clusters, crushed and similarly laid on with the shells, and edible snails without them, but most efficaciously by the bones of snakes with hare's rennet. These bones also, with the rennet of any quadruped, show a good result by the third day. Cantharides too are highly recommended, beaten up and applied with barley meal.

XLIII. For women's complaints the afterbirth of *Female complaints.* an ewe is of service, as I said when speaking of goats.[d] The dung too of sheep [e] has the same

[d] See XXVIII. § 256.
[e] The word *ovis* appears to be used when the sex must be female, and *pecus* when the sex of the sheep does not matter.

suffitu stranguriae maxume mulierum iuvantur.
gallinaceorum testes si subinde a conceptu edat
mulier, mares in utero fieri dicuntur. partus conceptos hystricum cinis potus continet, maturat
caninum lacte potum, evocat membrana e secundis
canum, si terram non attigerit, lumbis parturientium
124 tactis.[1] fimum murinum aqua pluvia dilutum mammas
mulierum a partu tumentes reficit. cinis irenace-
orum cum oleo perunctarum custodit partus contra
abortus. facilius enituntur quae . . .[2] anserinum
cum aquae duobus cyathis sorbuere, aut ex ventriculo
125 mustelino per genitale effluentes aquas. vermes
terreni inliti ne cervicis scapularumque nervi doleant
praestant. graves secundas pellunt in passo poti.
idem per se inpositi mammarum suppurationes con-
cocunt et aperiunt extrahuntque et ad cicatricem
perducunt. lac evocant poti cum mulso. inveniuntur et in gramine vermiculi qui adalligati collo
continent partum, detrahuntur autem sub partu,
alias eniti non patiuntur. cavendum et ne in terra
ponantur. conceptus quoque causa dantur in potu
126 quini aut septeni. cocleae in cibo sumptae ad-
celerant partum, item conceptum inpositae cum
croco. eaedem ex amylo et tragacantha inlitae profluvia sistunt. prosunt et purgationibus sumptae in
cibo et vulvam aversam corrigunt cum medulla
cervina ita ut uni cocleae denarii pondus addatur et

---

[1] tactis *Detlefsen, Mayhoff*: lactis *aut* potus lactis *codd.*
[2] *lacunam indicat Mayhoff*: cum VRE: adipem d T:
del. Detlefsen; serum *Brakman.*

---

[a] The *serum* (i.e. semen) of Brakman may be right.
[b] See *Index of Plants* in vol. VII.

medicinal uses. Fumigation with lobsters is of the
greatest help in strangury in women. If occasionally
after conception a woman eats the testicles of a cock,
males are said to be formed in the uterus. The
foetus is retained by taking in drink the ash of
porcupines, brought to maturity by drinking bitch's
milk, and withdrawn by the afterbirth of a bitch,
which must not touch the earth, laid on the loins of
the woman in childbed. Mouse dung diluted with
rain water reduces the breasts of women swollen after
childbirth. Rubbing the woman all over with the
ash of hedgehogs and oil prevents miscarriage. The
delivery of those is easier who have swallowed goose
. . .*a* with two cyathi of water, or the liquids that
flow from a weasel's uterus through its genitals.
Applying earth-worms prevents pains in the sinews
of neck and shoulders, and taken in raisin wine bring
away a sluggish afterbirth. These worms laid by
themselves on the breasts also mature suppurations
there, open them, draw out the pus, and make them
cicatrize. Taken with honey wine they stimulate
the flow of milk. There are also little worms found
in grass; these, tied round the neck as an amulet,
prevent a miscarriage, but they are taken off just
before the birth, otherwise they prevent delivery.
Care too must be taken not to lay them on the earth.
Further, to cause conception five or seven at a time
are given in drink. Snails taken in food hasten
delivery, and conception too if applied with saffron.
An application of snails in starch and tragacanth *b*
arrests fluxes. They are also good for menstruation
if taken in food, and correct with deer's marrow dis-
placements of the uterus; to one snail should be
added a denarius by weight of marrow and cyprus oil.

cypri. inflationes quoque vulvarum discutiunt exemptae testis tritae cum rosaceo. ad haec Asty-
127 palaeicae maxime eliguntur. alio modo Africanae binae tritae cum feni Graeci quod tribus digitis capiatur, addito melle coclearibus quattuor, inlinuntur alvo prius irino suco perunctae. sunt et minutae loricaeque [1] candidae cocleae passim oberrantes. hae arefactae sole in tegulis tusaeque in farinam miscentur lomento aequis partibus candoremque et levorem corpori adferunt. scabendi desideria tollunt minutae et latae cum polenta.
128 viperam mulier praegnans si transcenderit, abortum faciet, item amphisbaenam, mortuam dumtaxat, † nam vivam habentes in pyxide inpune transeunt; etiam si mortua sit atque adservata, partus faciles praestat; vel mortua mirum, si sine adservata transcenderit gravida, innoxium fieri, si protinus transcendat adservatam.† [2] anguis inveterati suffitu menstrua adiuvant.
129 XLIV. Anguium senectus adalligata lumbis faciliores partus facit, protinus a puerperio removenda. dant et in vino bibendam cum ture, aliter sumpta abortum facit. baculum quo angui rana excussa sit parturientes adiuvat, trixallidum cinis inlitus cum melle purgationes, item araneus qui filum deducit ex

---

[1] loricaeque VRdT: longaeque E *vulg. Detlefsen*: loricataeque *coni. Warmington.*
[2] *Sic codd.*: in *pro vel Mayhoff. Obelos ego addo loco, ut videtur, desperato.*

---

[a] The text and its explanation are so conjectural that I prefer to print the reading of the MSS. within daggers. I

Inflation too of the uterus is dispersed by snails taken out of their shells and beaten up with rose oil. For these purposes the most preferred are snails of Astypalaea. African snails are prepared in a different way; doses of two are beaten up with a three-finger pinch of fenugreek, four spoonfuls of honey added, and the whole applied after rubbing the abdomen with iris juice. There are also found straying everywhere small snails with a white corslet. Dried in the sun on tiles, crushed to powder, and mixed with an equal quantity of bean meal, these impart both whiteness and smoothness to the skin. The desire to scratch is removed by the small, broad snails with pearl barley. If a woman with child step across a viper she will miscarry; similarly if she cross an amphisbaena, a dead one at least, †but those that carry on their persons a live one in a box step across with impunity; even if it is a dead one and preserved it makes childbirth easy. In the case of a dead one, wonderful to relate, no harm is done should a pregnant woman cross it without a preserved one, if she at once crosses a preserved one†.[a] Fumigation with a dried snake assists menstruation.

XLIV. A snake's slough, tied to the loins as an amulet, makes childbirth easier, but it must be taken off immediately after delivery. They also give it in wine to be taken with frankincense; in any other way it causes miscarriage. A stick with which a frog has been shaken from a snake helps lying-in women, and the ash of the trixallis, applied with honey, helps menstruation, as does a spider that is

translate as a stop-gap the text of Mayhoff. See also Additional Note on p. 374.

alto. capi debet manu cava tritusque admoveri,
quod si redeuntem prenderit, inhibebit idem purga-
130 tiones. lapis aetites in aquilae repertus nido custodit
partus contra omnes abortuum insidias. penna
vulturina subiecta pedibus adiuvat parturientes.
ovum corvi cavendum gravidis constat, quoniam
transgressis abortum per os faciat. fimum accipitris
in mulso potum videtur fecundas facere. vulvarum
duritias, collectiones adeps anseris aut cygni [1] emollit.
131 XLV. Mammas a partu custodit adeps anseris cum
rosaceo et araneo. Phryges et Lycaones mammis
puerperio vexatis invenerunt otidum adipem utilem
esse. his quae vulva strangulentur et blattas inlinunt. ovorum perdicis putaminum cinis cadmiae
mixtus et cerae stantes mammas servat. putant et
ter circumductas ovo perdicis aut ortygis [2] non in-
clinari et, si sorbeantur eadem, fecunditatem facere,
lactis quoque copiam, cum anserino adipe perunctis
mammis dolores minuere, molas uteri rumpere,
scabiem vulvarum sedare, si cum cimice trito inlinantur.
132 XLVI. Vespertilionum sanguis psilotri vim habet,
sed alis puerorum inlitus non satis proficit nisi aerugo
vel cicutae semen postea inducatur. sic enim aut in

---

[1] cygni d r, *vulg.*, *Mayhoff*: ciconiae E, *Detlefsen*.
[2] ortygis *Brakman*: otidis *Detlefsen post Urlichs*: *om.*
codd.: *lacunam Sillig et Mayhoff*: anseris *coni. Mayhoff,*
Sereno collato.

---

[a] "The eagle stone." See XXXVI. § 149.
[b] An oxide of zinc.
[c] With the other conjectures, " bustard " or " goose."
[d] For these *molae* see VII. § 63 and X. § 184.

spinning a thread from a height. It should be caught in the hollow of the hand, crushed, and applied; but if it is caught as it ascends again, the same treatment will arrest menstruation. The stone aetites,[a] found in the eagle's nest, protects a foetus from all plots to cause abortion. A vulture's feather, placed under their feet, helps lying-in women. It is certain that pregnant women must avoid a raven's egg, since if they step over it they will miscarry through the mouth. A hawk's dung taken in honey wine seems to make women fertile. Indurations and abscesses of the uterus are softened by goose grease or by swan's grease.

XLV. The breasts after delivery are safeguarded by goose grease with rose oil and a spider's web. The Phrygians and Lycaonians have found that the fat of bustards is beneficial for teats disordered by childbirth. For uterine suffocation beetles also are applied. Ash of partridge egg-shells mixed with cadmia [b] and wax keeps the breasts firm. They also think that breasts do not droop if circles are traced round them three times with the egg of partridge or quail,[c] and that if this egg is swallowed it also produces fertility and an abundant supply of milk as well, that it lessens pains in the breasts if they are rubbed with it and goose grease, that it breaks up moles [d] in the uterus, and that uterine itch is relieved if it is applied with crushed bugs.[e]

XLVI. Bats' blood is a depilatory, but an application to the armpits of boys is not enough unless copper rust or hemlock seed is spread over it after- *Depilatories.*

[e] Probably *cimice* is a generic singular. The probable lacuna in this chapter is perhaps larger than one word, for the plural *eadem* has only the singular *ovum* to which to refer.

totum tolluntur pili aut non excedunt lanuginem.
idem et cerebro eorum profici putant—est autem
duplex, rubens † itaque † [1] et candidum—aliqui
133 sanguinem et iocur eiusdem admiscent. quidam in
tribus heminis olei discocunt viperam, exemptis ossibus psilotri vice utuntur evolsis prius pilis quos
renasci nolunt. fel irenacei psilotrum est, utique
mixto vespertilionis cerebro et lacte caprino, item per
se cinis. lacte canis primiparae [2] evolsis pilis vel
nondum natis perunctae partes alios non sufficiunt.
134 idem evenire traditur sanguine ricini evulsi cani,
item hirundinino sanguine vel felle, ovis formicarum.
supercilia denigrari muscis tritis tradunt, si vero
oculi nigri nascentium placeant, soricem praegnanti
edendum, capilli ne canescant vermium terrenorum
cinere praestari admixto oleo.
135 XLVII. Infantibus qui lacte concreto vexantur
praesidio est agninum coagulum ex aqua potum, aut
si hoc vitium coagulato lacte acciderit, discutitur
coagulo ex aceto dato. ad dentitionem cerebrum
pecoris utilissimum est. ossibus in canino fimo inventis adustio infantium quae vocatur siriasis adalligatis emendatur, ramex infantium lacertae viridis
admotae dormientibus morsu. postea harundini inligata [3] suspenditur [4] in fumo, traduntque pariter
136 cum expirante ea sanari infantem. coclearum saliva

---

[1] itaque *codd.*: utique *vulg., Detlefsen, Mayhoff, qui* atque
vel aliquando et *coni.*; *obelos addo.*

[2] primiparae *Mayhoff, qui* prius *addit*: primi partus *Detlefsen*: primi parae *aut*: -partus *codd.*

[3] inligata *Detlefsen*: alligata *Gelenius*: adalligatae *vulg.*:
inligant et *Mayhoff*: inligate (-ti d) *codd.*

[4] suspenditur *Gelenius, Detlefsen*: suspendunt *Mayhoff*:
suspenduntur *codd.*

wards; this treatment either removes the hair altogether or reduces it to down. They think that a bat's
brain is equally efficacious—this brain is double,
red and white [a]—some adding the bat's blood
and liver. Others in three heminae of oil thoroughly
boil a viper after taking out the bones, using the
decoction as a depilatory after first plucking out the
hairs they do not wish to grow again. The gall of
a hedgehog is a depilatory, especially when mixed
with a bat's brain and goat's milk, as is also the ash
by itself. Parts rubbed with the milk of a bitch with
her first litter, when the hairs have been plucked out
or not yet grown, do not grow hair again. The same
result is said to be produced by the blood of a tick
plucked from a dog, by the blood or gall of a swallow,
or by the eggs of ants. They say that eyebrows are
made black by crushed flies; if however it is desired
that the eyes of babies should be black, the expectant
mother must eat a shrewmouse; hair is prevented
from turning grey by the ash of earth-worms mixed
with oil.

XLVII. Babies that are troubled with curdled milk *The troubles*
have a preventative in lamb's rennet taken in water; *of babies.*
or if the trouble has occurred with milk already
curdled it is dispersed by this rennet given in vinegar.
For dentition the brain of a sheep is very beneficial.
The inflammation of babies called siriasis is cured by
the bones found in dog's dung worn as an amulet, and
hernia in babies by bringing a green lizard to bite
them when asleep. Afterwards they fasten the lizard
to a reed and hang it in smoke, and they say that as it
dies the baby recovers. The slime of snails applied

[a] This addition, which I treat as a parenthesis, seems point-
less.

inlita infantium oculis palpebras corrigit gignitque. ramicosis coclearum cinis cum ture ex ovi albo suco [1] inlitus per dies XXX medetur. inveniuntur in corniculis coclearum harenaceae duritiae, hae dentitionem facilem praestant adalligatae. coclearum inanium cinis cerae mixtus procidentium interan-137 eorum partes extremas prohibet. oportet autem cineri misceri saniem punctis emissam.[2] cerebrum viperae inligatum pellicula [3] dentitiones adiuvat. idem valent et grandissimi dentes serpentium. fimum corvi lana adalligatum infantium tussi medetur. vix est serio conplecti quaedam, non omittenda tamen, quia sunt prodita. ramici infantium lacerta mederi iubent. marem hanc prendi, id intellegi eo quod sub [4] cauda unam cavernam habeat, 138 id agendum ut per aureum vel argenteum clostrum [5] mordeat vitium, tum in calice novo inligatum [6] in fumo poni. urina infantium cohibetur muribus elixis in cibo datis: scarabaeorum cornua grandia denticulata adalligata iis amuleti naturam obtinent. 139 bovae capiti lapillum inesse tradunt, quem ab ea expui, si necem timeat, inopinantis praeciso capite exemptum adalligatumque mire prodesse dentitioni. item cerebrum eiusdem ad eundem usum adalligari iubent et limacis lapillum sive ossiculum; invenitur in dorso. magnifice iuvat et ovis cerebrum gingivis

---

[1] suco *Detlefsen, codd.*: specillo *Mayhoff.*
[2] emissam d, *Mayhoff, qui ante addit*: emissum E, *Detlefsen, qui cum* cerebrum *iungit.*
[3] pellicula d, *Mayhoff*: pelliculae RE, *Detlefsen.*
[4] eo quod sub *Detlefsen, Mayhoff*: *varia codd.*
[5] aureum vel argenteum clostrum *Mayhoff, qui* claustrum *scribit*: aurum et argentum et clostrum (closirum, dosirum) *codd.*: electrum *pro* clostrum *Warmington.*
[6] inligatum *Detlefsen*: inligatam *Mayhoff*: inligatur *codd.*

to the eyes of babies straightens the eyelashes and makes them grow. Hernia is cured by the ash of snails applied for thirty days with frankincense in white of egg.[a] There are found in the little horns of snails sandy grits; worn as an amulet these make dentition easy. The ash of snail shells mixed with wax checks procidence of the end of the bowel, but the ash should be mixed with the discharge that exudes when the snails are pricked. A viper's brain tied on with a piece of his skin helps dentition. The same effect have also the largest teeth of serpents. The dung of a raven attached with wool as an amulet cures babies' coughs. Certain details can scarcely be included as serious items, but I must not omit them, since they have been put on record. As a remedy for hernia in babies they recommend a lizard; there should be taken a male, which can be recognised by its having one vent beneath the tail. The necessary ritual is: that it must bite the lesion through a gold or silver barrier; then it must be fastened in an unused cup and placed in smoke. Incontinence of urine in babies is checked by giving in their food boiled mice. The tall, indented horns of the beetle, fastened to babies, serves as an amulet. In the head of the boa is said to be a little stone, which is spit out by it when in fear of violent death; they add that dentition is wonderfully aided if the creature's head is cut off unawares, the stone extracted and worn as an amulet. The brain too of the same creature they recommend to be worn for the same purpose, or the stone or little bone found on the back of a slug. A splendid help also is the brain of a ewe rubbed on the gums, as for the ears is

---

[a] With Mayhoff's reading : "applied with a probe, etc."

inlitum sicut aures adeps anserinus cum ocimi suco inpositus. sunt vermiculi in spinosis herbis asperi, lanuginosi, hos adalligatos protinus mederi tradunt infantibus, si quid ex cibo haereat.

140 XLVIII. Somnos adlicit oesypum cum murrae momento in vini cyathis duobus dilutum, vel cum adipe anserino et vino myrtite, avis cuculus leporina pelle adalligatus, ardiolae rostrum in pelle asinina fronti adalligatum. putant et per se rostrum effectus eiusdem esse vino collutum. e diverso somnum arcet vespertilionis caput aridum adalligatum.

141 XLIX. In urina virili enecata lacerta venerem eius qui fecerit cohibet. nam inter amatoria esse Magi dicunt. inhibent et cocleae, fimum columbinum cum oleo et vino potum. pulmonis vulturini dextrae partes venerem concitant viris adalligatae gruis pelle, item si lutea ex ovis quinque columbarum admixto adipis suilli denarii pondere ex melle sorbeantur, passeres in cibo vel ova eorum, gallinacei dexter

142 testis arietina pelle adalligatus. ibium cinere cum adipe anseris et irino perunctis, si conceptus[1] sit, partus contineri, contra inhiberi venerem pugnatoris galli testiculis anserino adipe inlitis adalligatisque pelle arietina tradunt, item cuiuscumque galli, si cum sanguine gallinacei lecto subiciantur. cogunt concipere invitas saetae ex cauda mulae,[2] si iunctis

143 evellantur, inter se conligatae in coitu. qui in

---

[1] conceptus sit *vulg.*, *Detlefsen* : conceptos *Mayhoff* : conceptus *codd.*

[2] mulae *codd.* : muli et mulae *coni. Mayhoff.*

---

[a] If *nam* is " for," *amatoria* would have to mean " antaphrodisiacs."

goose grease put in them with juice of ocimum. On prickly plants are grubs which are rough and downy. These worn by babies as an amulet are said to effect an immediate recovery when part of their food sticks in the throat.

XLVIII. Sleep is induced by wool grease with a morsel of myrrh diluted in two cyathi of wine, or else with goose grease and myrtle wine, by the cuckoo bird in a piece of hare's fur worn as an amulet, or by a heron's beak worn as an amulet on the forehead in a piece of ass's hide. It is thought too that the beak of the heron by itself rinsed in wine has the same effect. Sleep is kept away, on the contrary, by a dried bat's head worn as an amulet. *Remedies for sleep.*

XLIX. A lizard drowned in a man's urine is anta-phrodisiac to him who passed it, but *a* the Magi claim that it is a love-philtre. Antaphrodisiac too are snails, and pigeon's dung taken with oil and wine. Aphrodisiac for men are the right parts of a vulture's lung, worn as an amulet in a piece of crane's skin; aphrodisiac also are the yolks of five pigeons' eggs mixed with a denarius by weight of pig fat and swallowed in honey, sparrows or their eggs in food, or the right testicle of a cock worn as an amulet in a piece of ram's-skin. They say that rubbing with ibis ash, goose grease and iris oil prevent miscarriage when there has been conception; that desire on the contrary is inhibited if a fighting cock's testicles are rubbed with goose grease and worn as an amulet in a ram's skin, as it also is if with a cock's blood any cock's testicles are placed under the bed. Women unwilling to conceive are forced to do so by hairs from the tail of a she-mule, pulled out during the animal copulation and entwined during the human. *Aphrodi-siacs, etc.*

urinam canis suam egesserit dicitur ad venerem
pigrior fieri. mirum et de stelionis cinere, si verum
est, linamento involutum in sinistra manu venerem
stimulare, si transferatur in dextram, inhibere, item
vespertilionis sanguinem collectum flocco subposi-
tumque capiti mulierum libidinem movere aut
anseris linguam in cibo vel potione sumptam.

144   L. Phthiriasim et totius corporis pota membrana
senectutis anguium triduo necat, serum exempto
caseo potum cum exiguo sale. caseos, si cerebrum
mustelae coagulo addatur, negant corrumpi vetustate
aut a muribus attingi. eiusdem mustelae cinis si
detur in offa gallinaceis et columbinis, tutos esse a
mustelis. iumentorum urinae tormina vespertilione
adalligato finiuntur, verminatio ter circumlato mediis
palumbe. mirum dictu, palumbis emissus moritur
iumentumque liberatur confestim.

145   LI. Ebriosis ova noctuae per triduum data in vino
taedium eius adducunt. ebrietatem arcet pecudum
assus pulmo praesumptus. hirundinis rostri cinis
cum murra tritus et vino quod bibetur inspersus
securos praestabit a temulentia. invenit Orus
Assyriorum rex.

146   LII. Praeter haec sunt notabilia animalium ad
hoc volumen pertinentium: gromphena—avem in
Sardinia narrant grui similem, ignotam iam etiam

A man who passes his urine on a dog's is said to become less sexually active. A wonderful thing again (if it is true) is told about the ash of the spotted lizard: if wrapped in a linen cloth and held in the left hand it is aphrodisiac; if transferred to the right hand it is antaphrodisiac. Another wonder: the blood of a bat, collected on a flock of wool and placed under the head of women, moves them to lust, as does the tongue of a goose, taken either in food or in drink.

L. The lice of phthiriasis even of the whole body are destroyed in three days by taking in drink the cast slough of a snake, or by drinking, with a little salt, whey after the cheese has been taken out. They say that if the brain of a weasel is added to rennet, cheeses neither go rotten through age nor are touched by mice. If the ash too of a weasel is given to poultry or pigeons in their mash, they are said to be safe from weasels. Pains of draught animals in making urine are ended by a bat put on them as an amulet, and bots by a wood-pigeon carried three times round their middle. Wonderful to relate, the wood-pigeon on being set free dies, while the animal is at once freed from pain.

Lice, maggots, etc.

LI. The eggs of an owl, given for three days in wine to drunkards, produce distaste for it. Drunkenness is kept away by taking early the roasted lung of sheep. A swallow's beak reduced to ash, beaten up with myrrh, and sprinkled on the wine that will be drunk, will free drinkers from fear of becoming tipsy. This is a discovery of Orus, king of Assyria.

Drunkenness, etc.

LII. In addition to all this there are some notable things about the animals that belong to this Book: the gromphena, a bird spoken of in Sardinia as like a crane,

371

Sardis, ut existimo—in eadem provincia ophion, cervis tantum pilo similis nec alibi nascens. idem auctores, nomen habere [1] sirulugum, quod nec quale esset animal nec ubi nasceretur tradiderunt. fuisse quidem non dubito, cum et medicinae ex eo sint demonstratae. M. Cicero tradit animalia biuros vocari qui vites in Campania erodant.

147 LIII. Reliqua mirabilia ex his quae diximus. non latrari a cane membranam e secundis canis habentem aut leporis fimum vel pilos tenentem, in culicum genere muliones [2] non amplius quam uno die vivere, eosque qui arborarii pici rostrum habeant et mella eximant ab apibus non attingi, porcos sequi eos a
148 quibus cerebrum corvi acceperint in offa, pulverem in quo se mula volutaverit corpori inspersum mitigare ardores amoris. sorices fugare,[3] si unus castratus emittatur, anguina pelle et sale et farre et serpyllo contritis una deiectisque cum vino in fauces boum uva maturescente, toto anno eos valere, vel si hirundinum pulli tres tribus [4] offis dentur, pulvere e vestigio anguium collecto sparsas apes in alvos reverti,
149 arietis dextro teste praeligato oves tantum gigni, non lassescere in ullo labore qui nervos ex alis et cruribus gruis habeant, mulas non calcitrare cum vinum biberint. ungulas tantum mularum repertas, neque

---

[1] nomen habere E r, *Detlefsen* : nominavere R d(?) *vulg.*: *om. Mayhoff.*

[2] *Post* muliones *lacunam indicat Mayhoff.*

[3] fugare *codd.* : fugere *Mayhoff.*

[4] tres tribus *codd.* : terni ternis *Mayhoff.*

but now, I think, unknown even to the Sardinians. In the same province we have the *ophion*, a creature like deer only in its hair, and found nowhere else. The same authorities say that there is a creature called *sirulugum*, but they have not told us what kind of an animal it is or where it is found. I do not indeed doubt that it once existed, since even medicines from it have been prescribed. Marcus Cicero tells us that there are animals called *biuri* which gnaw the vines in Campania.

LIII. There are still some wonders in the animals that I have mentioned: that a dog does not bark at a person having on him the membrane from the after-birth of a bitch, or holding the dung or hair of a hare; included among gnats are *muliones*, which live only for a day; those taking honey from hives are not stung by the bees if they have on them the beak of a woodpecker; pigs follow those from whom they have received in their mash the brain of a raven; the dust in which a she-mule has wallowed, sprinkled on the body, lessens the fires of love. Shrew mice are put to flight if one of them is castrated and let go free; if a snake's skin, salt, emmer wheat, and wild thyme are pounded together and with wine poured down the throat of oxen when the grapes are ripening, they enjoy good health for a whole year, or if three young swallows are given at three meals in their mash; if dust is gathered from the track of a snake and sprinkled on bees, these return to their hives; if the right testicle of a ram is tied up he begets ewes only; those are not wearied by any toil who have on them sinews from the wings and legs of a crane; she-mules do not kick if they have drunk wine. The hoofs of she-mules are the

Wonders of animals.

aliam ullam materiam quae non perroderetur a veneno Stygis aquae, cum id dandum Alexandro Magno Antipater mitteret, memoria dignum est magna Aristotelis infamia excogitatum. nunc ad aquatilia praevertemur.

ADDITIONAL NOTE TO P. 361

Pliny, XXX, 128: *vel mortua mirum si sine adservata transcenderit gravida innoxium fieri si protinus transcendat adservatam.* A tentative effort towards a solution of this passage is given by Warmington as follows. The sentence began *vel mortuam mirum* but was continued, in erratic copying, by a wrongly written clause (a) *si sine adservata transcenderit gravida* which was then imperfectly corrected into another clause (b) *si protinus transcendat adservatam* written in the margin. This marginal correction (b) was later copied out in its right place while clause (a) was still retained. It is clause (a) which is really corrupt and superfluous, and it should be deleted; and the whole passage may then be read: *vel mortuam mirum innoxiam fieri si protinus adservatam transcendat gravida*: "Or a dead one, wonderful to relate, does no harm if a pregnant woman crosses it if it was preserved without delay." Warmington suggests that a scribe began writing *mirum si sine mora adservatam transcendat* instead of *si protinus a. t.* At some later stage the intruded word *mora* was omitted but *sine* was still left in and *adservatam* was made into an ablative *adservata*. Thus *si protinus transcendat adservatam* or *si protinus adservatam transcendat* seems likely to be right. Anyhow to retain both clauses (a) and (b) seems intolerable; and (a) is more wrong than (b).

only material discovered that is not rotted by the poisonous water of Styx,[a] a notable fact discovered by Aristotle, to his great infamy, when Antipater sent a draught of it to Alexander the Great. Now I will pass to things found in water.[b]

[a] A fountain in Arcadia.
[b] Practically the whole of this chapter is in indirect speech, to denote the scepticism of Pliny.

BOOK XXXI

LIBER XXXI

1 I. Aquatilium secuntur in medicina beneficia, opifice natura ne in illis quidem cessante et per undas fluctusque ac reciprocos aestus amniumque rapidos cursus inprobas exercente vires, nusquam potentia maiore, si verum fateri volumus, quippe hoc elemen-2 tum ceteris omnibus imperat. terras devorant aquae, flammas necant, scandunt in sublime et caelum quoque sibi vindicant ac nubium obtentu vitalem spiritum strangulant, quae causa fulmina elidit, ipso secum discordante mundo. quid esse mirabilius potest aquis in caelo stantibus ? at illae, ceu parum sit in tantam pervenire altitudinem, rapiunt eo secum piscium examina, saepe etiam lapides subeuntque portantes aliena pondera. 3 eaedem cadentes omnium terra enascentium causa [1] fiunt prorsus mirabili natura, si quis velit reputare, ut fruges gignantur, arbores fruticesque vivant, in caelum migrare aquas animamque etiam herbis vitalem inde deferre, victa confessione [2] omnes terrae

[1] causa] *Mayhoff* (*Appendix p.* 485) causae *coni.*

[2] victa confessione dTa r *vulg.*: confessione victa VR, *Sillig*: iusta confessione *Caesarius*, *Mayhoff*: confessione invita *Urlichs*.

---

[a] Or, as such things as salt are included, "creatures of the water."

[b] English allows the plural "waters," but not exactly in the sense of the Latin *aquae*. Here it is perhaps safer to use the singular in translating.

BOOK XXXI

I. There follow the medicinal benefits obtained from aquatic animals;[a] Nature the Creator is not idle even among them, but puts forth her tireless strength on waves, billows, ebb and flow of tides, and the rapid currents of rivers; and nowhere with greater might, if we will but admit the truth, seeing that this element is lord over all the others. Water [b] swallows up the land, destroys flames, climbs aloft claiming the sovereignty even of the sky, and by a blanket of clouds chokes the life-giving spirit, so forcing out thunderbolts, the world waging civil war with itself. What can be more wonderful than water seated [c] in the sky? But as though it were a little thing to reach this great height, water sucks up thither with itself shoals of fish, and often even stones, carrying up aloft a weight other than its own. This element also falls again to become the source of all things that spring from the earth. Right wonderful action this on the part of Nature, if one considers it: in order that crops may grow, and that trees and shrubs may live, water soars to the sky and brings down thence even to plants the breath of life, so we are forced [d] to admit that all the powers of earth too

Remedies from aquatic animals.

[c] Literally: "standing."
[d] "The admission being constrained" is perhaps possible Plinian Latin. Of the emendations that of Urlichs seems the best, giving much the same sense.

379

quoque vires aquarum esse beneficii. quapropter
ante omnia ipsarum potentiae[1] exempla ponemus.
cunctas enim enumerare quis mortalium queat?

4 II. Emicant benigne passimque in plurimis terris
alibi frigidae, alibi calidae, alibi iunctae, sicut in
Tarbellis Aquitanica gente et in Pyrenaeis montibus
tenui intervallo discernente, alibi tepidae egelidaeque,[2] auxilia morborum profitentes et e cunctis
animalibus hominum tantum causa erumpentes.
augent numerum deorum nominibus variis urbesque
condunt, sicut Puteolos in Campania, Statiellas in
Liguria. Sextias in Narbonensi provincia, nusquam
tamen largius quam in Baiano sinu nec pluribus
5 auxiliandi generibus, aliae sulpuris vi, aliae aluminis,
aliae salis, aliae nitri, aliae bituminis, nonnullae
etiam acida salsave mixtura. vapore ipso aliquae
prosunt tantaque est vis, ut balneas calefaciant ac
frigidam etiam in solis fervere cogant. quae in Baiano
Posidianae vocantur nomine accepto a Claudii
Caesaris liberto obsonia quoque percocunt. vaporant
et in mari ipso quae Licinii Crassi fuere, mediosque
inter fluctus existit aliquid valetudini salutare.

III. Iam generatim nervis prosunt pedibusve aut
6 coxendicibus, aliae luxatis fractisve, inaniunt alvos,
sanant vulnera. capiti, auribus privatim medentur,

---

[1] potentiae R *vulg.*, *Mayhoff*: potentia *ceteri codd.*,
Detlefsen.
[2] egelidaeque *codd.* (*aut* gelidaequae) *Detlefsen*: egelidae
atque *Mayhoff.*

---

[a] The word *vis* is hard to translate, as it sometimes com-
bines the sense of "power," "quality," and "magical

are part of the beneficence of water. Wherefore I shall first of all give examples of the might of water, for what mortal man could count them all?

II. Everywhere in many lands gush forth benefi- *Various* cent waters, here cold, there hot, there both, as *waters, and* among the Tarbelli, an Aquitanian tribe, and in the *qualities.* Pyrenees, with only a short distance separating the two, in some places tepid and lukewarm, promising relief to the sick and bursting forth to help only men of all the animals. Water adds to the number of the gods by its various names, and founds cities, such as Puteoli in Campania, Statiellae in Liguria, and Sextiae in the province of Narbonensis. Nowhere however is water more bountiful than in the Bay of Baiae, or with more variety of relief: some has the virtue *a* of sulphur, some of alum, some of salt, some of soda, some of bitumen, some are even acid and salt in combination; of some the mere steam is beneficial, of which the power *a* is so great that it heats baths and even makes cold water boil in the tubs. The water called Posidian in the region of Baiae, getting its name from a freedman of Claudius Caesar, cooks thoroughly even meat. In the sea itself too, steam rises from the water that belonged to Licinius Crassus, and there comes something valuable to health in the very midst of the billows.

III. To come now to the classes of water: some *Classes of* waters are good for sinews *b* or feet, or for sciatica; *water.* others for dislocations or fractures; they purge the bowels; heal wounds; are specific for head, or for

property." In § 3 *vires* seems to be, not " strength " but " powers."
b The Latin *nervus* includes tendons, ligaments, and nerves. It is used of all fibrous tissues or membranes.

oculis vero Ciceronianae. dignum [1] memoratu, villa
est ab Averno lacu Puteolos tendentibus inposita
litori, celebrata porticu ac nemore, quam vocabat
M. Cicero Academiam ab exemplo Athenarum; ibi
compositis voluminibus eiusdem nominis, in qua et
monumenta sibi instauraverat, ceu vero non in toto
7 terrarum orbe fecisset. huius in parte prima exiguo
post obitum ipsius Antistio Vetere possidente
eruperunt fontes calidi perquam salubres oculis,
celebrati carmine Laureae Tulli, qui fuit e libertis
eius, ut protinus noscatur etiam ministeriorum
haustus ex illa maiestate ingenii. ponam enim ipsum
carmen, ubique et non ibi tantum legi dignum.[2]

8   Quo tua, Romanae vindex clarissime linguae,
      silva loco melius surgere iussa viret
    atque Academiae celebratam nomine villam
      nunc reparat cultu sub potiore Vetus,
    hoc etiam apparent lymphae non ante repertae
      languida quae infuso lumina rore levant.
    nimirum locus ipse sui Ciceronis honori
      hoc dedit, hac fontes cum patefecit ope.
    ut, quoniam totum legitur sine fine per orbem,
      sint plures oculis quae medeantur aquae.

IV. In eadem Campaniae regione Sinuessanae
aquae sterilitatem feminarum et virorum insaniam
9 abolere produntur, V. in Aenaria insula calculosis
mederi, et quae vocatur Acidula ab Teano Sidicino

---

[1] dignum *Mayhoff*: dignae (*cum antecedentibus*) *Detlefsen*:
digno, dignu, digna *codd.*
[2] dignum *Brakman*: del. *Detlefsen*: queat *Mayhoff*, add.
ut *ante* ubique. *Vulg.* dignum ubique, et non ibi tantum
legi.

ears; while the Ciceronian are so for the eyes. It is worth while recording that there is a country seat on the coast as you go from Lake Avernus to Puteoli, with a famous portico and grove, which M. Cicero, copying Athens, called Academia. There he wrote the volumes called Academica, and in it he also erected memorials to himself, as though indeed he had not done so throughout the whole world. In the front part of this estate, when the owner was Antistius Vetus, a short time after Cicero's demise there burst out hot springs, very beneficial for eye complaints, which have been made famous by a poem of Tullius Laurea, who was one of Cicero's freedmen. From it we at once realize that even his servants drew inspiration from that mighty genius. For I will quote the actual poem, which deserves to be read, not only on this site, but everywhere.

" O famous champion of our Latin tongue, where grows with a fairer green the grove you bade rise, and the villa, honoured by the name of Academe, Vetus keeps in repair under a more careful tendance, here are also to be seen waters not revealed before, which with drops infused relieve wearied eyes. For indeed the site itself gave this gift as an honour to Cicero its master, when it disclosed springs with this healing power, so that, since he is read throughout the whole world, there may be more waters to give sight to eyes."

IV. In Campania too are the waters of Sinuessa, which are said to cure barrenness in women and insanity in men. V. The waters in the island of Aenaria are said to cure stone in the bladder, as does also the water called Acidula—it is a cold one—four

ĪĪĪĪ p.   haec frigida, item in Stabiano quae Dimidia
vocatur, et in Venafrano ex fonte Acidulo.   idem contingit in Velino lacu potantibus, item in Syriae fonte
iuxta Taurum montem auctor est M. Varro et in
Phrygiae Gallo flumine Callimachus.   sed ibi in
potando necessarius modus, ne lymphatos agat, quod
in Aethiopia accidere his qui e fonte Rubro biberint
Ctesias scribit.

10   VI. Iuxta Romam Albulae aquae volneribus
medentur, egelidae hae, sed Cutiliae in Sabinis
gelidissimae suctu quodam corpora invadunt, ut prope
morsus videri possit, aptissimae stomacho, nervis,
universo corpori.

VII. Thespiarum fons conceptus mulieribus repraesentat, item in Arcadia flumen Elatum.   custodit
autem fetum Linus fons in eadem Arcadia abortusque
fieri non patitur.   e diverso in Pyrrha flumen quod
Aphrodisium vocatur steriles facit.

11   VIII. Lacu Alphio vitiligines tolli Varro auctor est,
Titiumque praetura functum marmorei signi faciem
habuisse propter id vitium.   Cydnus Ciliciae amnis
podagricis medetur, sicut apparet epistula Cassi
Parmensis ad M. Antonium.   contra aquarum culpa
12 in Troezene omnium pedes vitia sentiunt.   Tungri
civitas Galliae fontem habet insignem plurimis bullis
stillantem, ferruginei saporis, quod ipsum non nisi in
fine potus intellegitur.   purgat hic corpora, tertianas
febres discutit calculorumque vitia.   eadem aqua igne

384

miles from Teanum Sidicinum, that at Stabiae called
Dimidia, and the water of Venafrum from the spring
Acidulus. The same result comes from drinking the
water of Lake Velia, also of the Syrian spring near
Mount Taurus, according to Marcus Varro, and of the
Phrygian river Gallus, according to Callimachus. But
here moderation is necessary in drinking lest it drive
people to madness, which Ctesias writes those suffer
from who drink of the Red Spring in Aethiopia.

VI. Near Rome the waters of Albula heal wounds.
These are lukewarm, but those of Cutilia of the
Sabines are very cold, penetrating the body with a
sort of suction, so that they might seem almost to
bite, being very healthful to the stomach, the sinews,
and the whole body.

VII. The spring at Thespiae causes women to conceive, as does the river Elatum in Arcadia, and the
spring Linus, also in Arcadia, guards the embryo and
prevents miscarriage. The river in Pyrrha, on the contrary, that is called Aphrodisium, causes barrenness.

VIII. The water of Lake Alphius removes
psoriasis, Varro tells us, adding that Titius, an
ex-praetor, as a result of this complaint had a face
like that of a marble statue. The Cydnus, a river of
Cilicia, cures gout, as appears from a letter of Cassius
of Parma to M. Antonius. On the other hand, it is
the fault of the water in Troezen that everyone there
suffers from diseases of the feet. The Tungri, a
state of Gaul, has a remarkable spring that sparkles
with innumerable bubbles, with a taste of iron rust,
which yet cannot be detected until the water has
been drunk. It is a purgative, and cures tertian
agues and stone in the bladder. This water also,
if fire is brought near it, becomes turbid, and

385

admoto turbida fit ac postremo rubescit. Leucogaei fontes inter Puteolos et Neapolim oculis et vulneribus medentur. Cicero in admirandis posuit Reatinis tantum paludibus ungulas iumentorum indurari.

13 IX. Eudicus in Hestiaeotide fontes duos tradit esse, Ceronam ex quo bibentes oves nigras fieri, Nelea ex quo albas, ex utroque varias, Theophrastus Thuriis Crathim candorem facere, Sybarim nigritiam

14 bubus ac pecori, X. quin et homines sentire differentiam eam; nam qui e Sybari bibant nigriores esse durioresque et crispo capillo, qui e Crathi candidos mollioresque ac porrecta coma. item in Macedonia qui velint sibi candida nasci ad Haliacmonem ducere, qui nigra aut fusca ad Axium. idem omnia fusca nasci quibusdam in locis dicit et fruges quoque, sicut in Messapis, at in Lusis Arcadiae quodam fonte mures terrestres vivere et conversari. Erythris Aleos amnis pilos gignit in corporibus.

15 XI. In Boeotia ad Trophonium deum iuxta flumen Hercynnum [1] e duobus fontibus alter memoriam alter oblivionem adfert, inde nominibus inventis.

XII. In Cilicia apud oppidum Cescum rivus fluit Nuus, ex quo bibentium subtiliores sensus fieri M. Varro tradit, at in Cea insula fontem esse quo hebetes fiant, Zamae in Africa ex quo canorae voces.

---

[1] Hercynnum *Sillig*: *varia codd.*

---

[a] The Greek names are referred to.
[b] The Greek *νοῦς* means "intelligence."

finally turns red. White Earth Springs, between Puteoli and Naples, is good for complaints of the eyes and for wounds. Cicero in his *Book of Marvels* alleges that only by marsh water of Reate are the hoofs of draught cattle hardened.

IX. Eudicus tells us that in Hestiaeotis are two springs: Cerona, which makes black the sheep that drink of it, and Neleus, which makes them white, while they are mottled if they drink of each. Theophrastus says that at Thurii the Crathis makes oxen and sheep white, and the Sybaris makes them black. X. He adds that men too are affected by this difference: that those who drink of the Sybaris are darker and more hardy, and with curly hair, while those who drink of the Crathis are fair, softer, and with straight hair. He also says that in Macedonia those who wish white young to be born lead their beasts to the Haliacmon, but to the Axius if they wish the young to be black or dark. The same authority adds that in certain places all produce grows to be dark, even grain and vegetables, as among the Messapii, and that in a certain spring at Lusi in Arcadia land mice live and dwell. At Erythrae the river Axios makes hair grow on the body.

XI. In Boeotia by the temple of Trophonius near the river Hercynnus are two springs; one brings remembrance, the other forgetfulness; hence the names [a] that have been given them.

XII. In Cilicia near the town Cescum flows the river Nuus.[b] Those that drink of it become, says Marcus Varro, of keener perception, but on the island of Cea there is a spring that makes men dull, and at Zama in Africa is one that gives the drinkers a tuneful voice.

387

16   XIII. Vinum taedio venire his qui ex Clitorio lacu
biberint ait Eudoxus, set Theopompus inebriari fontibus his quos diximus, Mucianus Andri e fonte Liberi
patris statis diebus septenis eius dei vinum fluere, si
auferatur e conspectu templi, sapore in aquam trans-
17 euente, XIV. Polyclitus ex¹ Lipari iuxta Solos
Ciliciae ungui, Theophrastus, hoc idem in Aethiopia
eiusdem nominis fonte, Lycos in Indis ² Oratis fontem
esse cuius aqua lucernae luceant. idem Ecbatanis
traditur. Theopompus in Scotussaeis lacum esse
18 dicit qui volneribus medeatur,³ XV. Iuba in Trogo-
dytis lacum Insanum malefica vi appellatum ter die
fieri amarum salsumque ac deinde dulcem, totiensque
et noctu, scatentem albis serpentibus vicenum cubi-
torum, idem in Arabia fontem exilire tanta vi ut
19 nullum non pondus inpactum respuat, XVI. Theophrastus Marsyae fontem in Phrygia ad Celaenarum
oppidum saxa egerere. non procul ab eo duo sunt
fontes Claeon et Gelon ab effectu Graecorum
nominum dicti. Cyzici fons Cupidinis vocatur ex
quo potantes amorem deponere Mucianus credit.
20   XVII. Crannone est fons calidus citra summum
fervorem, qui vino addito triduo calorem potionis

---

¹ ex Lipari *Detlefsen*: Lipari *Urlichs*: expleri *codd.*
² in Indis *Mayhoff*: Indis *Detlefsen*: varia *codd.*
³ medeatur *C. F. W. Müller*: medetur *codd.*

ᵃ Book II. § 230.
ᵇ " The oily river."
ᶜ For these people see Book VI. § 75.

XIII. Disgust at wine, says Eudoxus, comes upon those who have drunk of Lake Clitorius, but Theopompus says that drunkenness is caused by the springs that I have mentioned,[a] and Mucianus that at Andros, from the spring of Father Liber, on fixed seven-day festivals of this god, flows wine, but if its water is carried out of sight of the temple the taste turns to that of water. XIV. Polyclitus says that with the river Liparis [b] near Soli in Cilicia people are anointed, Theophrastus says this of a spring with the same name in Aethiopia, and Lycos that among the Oratae [c] of India is a spring the water of which keeps lamps burning bright. The same is said of one at Ecbatana. Theopompus says that among the people of Scotussa is a lake that heals wounds. Juba says that among the Trogodytae is a lake called Insanus,[d] so named from its evil character, for three times each day and three times each night it becomes bitter, and then again fresh, full of white serpents twenty cubits long; he also says that in Arabia is a spring that bursts forth with such violence that it throws out everything, no matter how heavy, that is heaved into it. XVI. Theophrastus tells us that a spring of Marsyas in Phygia, near the town of Celaenae, casts out rocks. Not far from it are two springs, named Claeon and Gelon, so called from the force of their Greek [e] names. A spring at Cyzicus is called Cupid's Spring; those who drink of it, Mucianus believes, lose their amorous desires.

XVII. In Crannon is a hot spring which just falls short of boiling, the water of which with wine added remains in vessels a hot drink for three days. There

[d] " The lake of Madness."
[e] " Weeping " and " Laughing."

389

custodit in vasis. sunt et Mattiaci in Germania
fontes calidi trans Rhenum, quorum haustus triduo
fervet, circa margines vero pumicem faciunt aquae.

21 XVIII. Quod si quis fide carere ex his aliqua
arbitratur, discat in nulla parte naturae maiora esse
miracula, quamquam inter initia operis abunde multa
rettulimus. Ctesias tradit Silan[1] vocari stagnum in
Indis in quo nihil innatet, omnia mergantur, Coelius
apud nos in Averno etiam folia subsidere, Varro aves
quae advolaverint emori. contra in Africae lacu
22 Apuscidamo omnia fluitant, nihil mergitur, item in
Siciliae fonte Phinthia, ut Apion tradit, et in Medo-
rum lacu puteoque Saturni. item fluvii[2] fons
Limyrae transire solet in loca vicina portendens
aliquid, mirumque quod cum piscibus transit. re-
sponsa ab his petunt incolae cibo, quem rapiunt
adnuentes, si vero eventum negent, caudis abigunt.
23 amnis Alcas in Bithynia Bryazum adluit—hoc est
templo et deo nomen—cuius gurgitem periuri negantur[3] pati[4] velut flammam urentem.[5] in Cantabria
fontes Tamarici in auguriis habentur. tres sunt
octonis pedibus distantes, in unum alveum coeunt
24 vasto amne. singuli siccantur duodenis diebus,

---

[1] Silan *Mayhoff* (*Strabo XV*. 1, 38): Siden r *Sillig.*
[2] puteoque Saturni. item fluvii *Mayhoff*: puteoque.
Saturni templum *Detlefsen*: temthuni r: themtumi V:
themtuni T: templum E.
[3] negantur VRdT *Hard.*, *Mayhoff*: necantur a *Detlefsen*:
notantur *Hermolaus Barbarus.*
[4] pati VRdT: parthi E: rapti *Detlefsen.*
[5] flammam urentem *codd.*: flamma urente *Detlefsen.*

---

[a] Wiesbaden. [b] See II. §§ 224 foll.

are also in Germany across the Rhine the hot springs of Mattiacum,[a] a draught from which is boiling hot for three days; around the borders indeed the water forms pumice.

XVIII. But if anybody thinks that some of these statements are incredible, he has to learn that in no sphere does Nature show greater marvels, although in the early parts of my work I have mentioned [b] plenty of examples. Ctesias tells us that there is in India standing water called Silas,[c] in which nothing floats but everything sinks to the bottom; Coelius says in our Avernus even leaves sink, and Varro that the birds that fly to it die. On the other hand, in the African lake Apuscidamus everything floats and nothing sinks; similarly in the Sicilian spring Phinthia, as Apion tells us, and among the Medes in the lake and well of Saturn. Again, the source of the river Limyra often crosses to neighbouring districts, indicating some portent, and a wonderful thing is that the fish cross with it. The inhabitants seek responses from them, offering food. To give a favourable answer the fish snap it up; but for an unfavourable one, they knock it away with their tails. The river Alcas in Bithynia flows by Bryazus—this is the name both of a god and of his temple—the current of which perjured persons are said to be unable to endure, as it burns like a flame. In Cantabria the springs of the Tamaris are supposed to be prophetic. Three in number they are eight feet apart, uniting in one channel to form a vast river. Each one dries up for periods of twelve, occasionally of twenty days, without the slightest trace of water,

The marvels of many waters.

---

[c] A reference to Strabo shows that Mayhoff's conjecture is correct, but Strabo calls the Silas a river.

aliquando vicenis,[1] citra suspicionem ullam aquae, cum sit vicinus illis fons sine intermissione largus. dirum est non profluere eos aspicere volentibus, sicut proxime Larcio Licinio legato pro praetore post septem dies accidit. In Iudaea rivus sabbatis omnibus siccatur.

25 XIX. E diverso miracula alia dira. Ctesias in Armenia fontem esse scribit, ex quo nigros pisces ilico mortem adferre in cibis quod et circa Danuvii exortum audivi, donec veniatur ad fontem alveo adpositum, ubi finitur id genus piscium ideoque ibi caput amnis eius intellegit fama. hoc idem et in 26 Lydia in stagno Nympharum tradunt. In Arcadia ad Pheneum aqua profluit e saxis Styx appellata, quae ilico necat, ut diximus, sed esse pisces parvos in ea tradit Theophrastus, letales et ipsos, quod non in 27 alio genere mortiferorum fontium. necari aquis Theopompus et in Thracia apud Cychros dicit, Lycos in Leontinis tertio die quam quis biberit, Varro ad Soracten in fonte, cuius sit latitudo quattuor pedum. sole oriente eum exundare ferventi similem, aves quae degustaverint iuxta mortuas iacere. namque et haec insidiosa conditio est quod quaedam etiam blandiuntur aspectu, ut ad Nonacrim Arcadiae, omnino enim nulla deterrent qualitate. hanc putant nimio frigore esse noxiam, utpote cum pro-28 fluens ipsa lapidescat. aliter circa Thessalica Tempe, quoniam virus omnibus terrori est, traduntque aena

---

[1] singuli siccantur duodenis diebus, aliquando vicenis *Mayhoff*: siccantur duodecies singulis diebus, aliquando vicies *Detlefsen*: *varia codd.*

---

[a] Perhaps " black."   [b] Book II. § 231.

although there is a copious spring near them that never dries up. It is an evil portent if those wishing to look at them find them not flowing, as recently Larcius Licinius, a legate pro-praetore discovered after seven days. In Judaea is a stream that dries up every Sabbath.

XIX. On the other hand some other marvels are deadly. Ctesias writes that in Armenia is a spring in which are dark *a* fish that, eaten as food, bring instant death, as I have heard do the fish also from the water around the rising of the Danube, until a spring is reached close to the main channel, where the fish of this sort go no further. At this point, therefore, report says is the real source of that river. They tell us that this same phenomenon occurs in Lydia in the marsh of the Nymphs. In Arcadia near the Pheneus there flows from the rocks a stream called Styx, which I have said *b* proves instantly fatal to life, but Theophrastus tells us that in it are small fish equally deadly; no other kind of poisonous spring is like this. Theopompus also says that near Cychri in Thrace are deadly waters, Lycos that at Leontini is water that kills on the third day after drinking, and Varro that on Soracte is poisonous water in a spring four feet wide. At sunrise, he adds, this bubbles out as though it boiled, and birds that have tasted it lie dead close by. For certain waters have also this insidious property, that the very prospect is attractive; as at Nonacris in Arcadia, which has nothing at all about it to serve as a warning. They think that this water harms by its excessive cold, seeing that as it flows it itself turns to stone. It is otherwise around Tempe in Thessaly, for its poison is a terror to everyone, and they tell us that by the

Deadly waters.

PLINY: NATURAL HISTORY

etiam ac ferrum erodi illa aqua. profluit, ut indicavimus, brevi spatio, mirumque siliqua silvestris
amplecti radicibus fontem eum dicitur semper florens
purpura. et quaedam sui generis herba in labris
fontis viret. In Macedonia, non procul Euripidis
poetae sepulchro, duo rivi confluunt. alter salu-
berrimi potus, alter mortiferi.

29   XX. In Perperenis fons est quamcumque rigat
lapideam faciens terram, item calidae aquae in
Euboeae Adepso. nam quae [1] adit [2] rivus saxa in
altitudinem crescunt. in Eurymenis deiectae coro-
nae in fontem lapideae fiunt. in Colossis flumen est
quo lateres coniecti lapidei extrahuntur. in Scyre-
tico metallo arbores quaecumque flumine adluuntur
30 saxeae fiunt cum ramis. destillantes quoque guttae
lapide durescunt in antris, conchatis ideo,[3] Miezae in
Macedonia etiam pendentes in ipsis camaris, at in
Corinthio [4] cum cecidere, in quibusdam speluncis
utroque modo, columnasque faciunt, ut in Phausia
Cherrhonesi adversae Rhodo in antro magno etiam
discolori aspectu. et hactenus contenti simus exemplis.

31   XXI. Quaeritur inter medicos cuius generis aquae
sint utilissimae. stagnantes pigrasque merito dam-

[1] quae E *Detlefsen, Mayhoff*: qua *plerique codd., Hard.*
[2] adit E *Mayhoff*: cadit *plerique codd. Hard.*: alluit *vulg.*:
adluit *Detlefsen.*
[3] conchatis *Mayhoff, coll. XI.* § 270: coricis *codd.*: Coryciis
vulg.; *fortasse* ideo *ex* lapide *est ortum.*
[4] Corinthio R *Ianus*: Corintio VdTf: coricio E: Corycio
Sillig, Mayhoff.

---

[a] Book IV. § 31.
[b] A *locus adhuc corruptus* says Mayhoff. I adopt his conjecture with certain doubts, for unless we discard in the next

394

water there even bronze and iron are corroded. It flows, as I have pointed out,[a] for only a short distance, and a marvellous thing is related of this spring: it is embraced by the roots of a wild carob always bearing purple blossom. And a unique kind of herb flourishes on the margins of the spring. In Macedonia, not far from the tomb of the poet Euripides, two streams join, one very wholesome to drink, the other a deadly poison.

XX. At Perperena is a spring that turns to stone whatever land it irrigates, as do also the hot waters at Aedepsus in Euboea, for, whatever rocks the stream reaches increase in height. At Eurymenae chaplets, thrown into a spring, turn to stone. At Colossae is a river, and bricks when cast into it are of stone when taken out. In Scyros in the mine all the trees watered by the river are turned to rock, branches and all. Drops too dripping from the stone harden in certain caves, and hence these are concave in shape.[b] But at Mieza in Macedonia the drops actually hang from the arched roofs, while in the Corinthian cave they petrify after falling; in certain caverns the stone forms in both ways and makes pillars, as at Phausia in the Chersonesus opposite to Rhodes in a huge cave, where the pillars are actually of different colours to look at. These examples must be enough for the present. *Petrifying waters, stalactites and stalagmites.*

XXI. It is a question debated by the physicians what kinds of water are most beneficial. They *The most beneficial waters.*

sentence a well attested reading *Corinthio* (*-tio*) there will be no reference to the famous Corycian cavern. It seems just possible that an absent-minded scribe repeated *lapide* (or part of it) after *Coryciis*, and that the vulgate, which omits *ideo*, is correct, or nearly so.

nant, utiliores quae profluunt existimantes, cursu
enim percussuque ipso extenuari atque proficere,
eoque miror cisternarum ab aliquis maxime probari.
sed hi rationem adferunt, quoniam levissima sit
imbrium, ut quae subire potuerit ac pendere in aere.
32 ideo et nives praeferunt, nivibusque etiam glaciem
velut ad infinitum coacta subtilitate. leviora enim
haec esse et glaciem multo leviorem aqua. horum sen-
tentiam refelli interest vitae. in primis enim levitas
illa deprehendi aliter quam sensu vix potest, nullo
paene momento ponderis aquis inter se distantibus.
nec levitatis in pluvia aqua argumentum est subisse
eam in caelum, cum etiam lapides subire appareat
cadensque inficiatur halitu terrae, quo fit ut pluviae
aquae sordium plurimum inesse sentiatur citissime-
33 que ideo calefiat aqua pluvia. nivem quidem
glaciemque subtilissimum elementi eius videri miror
adposito grandinum argumento, e quibus pestilentis-
simum potum esse convenit. nec vero pauci inter
ipsos e contrario ex gelu ac nivibus insaluberrimos
potus praedicant, quoniam exactum sit inde quod
tenuissimum fuerit. minui certe liquorem omnem
congelatione deprehenditur et rore nimio scabiem
fieri, pruina uredinem, cognatis et nivis causis.
34 pluvias quidem aquas celerrime putrescere convenit

<hr>

[a] The opposite is the truth.

rightly condemn stagnant and sluggish waters, holding that running water is more beneficial, as it is made finer and more healthy by the mere agitation of the current. For this reason I am surprised that some physicians recommend highly water from cisterns. But these physicians put forward a reason; the lightest water, they say, is rain-water, seeing that it has been able to rise and to be suspended in the atmosphere. Therefore they also prefer snow and ice even more than snow, as though its texture were rarefied to the utmost; for, they say, snow and ice are lighter than water, and ice much lighter. To refute this view is a matter that is important to all men. For first of all, this lightness of water can be discovered with difficulty except by sensation, as the kinds of water differ practically nothing in weight. Nor is it proof of the lightness of rain water that it rose to the sky, since even stones are seen to do the same, and as it falls it is infected with exhalations from the earth. Hence it comes about that rain-water is found to be full of dirt, for which reason this water becomes hot very quickly. That snow indeed and ice should be considered the finest form of that element makes me wonder, when I have before me the evidence of hailstones, to drink the water of which it is agreed is most unwholesome. Not a few physicians however themselves maintain that hail and snow on the contrary make very unhealthy drink, since there has been taken from it what was its thinnest part. Certainly it is found that every liquid becomes smaller when frozen,[a] that too much dew brings blight, and hoar frost blast, effects caused by snow also being akin. Rain-water, it is agreed, becomes putrid very quickly, and it is the worst

minimeque durare in navigatione. Epigenes autem
aquam quae septies putrefracta purgata sit tradit[1]
amplius non putrescere. nam cisternas etiam medici
confitentur inutiles alvo duritia faucibusque, etiam
limi non aliis inesse plus aut animalium quae faciunt
35 taedium. at iidem[2] confitendum habent nec statim
amnium utilissimas esse, sicuti nec torrentium ullius,
lacusque plurimos salubres. quaenam igitur et cuius
generis aptissimae? aliae alibi. Parthorum reges
ex Choaspe et Eulaeo tantum bibunt, hae quamvis in
longinqua comitantur illos. sed horum placere non
quia sint amnes apparet, quoniam neque e Tigri neque
Euphrate, neque e multis aliis bibunt.

36   XXII. Limus aquarum vitium est. si tamen idem
amnis anguillis scateat, salubritatis indicium habetur,
sicuti frigoris taeneas in fonte gigni. ante omnia
autem damnantur amarae et quae sorbentem statim
implent, quod evenit Trozene. nam nitrosas atque
salmacidas in desertis Rubrum mare petentes addita
polenta utiles intra duas horas faciunt ipsaque vescuntur polenta. damnantur in primis quae fonte
caenum faciunt quaeque malum colorem bibentibus,
refert et si vasa aerea inficiunt aut si legumina tarde
percocunt, si liquatae lentiter[3] terram relinquunt
37 decoctaeque crassis obducunt vasa crustis. est
etiamnum vitium non fetidae modo verum omnino
quicquam resipientis, iucundum sit illud licet gratum-

---

[1] tradit *coni. Mayhoff*: perhibet R (?) *Detlefsen*, " *contra
Plinii usum* " (*Mayhoff*).
[2] at iidem *coni. Mayhoff*, item *scribit*; *om. codd. et Detlefsen*.
[3] lente *coni. Warmington*.

water to stand a voyage. Epigenes, however, says
that water which has become putrid and been purified
seven times becomes putrid no more. But cistern
water even physicians admit is harmful to the bowels
and throat because of its hardness, and no other
water contains more slime or disgusting insects. Yet
it must be admitted, they hold, that river water is not
ipso facto the most wholesome, nor yet that of any
torrent whatsoever, while there are very many lakes
that are wholesome. What water then, and of what
kind, is the best? It varies with the locality. The
kings of Parthia drink only of the Choaspes and the
Eulaeus; water from these rivers is taken with them
even into distant regions. But it is clear that
the water of these rivers does not find favour just
because they are rivers, for the kings do not drink
from the Tigris, Euphrates, or many other rivers.

XXII. Slime in water is bad. If however the same
river is full of eels, it is held to be a sign of wholesomeness, as it is of coldness for worms to breed in a
spring. But before all are condemned bitter waters,
and those that give a full feeling immediately after
drinking, as does the water at Troezen. But the
nitrous and salty-acid streams that in the desert
flow to the Red Sea are made sweet within two hours
if pearl barley is added, and the barley itself they
eat. Especially are condemned waters that have mud
at their source, and those that give a bad colour
to those who drink of them. It also makes a difference
if water stains bronze vessels, or if it cooks greens
slowly, if when gently filtered out it leaves a sediment
of earth, or when boiled thickly encrusts the vessel.
Not only too is fetid water bad, but also that which
tastes of anything at all, though the taste may be

que et ut saepe ad viciniam lactis accedens. aquam
salubrem aeris quam simillimam esse oportet. unus
in toto orbe traditur fons aquae iucunde olentis in
Mesopotamia Chabura. fabulae rationem adferunt,
quoniam eo Iuno perfusa sit. de cetero aquarum
salubrium sapor odorve nullus esse debet.

38   XXIII. Quidam statera iudicant de salubritate,
frustrante diligentia, quando perrarum est ut levior
sit aliqua. certior subtilitas inter pares meliorem
esse quae calefiat refrigereturque celerius. quin. et
haustam vasis, † ne manus pendeant,[1] depositisque †
in humum tepescere adfirmant. ex quonam ergo
genere maxime probabilis continget? puteis nimi-
rum, ut in oppidis video constare, sed his quibus et
exercitationis ratio crebro haustu contingit et illa
39   tenuitas colante terra. salubritati haec satis sunt.
frigori et opacitas necessaria utque caelum videant.
super omnia una observatio—eadem et ad perennita-
tem pertinet—ut illa e vado exiliat vena, non e lateribus. nam ut tactu gelida sit etiam arte contingit, si
expressa in altum aut e sublimi deiecta verberatum
corripiat aera. in natando quidem spiritum con-
40   tinentibus frigidior sentitur eadem. Neronis principis

---

[1] ne manus pendeant *codd.*: ne manus suspendant *Detlefsen*:
ne manu pendeant *Mayhoff, qui post* vasis *add.* portatis.

---

a See Additional Note H (p. 567).

pleasant and agreeable, or, as often happens, approaching that of milk. Wholesome water ought to be very like air. In the whole world one spring of water only is said to have a pleasant smell, and that is at Chabura in Mesopotamia; a reason is sought in the legend that with it Juno was bathed. Apart from this wholesome water should have no sort of taste or smell.

XXIII. Some judge the wholesomeness of water by means of the balance. This is wasted carefulness, for it is very rare for one water to be lighter than another. A more reliable and a delicate test is that, other things being equal, a water is better that becomes warm and cool more quickly. Moreover we are told that if drawn in vessels [without being weighed, *or* without being warmed by the hand] *a* and placed on the ground, the better water becomes warm. From what source then shall we obtain the most commendable water? From wells surely, as I see they are generally used in towns, but they should be those the water of which by frequent withdrawals is kept in constant motion, and those where due thinness is obtained by filtering through the earth. For wholesomeness so much suffices; for coolness both shade is necessary and that the well should be open to the air. One point above all must be observed— and this is also important for a continuous flow—well water should issue from the bottom, not the sides. But coolness to the touch can also be obtained artificially, if the water is forced aloft or let fall from a height, beating and absorbing the air. In swimming indeed the same water is felt to be cooler by those who hold their breath. It was a discovery of the Emperor Nero to boil water and cool it in a glass

inventum est decoquere aquam vitroque demissam in
nives refrigerare. ita voluptas frigoris contingit sine
vitiis nivis. omnem utique decoctam utiliorem esse
convenit, item calefactam magis refrigerari, subti-
lissimo invento. vitiosae aquae remedium est, si
decoquatur ad dimidias partes. aqua frigida ingesta
sistitur sanguis. aestus in balneis arcetur, si quis ore
teneat. quae sint haustu frigidissimae non perinde
et tactu esse, alternante hoc bono, multi familiari
exemplo colligunt.

41 XXIV. Clarissima aquarum omnium in toto orbe
frigoris salubritatisque palma praeconio urbis Marcia
est inter reliqua deum munera urbi tributa. vocabatur haec quondam Aufeia, fons autem ipse Pitonia.
oritur in ultimis montibus Paelignorum, transit Mar-
sos et Fucinum lacum, Romam non dubie petens.
mox in specus mersa in Tiburtina se aperit novem
milibus passuum fornicibus structis perducta. primus
eam in urbem ducere auspicatus est Ancus Marcius
unus e regibus, postea Q. Marcius Rex in praetura,
rursusque restituit M. Agrippa.

42 XXV. Idem et Virginem adduxit ab octavi lapidis
diverticulo duo milia passuum Praenestina via. iuxta
est Herculaneus rivus, quem refugiens Virginis
nomen obtinuit. horum amnium comparatione
differentia supra dicta deprehenditur, cum quantum

vessel by thrusting it into snow. In this way is obtained a pleasant coolness without the injurious qualities of snow. At any rate it is agreed that all water is more serviceable when boiled, and that water which has been heated can be cooled to a greater degree—a most clever discovery. It purifies bad water to boil it down to one half. Cold water taken internally checks bleeding, and to hold it in the mouth prevents overheating in the bath. Water that is very cold to swallow is not always so to the touch; this good quality alternates,[a] as many find out by personal experience.

XXIV. The first prize for the coolest and most wholesome water in the whole world has been awarded by the voice of Rome to the Aqua Marcia, one of the gods' gifts to our city. This was once called the Aqua Aufeia, and the source itself Aqua Pitonia. It rises at the extreme end of the Paelignian range, crosses the country of the Marsi and the Fucine lake, plainly making straight for Rome. Next it sinks into the underground caves near Tibur, reappearing and completing its journey of nine more miles along an aqueduct. The first to begin the bringing of this water to Rome was one of the kings, Ancus Marcius; later, repairs were carried out by Quintus Marcius Rex in his praetorship, and again by Marcus Agrippa.

XXV. The same Agrippa also brought the Virgin Water to Rome from the bye-road, eight miles away, that extends two miles along the road to Praeneste. Nearby is the stream of Hercules, and because the Virgin Water runs away from this it was so named. A comparison of these rivers illustrates the difference

---

[a] We might say: " and *vice versa*."

Virgo tactu praestet, tantum praestet Marcia haustu,
quamquam utriusque iam pridem urbi perit voluptas,
ambitione avaritiaque in villas ac suburbana detor-
quentibus publicam salutem.

43 XXVI. Non ab re sit quaerendi aquas iunxisse
rationem. repperiuntur in convallibus maxime et
quodam convexitatis cardine aut montium radicibus.
multi septentrionales ubique partes aquosas existimavere, qua in re varietatem naturae aperuisse con-
veniat. in Hyrcanis montibus a meridiano latere non
pluit, ideo silvigeri ab aquilonis tantum parte sunt.
at Olympus, Ossa, Parnasus, Appenninus, Alpes
undique vestiuntur amnibusque perfunduntur, aliqui
ab austro, sicut in Creta Albi montes. nihil ergo in
his perpetuae observationis iudicabitur.

44 XXVII. Aquarum sunt notae iuncus [1] et herba de
qua dictum est multumque alicui loco pectore incu-
bans rana. salix enim erratica et alnus aut vitex aut
harundo aut hedera sponte proveniunt et conrivatione
aquae pluviae in locum humiliorem e superioribus
defluentis, augurio fallaci, certiore multo nebulosa
exhalatione ante ortum solis longius intuentibns,
quod quidam ex edito speculantur proni terram
45 adtingente mento. est et peculiaris aestimatio
peritis tantum nota, quam ferventissimo aestu secuntur dieique horis ardentissimis, qualis ex quoque loco
repercussus splendeat. nam si terra sitiente umidior

[1] *Post* iuncus *add.* aut harundo *codd. Cf. infra.*

---

[a] See § 40.
[b] This is bechion (tussilago); see XXVI. § 30.

mentioned above; [a] for the Aqua Marcia is as much superior to swallow as the Virgin is cool to touch. And yet Rome has long since lost the delights of each, for love of display and greed have diverted these means of public health to country seats and suburbs.

XXVI. It would be pertinent to add the method of searching for water. It is found mostly in enclosed valleys, and what may be called the hinge of converging slopes, or at the foot of mountains. Many have thought that everywhere the northern are the watery slopes. On this matter it would be well to point out the variableness of Nature. In the Hyrcanian mountains it does not rain on the southern slope, and so only on the north side are there woods. But Olympus, Ossa, Parnassus, the Apennines, and the Alps, are everywhere covered with trees and watered by rivers; others are so only on the south side, as are the White Mountains in Crete. So in this matter there will be no unvarying rule to follow. *Water-finding.*

XXVII. Signs of the presence of water are rushes, the plant about which I have spoken,[b] and frogs squatting on their chest in great numbers for any one place. For wild willow, alder, vitex, reed, or ivy, which grow spontaneously and where there is a settling of rain-water flowing from higher regions to one lower down, are deceptive indications; one much more reliable is a misty steam, visible from a distance before sunrise, for which some water-finders watch from a height, lying prone with their chin touching the earth. There is also a special sign, known only to experts, which they look for in the hottest season and in the most blazing heat of the day, the nature of the reflection that shines from each locality. For if one spot looks moister while the earth around is

46 est ille, indubitata spes promittitur. sed tanta
oculorum intentione opus est ut indolescant. quod
fugientes ad alia experimenta decurrunt, loco in
altitudinum pedum quinque defosso ollisque e figlino
opere crudis aut peruncta pelvi aerea,[1] cooperto,[2]
lucernaque ardente concamarata frondibus, dein
terra, si figlinum umidum ruptumve, aut in aere
sudor vel lucerna sine defectu olei restincta aut etiam
vellus lanae madidum repperiatur, non dubie
promittunt aquas. quidam et igni prius excocunt
locum tanto efficaciore vasorum argumento.

47 XXVIII. Terra vero ipsa promittit candicantibus
maculis aut tota glauci coloris. in nigra enim
scaturigines non fere sunt perennes. figularis creta
semper adimit spes, nec amplius puteum fodiunt
coria terrae observantes, ut a nigra descendat ordo
48 supra dictus. aqua semper dulcis in argillosa terra,
frigidior in tofo. namque et hic probatur, dulces
enim levissimasque facit et colando continet sordes.
sabulum exiles limosasque promittit, glarea incertas
venas, sed boni saporis, sabulum masculum et harena
carbunculus certas stabilesque et salubres, rubra
saxa optimas speique certissimae, radices montium
saxosae et silex hoc amplius rigentes. oportet autem

---

[1] *Post* aerea *add.* lanae vellere *Mayhoff.*
[2] *Post* terra *trans.* cooperto *Detlefsen.*

---

[a] Mayhoff adds *lanae vellere* after *aerea,* comparing passages
in Vitruvius, Palladius, and *Geoponica.* The asyndeton is
awkward, and perhaps Pliny omitted to mention the wool in
his first list, and when he came across it again in the second
list, did not think it necessary for the sense to go back and
add it to the previous clause.

parching, that is an infallible sign. But so great is the necessary strain on the eyes that pain results. To avoid this strain they have recourse to other tests. They dig a hole to the depth of five feet, covering it with jars of unbaked potters' clay, or else with a well-oiled bronze basin, and also a burning lamp arched over with foliage and earth on top; if the clay is found to be wet or broken, or if moisture covers the bronze, or the lamp goes out without any failure of oil, or perchance a flock of wool is wet,[a] then the finding of water is assured. Some also light a fire first and dry the hole, making yet more conclusive the evidence of the vessels.

XXVIII. The earth however itself guarantees water by white spots or by being green all over. For in black earth the springs are generally not permanent. Potters' clay always dashes hopes of water, and further well-digging ceases when it is observed that the earth's *strata* begin with black and go down in the order given above.[b] Water in clay is always sweet, but cooler in tufa. For tufa too is commended, for it makes water sweet and very light; acting as a strainer it keeps back any dirt. Loam [c] indicates scanty trickles with slime, gravel intermittent springs but of a good flavour, male loam [d] or carbunculus-sand [e] continuous streams, steady and wholesome; red rock points to the certain presence of excellent water; the rocky bases of mountains, or flint, point to the same kind of water, with great

---

[b] Apparently black, white, green.

[c] *Sabulum,* apparently soil containing coarse sand and clay.

[d] *Sabulum masculum* was coarse *sabulum.*

[e] See Varro I. 9, 2; earth so scorched by the sun that roots are charred.

fodientibus umidiores adsidue respondere glaebas
49 faciliusque ferramenta descendere. depressis puteis
sulpurata vel aluminosa occurrentia putearios necant.
experimentum huius periculi est demissa ardens
lucerna si extinguitur, tunc secundum puteum dextra
ac sinistra fodiuntur aestuaria quae graviorem illum
halitum recipiant. fit et sine his vitiis altitudine
ipsa gravior aer quem emendant adsiduo linteorum
iactatu eventilando. cum ad aquam ventum est, sine
50 harenato opus surgit ne venae obstruantur. quaedam aquae vere statim incipiente frigidiores sunt,
quarum non in alto origo est—hibernis enim constant
imbribus—quaedam a canis ortu, sicut in Macedoniae
Pella utrumque. ante oppidum enim incipiente
aestate frigida est palustris, dein maximo aestu in
excelsioribus oppidi riget. hoc et in Chio evenit
simili ratione portus et oppidi. Athenis Enneacrunos
nimbosa aestate frigidior est quam puteus in Iovis
horto, at ille siccitatibus riget. maxime autem putei
circa arcturum non ipsa aestate deficiunt, omnesque
quatriduo eo subsidunt, iam vero multi hieme tota, ut
51 circa Olynthum, vere primum aquis redeuntibus. in
Sicilia quidem circa Messanam et Mylas hieme in
totum inarescunt fontes, ipsa aestate exundant am-
nemque faciunt. Apolloniae in Ponto fons iuxta

---

coolness in addition. But as the diggers go deeper, the clods should prove continually moister, and the spades cut down more easily. When wells have been sunk deep, the well-diggers are killed if they meet with sulphurous or aluminous fumes. A test for this danger is to let down a lighted lamp and see if it goes out. If it does, vent-holes are sunk at the side of the well, on the right and on the left, to take off the oppressive gas. Apart from these injurious substances, mere depth makes the air oppressive; it is dissipated by continuous fanning with linen cloths. When water has been reached, walls are built from the bottom no cement being used lest the springs be dammed up. Some water, the source of which is not at a height, is cooler right from the beginning of spring—for it is made up of winter rain—some is cooler after the rising of the Dog-star *a*; in Macedonia at Pella are both kinds. For before the town there is a marsh stream that is cold at the beginning of summer; then in the higher parts of the town the water is very cold even in the height of summer. A similar phenomenon occurs in Chios also, the relative position of harbour and town being the same. At Athens, Enneacrunos in a cloudy summer is cooler than the well in the Garden of Juppiter, while this latter is very cold during summer droughts. Wells however generally run dry about Arcturus,*b* not in the actual summer, and all sink low during the four days of its rising. Moreover many wells fail throughout the winter, as those around Olynthus, the water returning first in the spring. In Sicily indeed, in the region of Messana and Mylae, springs in winter dry up altogether, but in the actual summer overflow and form rivers. At Apollonia in Pontus a spring

Wells and well-digging.

Cool waters.

409

mare aestate tantum superfluit et maxime circa canis
ortum, parcius, si frigidior sit aestas. quaedam
terrae imbribus sicciores fiunt, velut in Narniensi
agro, quod admirandis suis inseruit M. Cicero,
siccitate lutum fieri prodens, imbre pulverem.

52 XXIX. Omnis aqua hieme dulcior est, aestate
minus, autumno minime, minusque per siccitates.
neque aequalis amnium plerumque gustus est magna
alvei differentia. quippe tales sunt aquae qualis
terra per quam fluunt qualesve herbarum quas lavant
suci. ergo idem amnes parte aliqua repperiuntur
insalubres. mutant saporem et influentes rivi, ut
Borysthenen, victique diluuntur. aliqui vero et
imbre mutantur. ter accidit in Bosporo ut salsi
deciderent necarentque frumenta, totiens et Nili
rigua pluviae amara fecere magna pestilentia Aegypti.

53 XXX. Nascuntur fontes decisis plerumque silvis,
quos arborum alimenta consumebant, sicut in Haemo
obsidente Gallos Cassandro, cum valli gratia silvas
cecidissent. plerumque vero damnosi torrentes con-
rivantur detracta collibus silva continere nimbos ac
digerere consueta. et coli moverique terram callum-
que summae cutis solvi aquarum interest. proditur
certe in Creta expugnato oppido quod vocabatur
Arcadia cessasse fontes amnesque qui in eo situ multi
erant rursus condito post sex annos emersisse, ut
quaeque coepissent partes coli.

---

a Or: " disperse ".

near the sea is flooded only in summer, and especially about the rising of the Dog-star, but less so if the summer is colder than usual. Certain lands become drier in rainy weather, as the region of Narnia; Marcus Cicero included this in his *Marvels*, saying that drought brings mud, and rain dust.

XXIX. All water is sweeter in winter, in summer *Varieties of* less so, in autumn least, and less during droughts. *water.* The taste of rivers is usually variable, owing to the great difference in river beds. For waters vary with the land over which they flow, and with the juices of the plants they wash. Therefore the same rivers are found in some parts to be unwholesome. Tributaries too alter the flavour of a river, as do those of the Borysthenes, and being absorbed are diluted. Some rivers indeed are also changed by rain. Three times it has happened in the Bosphorus that salt rains fell and ruined the crops, and three times rains have made bitter the inundations of the Nile, a great plague for Egypt.

XXX. Springs arise often when woods have been *Various* cut down, being used up before as sustenance for the *phenomena* trees; this happened when Cassander was besieging *of water.* the Gauls after the woods on Mount Haemus had been felled by them to make a rampart. Often indeed devastating torrents unite when from hills has been cut away the wood that used to hold the rains and absorb *a* them. It also improves the water supply for the earth to be dug and tilled, and for the hard surface crust to be broken up. It is at any rate reported that in Crete, when a town called Arcadia had been stormed, the many springs and rivers of that region went dry, and six years afterwards, when the town was rebuilt, they reappeared, as each piece of land

411

54 Terrae quoque motus profundunt sorbentque aquas,
sicut circa Pheneum Arcadiae quinquies accidisse
constat. sic et in Coryco monte amnis erupit posteaque [1] coeptus est coli. illa mutatio mira, cuius causa
nulla evidens apparet, sicut in Magnesia e calida facta
frigida, salis non mutato sapore, et in Caria, ubi
Neptuni templum est, amnis qui fuerat ante dulcis
55 mutatus in salem est. et illa miraculi plena,
Arethusam Syracusis fimum redolere per Olympia,
verique simile, quoniam Alpheus in eam insulam sub
maria permeet. Rhodiorum fons in Cherroneso nono
anno purgamenta egerit. mutantur et colores
aquarum, sicut Babylone lacus aestate rubras habet
56 diebus undecim. et Borysthenes statis [2] temporibus
caeruleus fertur, quamquam omnium aquarum
tenuissimus, ideoque innatans Hypani, in quo et
illud mirabile, austris flantibus superiorem Hypanim
fieri. sed tenuitatis argumentum et aliud est quod
nullum halitum, non modo nebulam emittit. qui
volunt diligentes circa haec videri dicunt aquas
graviores post brumam fieri.
57 XXXI. Ceterum a fonte duci fictilibus tubis utilissimum est crassitudine binum digitorum, commissuris
pyxidatis ita ut superior intret, calce viva ex oleo

---

[1] posteaque *codd.*: posteaquam *cod. a vulg., Detlefsen.*
[2] statis *Mayhoff ex Athen. II.* 16: aestatis *codd., Detlefsen.*

---

[a] With the reading *posteaquam*: "after it came under
cultivation."
[b] The MSS. reading: "in summer time." Perhaps *aestatis*
because a scribe had just written *aestate*.

came under cultivation. Earthquakes too make water break out or swallow it up, for example, as is well known, around Pheneus in Arcadia this has happened five times. Thus too on Mount Corycus a river burst out, but afterwards *a* came to be tilled ground. Any change is startling when no obvious reason for it is to be seen. In Magnesia for instance hot water became cold but its salty flavour remained unaltered; while in Caria, where the temple of Neptune is, a river which before had been sweet was changed to salt. The following phenomena too are very wonderful: the Arethusa at Syracuse smells of dung during the Olympian games, a likely thing, for the Alpheus crosses to that island under the bed of the seas. A spring in the Rhodian Chersonesus pours out refuse every ninth year. The colour too of water changes, for example at Babylon a lake in summer has red water for eleven days, and the Borysthenes at fixed intervals *b* flows*c* with a blue colour, although of all waters it is the thinnest, and for that reason flows above the Hypanis. Wherein is another marvel: when south winds blow the Hypanis goes above. But other evidence for the thinness of the Borysthenes is that it gives out no exhalation, not to say no mist. Those who wish to be thought careful enquirers into these matters say that water becomes heavier after the winter solstice.

XXXI. For the rest, the best way for water to be brought from a spring is in earthenware pipes two fingers *d* thick, the joints boxed together so that the upper pipe fits into the lower, and smoothed with quicklime and oil. The gradient of the water should

Water pipes.

---

c It is less likely that *fertur* means " is said (to be)."
d The *digitus* was about one inch.

levigatis. libramentum aquae in centenos pedes
sicilici minimum erit, si cuniculo veniet, in binos actus
lumina esse debebunt. quam surgere in sublime
opus fuerit plumbo veniat. subit altitudinem exortus
sui. si longiore tractu veniet, subeat crebro descend-
58 atque, ne libramenta pereant. fistulas denum pedum
longitudinis esse legitimum est et si quinariae erunt
sexagena pondo pendere, si octonariae centena, si
denariae centena vicena, ac deinde ad has portiones.
denaria appellatur cuius lamnae latitudo, antequam
curvetur, digitorum decem est, dimidioque eius
quinaria. in anfractu omni collis quinariam fieri, ubi
dometur impetus, necessarium est, item castella,
prout res exigit.
59   XXXII. Homerum calidorum fontium mentionem
non fecisse demiror, cum alioqui lavari calida frequenter induceret, videlicet quia medicina tunc non
erat haec quae nunc aquarum perfugio utitur. est
autem utilis sulpurata nervis, aluminata paralyticis
aut simili modo solutis, bituminata aut nitrosa, qualis
60 Cutilia est, bibendo atque purgationibus. plerique
in gloria ducunt plurimis horis perpeti calorem earum,
quod est inimicissimum, namque paulo diutius quam
balineis uti oportet, ac postea frigida dulci, nec sine
oleo discedentes, quod vulgus alienum arbitratur,
idcirco non alibi corporibus magis obnoxiis, quippe et
vastitate odoris capita replentur et frigore infestantur
sudantia, reliqua corporum parte mersa. similis

---

^a The *actus* was 120 feet long.
^b I.e. of sulphur.

be at least a quarter of an inch every hundred feet; should it come in a tunnel, there must be vent holes every two *actus*.[a] When water is required to form a jet, it should come in lead pipes. Water rises as high as its source. If it comes from a long distance, the pipe should frequently go up and down, so that no momentum may be lost. The usual length for a piece of piping is ten feet; five-finger lengths should weigh 60 pounds, eight-finger lengths 100 pounds, ten-finger lengths 120 pounds, and so on in proportion. A ten-finger pipe is so called when the breadth of the strip before bending is ten fingers, and one half as large a five-finger pipe. At every bend of a hill where the momentum must be controlled, it is necessary to use a five-finger pipe; reservoirs must be made according as circumstances require.

XXXII. I wonder that Homer made no mention *Hot and* of hot springs, and that though he frequently speaks *medicinal* of hot baths, the reason being that modern hydro- *springs.* pathic treatment was not then a part of medicine. Sulphur waters, however, are good for the sinews, alum waters for paralysis and similar cases of collapse, waters containing bitumen and soda, such as that of Cutilia, are good for drinking and as a purge. Many people make a matter of boasting the great number of hours they can endure the heat of these sulphur waters—a very injurious practice, for one should remain in them a little longer than in the bath, afterwards rinse in cool, fresh water, and not go away without a rubbing with oil. The common people find these details irksome, and so there is no greater risk to health than this treatment, because an overpowering smell [b] goes to the head, which sweats and is seized with chill, while the rest of the body is im-

error, quam plurimo potu gloriantur. vidique iam
turgidos bibendo in tantum ut anuli integerentur
cute, cum reddi non posset hausta multitudo aquae.
nec hoc ergo fieri convenit sine crebro salis gustu.
61 utuntur et caeno fontium ipsorum utiliter, sed ita si
inlitum sole inarescat. nec vero omnes quae sint
calidae medicatas esse credendum, sicut in Segesta
Siciliae, Larisa Troade,[1] Magnesia, Melo, Lipara.
nec decolor species aeris argentive, ut multi existima-
verunt, medicaminum argumentum est, quando nihil
eorum in Patavinis fontibus, ne odoris quidem
differentia aliqua deprehenditur.
62   XXXIII. Medendi modus idem et in marinis erit
quae calefiunt ad nervorum dolores, feruminanda a
fracturis ossa contusa, item corpora siccanda, qua de
causa et frigido mari utuntur. praeterea est alius
usus multiplex, principalis vero navigandi phthisi
adfectis, ut diximus, aut sanguine egesto, sicut
proxime Annaeum Gallionem fecisse post consula-
63 tum meminimus. neque enim Aegyptus propter se
petitur, sed propter longinquitatem navigandi. quin
et vomitiones ipsae instabili volutatione commotae
plurimis morbis capitis, oculorum, pectoris medentur
omnibusque propter quae helleborum bibitur. aquam
vero maris per se efficaciorem discutiendis tumoribus
putant medici, si illa decoquatur hordeacia farina, ad

---

[1] *Inter* Larisa *et* Troade *comma multi edd.*

---

[a] See XXIV. § 28 and XXVIII. § 54.

mersed. Those make a like mistake who boast of the great quantity they can drink. I have seen some already swollen with drinking to such an extent that their rings were covered by skin, since they could not void the vast amount of water they had swallowed. So it is not good to drink these waters without a frequent taste of salt. The mud too of medicinal springs is used with advantage, but the application should be dried in the sun. We must not think, however, that all hot waters are medicinal; for there are those at Segesta in Sicily, at Larisa in the Troad, at Magnesia, in Melos and Lipara. Nor is the discoloration of bronze or silver a proof, as many have thought, of medicinal properties, since there are none in the springs of Patavium. Between medicinal and other water there is not even a difference of smell to be detected.

XXXIII. The same method of treatment will also apply to sea water, which is used hot for pains in the sinews, for joining fractured bones, and for bruised bones; also for drying the body, in which treatment cold sea water is also employed. There are besides many other uses, the chief however being a sea voyage for those attacked by consumption, as I have said,[a] and for haemoptysis, such as quite recently within our memory was taken by Annaeus Gallio after his consulship. Egypt is not chosen for its own sake, but because of the length of the voyage. Moreover the mere sea-sickness caused by rolling and pitching are good for very many ailments of the head, eyes, and chest, as well as for all complaints for which hellebore is given. Sea water indeed by itself physicians think to be more efficacious for dispersing tumours, if with it a decoction is made of barley meal for parotid swell-

Medicinal uses of sea water.

417

parotidas. emplastris etiam, maxime albis et malag-
64 matis miscent, prodest et infusa crebro ictu. bibitur
quoque, quamvis non sine iniuria stomachi, ad purganda corpora bilemque atram aut sanguinem con-
cretum reddendum alterutra parte. quidam et in
quartanis dedere eam bidendam et in tenesmis
articulariisque morbis adservatam in hoc, vetustate
virus deponentem, aliqui decoctam, omnes ex alto
haustam nullaque dulcium mixtura corruptam, in quo
usu praecedere vomitum volunt. tunc quoque
65 acetum aut vinum ea aqua miscent. qui puram
dedere raphanos supermandi ex mulso aceto iubent,
ut ad vomitiones revocent. clysteribus quoque
marinam infundunt tepefactum. testium quidem
tumorem fovendo non aliud praeferunt, item pernio-
num vitio ante ulcera, simili modo pruritibus, psoris et
lichenum curationi. lendes quoque et taetra capitis
animalia hac curantur. et liventia reducit eadem
ad colorem.[1] in quibus curationibus post marinam
aceto calido fovere plurimum prodest. quin et ad
ictus venenatos salutaris intellegitur, ut phalangi-
orum et scorpionum, et ptyade aspide respersis,
66 calida autem in his adsumitur. suffitur eadem cum
aceto capitis doloribus. tormina quoque et choleras
calida infusa clysteribus sedant. difficilius per-
frigescunt marina calefacta. mammas sororientes,
praecordia maciemque corporis piscinae maris corrigunt, aurium gravitatem, capitis dolores cum aceto
ferventium vapor. rubiginem ferro marinae celer-

---

[1] colorem *Mayhoff*: colores *codd., contra Plinii usum.*

---

[a] White plasters were made with *cerussa,* white lead. See
Celsus V. 19, 2.

ings. It is also an ingredient of plasters, especially white plasters,[a] and poultices. It is beneficially used too when poured over in frequent douches. It is also drunk, though not without harm to the stomach, for purging the body and for getting rid of black bile or clotted blood by vomit or stool. Some have also given it to be drunk in quartan agues, in tenesmus, and for diseased joints, keeping it for this purpose, for age takes away its injurious qualities. Some boil it; all draw it up out at sea, use it unspoiled by any addition of fresh water, and in using this remedy prefer that an emetic should precede the draught. Then also they mix with the water vinegar or wine. Those who have given it pure, recommend to eat afterwards radishes with oxymel to provoke further vomiting. Sea water warmed is also injected as an enema. Nothing is preferred to it for fomenting swollen testicles, or for bad chilblains before ulceration; similarly for itching, psoriasis, and the treatment of lichen. Nits too and foul vermin on the head are treated with sea water. It also restores the natural colour to livid patches. In this treatment it is of very great advantage to foment with hot vinegar after the sea water. It is moreover known to be healing for poisonous stings, as of spiders and scorpions, and for persons wetted by the spittle of the asp *ptyas*, but for these purposes it is employed hot. Steam from sea water and vinegar is beneficial for headaches. Colic too and cholera are relieved by warm enemas of sea water. Things warmed by it are harder to cool thoroughly. Swollen breasts, the viscera, and emaciation, are rectified by sea baths, deafness and headache by the vapour of boiling sea water and vinegar. Sea water removes

rime exterunt, pecorum quoque scabiem sanant
lanasque emolliunt.

67 XXXIV. Nec ignoro haec mediterraneis super-
vacua videri posse. verum et hoc cura providit inventa ratione qua sibi quisque aquam maris faceret.
illud in ea ratione mirum, si plus quam sextarius salis
in quattuor sextarios aquae mergatur, vinci aquam
salemque non liquari. cetero sextarius salis cum
quattuor aquae sextariis salsissimi maris vim et
naturam implet. moderatissimum autem putant
supra dictam aquae mensuram octonis cyathis salis
temperari, quoniam ita et nervos excalefaciat et
corpus non exasperet.

68 XXXV. Inveteratur et quod vocant thalassomeli
aequis portionibus maris, mellis, imbris. ex alto et
ad hunc usum advehunt fictilique vaso et picato condunt. prodest ad purgationes maxime sine stomachi
vexatione et sapore grato et odore.

69 XXXVI. Hydromeli quoque ex imbre puro cum
melle temperabatur quondam, quod daretur adpe-
tentibus vini aegris veluti innocentiore potu, damna-
tum iam multis annis, isdem vitiis quibus vinum nec
isdem utilitatibus.

70 XXXVII. Quia saepe navigantes defectu aquae
dulcis laborant, haec quoque subsidia demonstrabimus.
expansa circa navem vellera madescunt accepto halitu
maris, quibus dulcis umor exprimitur, item demissae
reticulis in mare concavae ex cera pilae vel vasa

ᵃ It is hard to reconcile this remark with the many prescriptions containing *hydromeli* (*aqua mulsa*) in Pliny. Perhaps there is a reference here to a particular kind of hydromel.

very quickly rust from iron, heals too scab on sheep, and softens wool.

XXXIV. I am well aware that to inland dwellers these remarks may appear superfluous, but research has provided for this also by discovering a method whereby every man may make sea water for himself. In this method there is one strange feature: if more than a sextarius of salt is dropped into four sextarii of water, the water is overpowered, and the salt does not dissolve. However, a sextarius of salt and four sextarii of water give the strength and properties of the saltest sea. But it is thought that the most reasonable proportion is to compound the measure of water given above with eight cyathi of salt. This mixture warms the sinews without chafing the skin.

XXXV. What is called thalassomeli is a mixture, *Thalasso-* kept till old, of sea water, honey, and rain water in *meli.* equal proportions. For this purpose too the water is brought from out at sea, and the mixture is stored in an earthenware vessel lined with pitch. It is good especially for purges, does not disturb the stomach, and has a pleasant flavour and smell.

XXXVI. Hydromel too is a mixture once prepared *Hydromel.* from pure rainwater and honey, to be given as a less injurious drink to patients who craved for wine. It has been condemned now for many years [a] as having all the faults of wine with none of its advantages.

XXXVII. Because those at sea often suffer from *Fresh water* the failure of fresh water, I shall describe ways of *from the sea.* meeting this difficulty. If spread around a ship, fleeces become moist by absorption of evaporated sea water, and from them can be squeezed water which is fresh. Again, hollow wax balls, let down into the sea in nets, or empty vessels with their

inania opturata dulcem intra se colligunt umorem.
nam in terra marina aqua argilla percolata dulcescit.
71 luxata corpora et hominum et quadrupedum natando
in cuius libeat generis aqua facillime in artus redeunt.
est et in metu peregrinantium ut temptent vali-
tudinem aquae ignotae. hoc cavent e balneis
egressi statim frigidam suspectam hauriendo.

72 XXXVIII. Muscus qui in aqua fuerit podagris inlitus prodest, item oleo admixto talorum dolori tumo-
rique. spuma aquae adfrictu verrucas tollit, nec non
harena litorum maris, praecipue tenuis et sole candens, in medicina est siccandis corporibus coopertis
hydropicorum aut rheumatismos sentientium. et
hactenus de aquis, nunc de aquatilibus. ordiemur
autem ut in reliquis a principalibus eorum quae sunt
salsa ac spongea.

73 XXXIX. Sal omnis aut fit aut gignitur, utrumque
pluribus modis, sed causa gemina, coacto umore vel
siccato. siccatur in lacu Tarentino aestivis solibus,
totumque stagnum in salem abit, modicum alioqui,
altitudine genua non excedens, item in Sicilia in lacu
qui Cocanicus vocatur et alio iuxta Gelam. horum
extremitates tantum inarescunt, sicut in Phrygia,
Cappadocia, Aspendi, ubi largius coquitur et usque
ad medium. aliud etiam in eo mirabile quod tantundem nocte subvenit quantum die auferas. omnis e
74 stagnis sal minutus atque non glaeba est. aliud
genus ex aquis maris sponte gignitur spuma in
extremis litoribus ac scopulis relicta. hic omnis rore

mouth sealed, collect fresh water inside. But on land sea water is made fresh by filtering through clay. Dislocated limbs of both man and quadrupeds are very easily re-set by swimming in any kind of water. Travellers too are sometimes afraid lest unknown water should endanger their health. A precaution against this danger is to drink the suspected water cold immediately on leaving the bath.

XXXVIII. An application of moss that has grown *Moss as a* in water is good for gout, and mixed with oil for pain- *cure.* ful and swollen ankles. Rubbing with foam of water removes warts, as does also sand of the sea shores, especially fine sand whitened by the sun; it is used in medicine as a covering for drying the bodies of patients suffering from dropsy or catarrhs. So much for waters; now for the products of water. I shall begin, as elsewhere, with the chief of them, that is, with salts and sponge.

XXXIX. All salt is artificial or native; each is *Salt,* formed in several ways, but there are two agencies, *artificial* *and native.* condensation or drying up of water. It is dried out of the Tarentine lake by summer sun, when the whole pool turns into salt, although it is always shallow, never exceeding knee height, likewise in Sicily from a lake, called Cocanicus, and from another near Gela. Of these the edges only dry up; in Phrygia, Cappadocia, and at Aspendus, the evaporation is wider, in fact right to the centre. There is yet another wonderful thing about it: the same amount is restored during the night as is taken away during the day. All salt from pools is fine powder, and not in blocks. Another kind produced from sea water spontaneously is foam left on the edge of the shore and on rocks. All this is condensation from drift,

densatur, et est acrior qui in scopulis invenitur. sunt
etiamnum naturales differentiae tres. namque in
Bactris duo lacus vasti, alter ad Scythas versus alter
ad Arios, sale exaestuant, sicut ad Citium in Cypro
et circa Memphin extrahunt e lacu, dein sole siccant.
75 sed et summa fluminum densantur in salem amne
reliquo veluti sub gelu fluente, ut apud Caspias portas
quae salis flumina appellantur, item circa Mardos et
Armenios. praeterea et apud Bactros amnis Ochus
et Oxus ex adpositis montibus deferunt salis ramenta.
76 sunt et in Africa lacus, et quidem turbidi, salem
ferentes. ferunt quidem et calidi fontes, sicut
Pagasaei. et hactenus habent se genera ex aquis
77 sponte provenientia. sunt et montes nativi salis, ut
Indis Oromenus, in quo lapicidinarum modo caeditur
renascens, maiusque regum vectigal ex eo est quam
ex auro atque margaritis. effoditur e terra, ut palam
est umore densato, in Cappadocia. ibi quidem caeditur specularium lapidum modo. pondus magnum
78 glaebis quas micas vulgus appellat. Gerris Arabiae
oppido muros domosque e massis salis faciunt aqua
feruminantes. invenit et iuxta Pelusium Ptolo-
maeus rex, cum castra faceret. quo exemplo postea
inter Aegyptum et Arabiam etiam squalentibus locis
coeptus est inveniri detractis harenis, qualiter et per
Africae sitientia usque ad Hammonis oraculum, is
79 quidem crescens cum luna noctibus. nam et Cyre-

and that found on rocks has the sharper taste. There are also three different kinds of native salt; for in Bactra are two vast lakes, one facing the Scythians, the other the Arii, which exude salt, while at Citium in Cyprus and around Memphis salt is taken out of a lake and then dried in the sun. But the surface too of rivers may condense into salt, the rest of the stream flowing as it were under ice, as near the Caspian Gates are what are called "rivers of salt," also around the Mardi and the Armenians. Moreover, in Bactria too the rivers Ochus and Oxus bring down scrapings of salt from nearby mountains. There are also lakes in Africa, and that muddy ones, which carry salt. Indeed hot springs too carry it, such as those at Pagasae. So much for the different kinds of salt which come, as natural products, from waters. There are also mountains of natural salt, *Block salt.* such as Oromenus in India, where it is cut out like blocks of stone from a quarry, and ever replaces itself, bringing greater revenues to the rajahs than those from gold and pearls. It is also dug out of the earth in Cappadocia, being evidently formed by condensation of moisture. Here indeed it is split into sheets like mica; the blocks are very heavy, nicknamed by the people "grains." At Gerra, a town of Arabia, the walls and houses are made of blocks of salt cemented with water. Near Pelusium too King Ptolemy found salt when he was making a camp. This led afterwards to the discovery of salt by digging away the sand even in the rough tracts between Egypt and Arabia, as it is also found as far as the oracle of Hammon through the parched deserts of Africa, where at night it increases as the moon waxes. But the region of Cyrenaica too is

425

naici tractus nobilitantur Hammoniaco et ipso, quia
sub harenis inveniatur, appellato. similis est colore
alumini quod schiston vocant, longis glaebis neque
perlucidis, ingrato sapore, sed medicinae utilis. probatur quam maxime perspicuus, rectis scissuris.
insigne de eo proditur quod levissimus intra specus
suos in lucem universam prolatus vix credibili pondere
ingravescat. causa evidens, cuniculorum spiritu
madido sic adiuvante molientes ut adiuvant aquae.
adulteratur Siculo quem Cocanicum appellavimus,
80 nec non et Cyprio mire simili. in Hispania quoque
citeriore Egelestae caeditur glaebis paene trans-
lucentibus cui iam pridem palma a plerisque medicis
inter omnia salis genera perhibetur. omnis locus in
quo repperitur sal sterilis est nihilque gignit. et in
81 totum sponte nascens intra haec est. facticii varia
genera, volgaris plurimusque in salinis mari adfuso
non sine aquae [1] dulcis [2] riguis, sed imbre maxime
iuvante ac super omnia sole multo,[3] aliter non
inarescens. Africa circa Uticam construit acervos
salis ad collium speciem, qui ubi sole lunaque induru-
ere, nullo umore liquescunt vixque etiam ferro cae-
duntur. fit tamen et in Creta sine riguis mare in
salinas infundentibus et circa Aegyptum ipso mari

[1] aquae d : aquis VRE, *Mayhoff.*
[2] dulcis *codd.* : dulcibus *Mayhoff.*
[3] *Post* multo *in* VR que : *Mayhoff* multo assiduoque *coni.*,
multo altoque *Brakman.*

---

[a] This salt consists of chlorides of sodium, calcium, and
magnesium. The Greek for " sand " is ἄμμος.
[b] I.e. " cleft."
[c] See § 73.
[d] Brakman's *alto* would mean "overhead." Mayhoff also
conjectures *lunaque*, as just below.

famous for Hammoniac salt, itself so called because
it is found under the sand.[a] It is in colour like the
alum called *schiston*,[b] consisting of long opaque slabs,
of an unpleasant flavour, but useful in medicine.
That is most valued which is most transparent and
splits into straight flakes. A remarkable feature is
reported of it: of very little weight in its underground
pits, when brought into the light of day it becomes
incredibly heavy. The reason is obvious; the damp
breath of the pits helps the workers by supporting
the weight as does water. It is adulterated by the
Sicilian salt I have said[c] comes from the lake
Cocanicus, as well as by Cyprian salt, which is wonderfully like it. In Hither Spain too at Egelesta salt is
cut into almost transparent blocks; to this for some
time past most physicians have given the first place
among all kinds of salt. Every region in which salt
is found is barren, and nothing will grow there. To
speak generally, these remarks about the various
kinds of native salt are comprehensive. Of artificial *Artificial*
salt there are various kinds. The usual one, and the *salts.*
most plentiful, is made in salt pools by running into
them sea water not without streams of fresh water,
but rain helps very much, and above all much ⟨warm⟩[d]
sunshine, without which it does not dry out. In
Africa around Utica are formed heaps of salt like
hills; when they have hardened under sun and moon,
they are not melted by any moisture, and even iron
cuts them with difficulty. It is also however made
in Crete without fresh water[e] by letting the sea flow
into the pools, and around Egypt by the sea itself,

[e] K. C. Bailey in *Hermathena* for 1926 points out that fresh
water could be profitably used only for washing salt already
obtained by evaporation.

influente in solum, ut credo, Nilo sucosum. fit et
82 puteis in salinas ingestis. prima densatio Babylone
in bitumen liquidum cogitur oleo simile, quo et in
lucernis utuntur. hoc detracto subest sal. et in
Cappadocia e puteis ac fonte aquam in salinas in-
gerunt. in Chaonia excocunt aquam ex fonte re-
frigerandoque salem faciunt inertem nec candidum.
Galliae Germaniaeque ardentibus lignis aquam
salsam infundunt.

83   XL. Hispaniae quadam sui parte e puteis hauriunt
muriam appellantes. illi quidem et lignum referre
arbitrantur. quercus optima, ut quae per se cinere
sincero vim salis reddat, alibi corylus laudatur. ita
infuso liquore salso arbor[1] etiam in salem vertitur.
quicumque ligno confit sal niger est. apud Theo-
phrastum invenio Umbros harundinis et iunci cinerem
decoquere aqua solitos donec exiguum superesset
umoris. quin et e muria salsamentorum recoquitur
iterumque consumpto liquore ad naturam suam redit,
vulgo e menis iucundissimus.

84   XLI. Marinorum maxume laudatur Cyprius a
Salamine, de stagnis Tarentinus ac Phrygius qui
Tattaeus vocatur. hi duo oculis utiles. e Cappa-
docia qui in laterculis adfertur cutis nitorem dicitur

---

[1] arbor E *Detlefsen, Mayhoff*: carbo *ceteri codd., vulg.*

---

[a] Mayhoff takes this sentence as part of the last. It may
be a parenthesis.
[b] The well attested *carbo* makes good sense, and it bears a
strong resemblance to *arbor*. The former is obviously an
easier reading, so perhaps Detlefsen and Mayhoff have chosen
the harder.

which penetrates the soil, soaked as I believe it is,
by the Nile. Salt is also made by pouring water
from wells into salt pools. At Babylon the first condensation solidifies into a liquid bitumen like oil,
which is also used in lamps. When this is taken
away, salt is underneath. In Cappadocia too they
bring water into salt pools from wells and a spring.
In Chaonia there is a spring, from which they boil
water, and on cooling obtain a salt that is insipid and
not white. In the provinces of Gaul and Germany
they pour salt water on burning logs. XL. (In one
part of the provinces of Spain they draw the brine
from wells and call it *muria.*[a]) The former indeed
think that the wood used also makes a difference.
The best is oak, for its pure ash by itself has the
properties of salt; in some places hazel finds favour.
So when brine is poured on it even wood [b] turns into
salt. Whenever wood is used in its making salt is
dark. I find in Theophrastus that the Umbrians
were wont to boil down in water the ash of reeds and
rushes, until only a very little liquid remained.
Moreover, from the liquor of salted foods salt is
recovered by reboiling, and when evaporation is
complete its saline character is regained. It is
generally thought that the salt obtained from
sardine brine is the most pleasant.

XLI. Of sea salt the most in favour comes from *Salts from various localities.*
Salamis in Cyprus, of pool salt that from Tarentum
and that from Phrygia which is called Tattaean.
The last two are useful for the eyes. The salt
imported from Cappadocia in little bricks [c] is said to
impart a gloss to the skin. But the salt I have said

<hr />

[c] Littré has: " dans des vaisseaux de brique."

facere. magis tamen extendit is quem Citium
appellavimus, itaque a partu ventrem eo cum melan-
85 thio inlinunt. salissimus sal qui siccissimus, suavissi-
mus omnium Tarentinus atque candidissimus est,[1] de
cetero fragilis qui maxime candidus. pluvia dulcescit
omnis, suaviorem tamen rores faciunt, sed copiosum
aquilonis flatus. austro non nascitur. flos salis non
fit nisi aquilonibus. in igni nec crepitat nec exilit
Tragasaeus neque Acanthius ab oppido appellatus,
86 nec ullius spuma aut[2] ramenta aut tenuis.[3] Agri-
gentinus ignium patiens ex aqua exilit.[4] sunt
et colorum differentiae. rubet Memphi, rufus est
circa Oxum, Centuripis purpureus, circa Gelam in
eadem Sicilia tanti splendoris ut imaginem recipiat.
in Cappadocia crocinus effoditur, tralucidus et
odoratissimus. ad medicinae usus antiqui Taren-
tinum maxime laudabant, ab hoc quemcumque e
marinis, ex eo genere spumeum praecipue, iumentorum vero et boum oculis Tragasaeum et Baeticum.
87 ad opsonium et cibum utilior quisquis facile liquescit,
item umidior, minorem enim amaritudinem habent,
ut Atticus et Euboicus. servandis carnibus aptior
acer et siccus, ut Megaricus. conditur etiam odori-
bus additis et pulmentarii vicem implet, excitans
aviditatem invitansque in omnibus cibis ita, ut sit

---

[1] est *Urlichs, Detlefsen*: set *Mayhoff*: et *codd.*

[2] aut at Er: aut ab *Detlefsen*: *om.* at *ceteri codd.*

[3] ramenta aut tenuis *ego*: ramento tenuis *Detlefsen*: ramen-
tum tenuius *Mayhoff*: ramento aut tenuis *codd.*

[4] ignium patiens ex aqua exilit *Detlefsen, Mayhoff, codd.*
ignis impatiens est atque exilit *K. C. Bailey.*

---

[a] See § 74.

[b] See XIII. § 14 and XXXI. § 90.

[c] Tragasa and Acanthus.

comes from Citium *a* smooths the skin better, and so after child-birth it is applied with melanthium to the abdomen. The saltest salt is the driest, the most agreeable and whitest of all is the Tarentine; for the rest, it is the whitest that is the most friable. All salt is made sweet by rain water, more agreeable, however, by dew, but plentiful by gusts of north wind. It does not form under a south wind. Flower of salt *b* forms only with north winds. Tragasaean salt and Acanthian, so named after towns,*c* neither crackles nor sputters in a fire, nor does froth *d* of any salt, or scrapings, or powder. Salt of Agrigentum submits to fire and sputters in water.*e* The colour too of salt varies: blushing red at Memphis, tawny red near the Oxus, purple at Centuripae, it is of such brightness near Gela (also in Sicily) that it reflects an image. In Cappadocia salt is quarried of a saffron colour, transparent, and very fragrant. For medicinal purposes the ancients used to favour most highly Tarentine salt, next, all kinds of sea salt, and of these especially that from foam, while for the eyes of draught animals and cattle salt of Tragasa and Baetica. To season meats and foods the most useful *Other* one melts easily and is rather moist, for it is less *varieties.* bitter, such as that of Attica and Euboea. For preserving meat the more suitable salt is sharp and dry, like that of Megara. A conserve too is made with fragrant additions, which is used as a relish, creating and sharpening an appetite for every kind

d See § 74.
e K. C. Bailey's emendation in *Hermathena* 1926 is contrary to passages in Isodore (16. 2. 4 and 14. 6. 34), Solinus (*Polyist.* 5. 18), and Augustine (*De Civ. Dei* 21. 5). He suggests that either "Agrigentum salt" was lime, or that a mistake occurred in Pliny's MSS. very early.

peculiaris ex eo intellectus inter innumera condi-
88 menta ciborum item in mandendo quaesitus garo.[1]
quin et pecudes armentaque et iumenta sale maxime
sollicitantur ad pastus multum largiore lacte multoque gratiore etiam in caseo dote. ergo, Hercules,
vita humanior sine sale non quit degere,[2] adeoque
necessarium elementum est uti transierit intellectus
ad voluptates animi quoque nimias.[3] sales appellantur, omnisque vitae lepos et summa hilaritas
laborumque requies non alio magis vocabulo constat.
89 honoribus etiam militiaeque interponitur salariis
inde dictis magna apud antiquos auctoritate, sicut
apparet ex nomine Salariae viae, quoniam illa salem
in Sabinos portari convenerat. Ancus Marcius rex
salis modios v̄ɪ in congiario dedit populis et salinas
primus instituit. Varro etiam pulmentarii vice usos
veteres auctor est, et salem cum pane esitasse eos
proverbio apparet. maxime tamen in sacris intellegitur auctoritas, quando nulla conficiuntur sine mola
salsa.
90   XLII. Salinarum sinceritas summam fecit suam
differentiam quandam favillam salis quae levissima
ex eo est et candidissima. appellatur et flos salis in
totum diversa res umidiorisque naturae et crocei
coloris aut rufi, veluti rubigo salis, odore quoque

---

[1] item in mandendo quaesitus garo *Mayhoff*: ciborum in
mandendo quaesitus garo *Detlefsen*: item E² a: ita E¹:
iterum *multi codd.*: " *locus adhuc corruptus* " (*Mayhoff*).
[2] degere *codd. et edd.*: degi *coni. Mayhoff*: degier *coni.
Brakman.*
[3] nimias *ego*: eximias *Mayhoff*: nimia *codd.*: del. *Detlefsen.*

---

[a] The exact text is very uncertain, but the general sense is
clear.

of food, so that in innumerable seasonings it is the taste of salt that predominates, and it is looked for when we eat garum.*a* Moreover sheep, cattle, and draught animals are encouraged to pasture in particular by salt; the supply of milk is much more copious, and there is even a far more pleasing quality in the cheese. Therefore, Heaven knows, a civilized life is impossible without salt, and so necessary is this basic substance that its name is applied metaphorically even to intense mental pleasures. We call them *sales* (wit); all the humour of life, its supreme joyousness, and relaxation after toil, are expressed by this word more than by any other. It has a place in magistracies also and on service abroad, from which comes the term " salary " (salt money); it had great importance among the men of old, as is clear from the name of the Salarian Way, since by it, according to agreement, salt was imported to the Sabines. King Ancus Marcius gave a largess to the people of 6,000 bushels of salt, and was the first to construct salt pools. Varro too is our authority that the men of old used salt as a relish, and that they ate salt with their bread is clear from a proverb.*b* But the clearest proof of its importance lies in the fact that no sacrifice is carried out without the *mola salsa* (salted meal).

XLII. Salt-pools have reached their highest degree of purity in what may be called embers of salt, which is the lightest and whitest of its kind. " Flower of salt " is also a name given to an entirely different thing, with a moister nature and a saffron or red colour, a kind of salt rust; it has an unpleasant smell,

b We do not know the proverb referred to, but several suitable ones suggest themselves.

Value of salt.

ingrato ceu gari dissentiens a sale, non modo a spuma.
Aegyptus invenit, videturque Nilo deferri. et fonti-
91 bus tamen quibusdam innatat. optimum ex eo quod
olei quandam pinguitudinem reddit. est enim
etiam in sale pinguitudo, quod miremur. adulteratur
autem tinguiturque rubrica aut plerumque testa trita,
qui fucus aqua deprehenditur diluente facticium
colorem, cum verus ille non nisi oleo resolvatur et un-
guentarii propter colorem eo maxime utantur,
canitia in vasis summa est, media vero pars umidior.
92 ut diximus. floris natura aspera, excalfactoria,
stomacho inutilis, sudorem ciet, alvum solvit in vino
et aqua, acopis et zmecticis utilis. detrahit et ex
palpebris pilos. ima faecis concutiuntur, ut croci
color redeat. praeter haec etiamnum appellatur in
salinis salsugo, ab aliis salsilago, tota liquida, a
marina aqua salsiore vi distans.
93   XLIII. Aliud etiamnum liquoris exquisiti genus,
quod garum vocavere, intestinis piscium ceterisque
quae abicienda essent sale maceratis, ut sit illa putres-
centium sanies. hoc olim conficiebatur ex pisce
quem Graeci garon vocabant, capite eius usto suffito
extrahi secundas monstrantes. nunc e scombro pisce
94 laudatissimum in Carthaginis Spartariae cetariis—
sociorum id appellatur—singulis milibus nummum

ª See § 90. This whole chapter is confused. The first
sentence does not contain the term *flos salis*, although the *et*
of the second sentence implies that it does. This white salt is
apparently referred to in *canitia . . . diximus*, a sentence
placed in the middle of a description of a saffron or red salt.
It seems hopeless to attempt to emend, and the faulty structure may be due to Pliny himself. The sentence *canitia . . .
diximus* is probably an interpolation, and in any case hard to
understand.

like that of garum, and is different from salt, not only
from foam salt. Egypt discovered it, and it appears
to be brought down by the Nile. It also however
floats on the surface of certain springs. The best
kind of it yields a sort of oily fat, for there is, surprising as it may seem, a fat even in salt. It is
adulterated too and coloured by red ochre, or usually
by ground crockery; this sham is detected by water,
which washes out the artificial colour, while the
genuine is only removed by oil, and perfumers use it
very commonly because of its colour. In vessels the
whiteness is seen on the surface, but the inner
part, as I have said,[a] is moister. The nature of
flower of salt is acrid, heating, bad for the stomach,
sudorific, aperient when taken in wine and water, and
useful for anodynes and detergents. It also removes
hair from eye-lids. The sediment is shaken up in
order to restore the saffron colour. Besides these
salines there is also what is called at the salt-pools
salpugo, or sometimes *salsilago*. It is entirely liquid,
differing from sea brine by its more salty character.

XLIII. There is yet another kind of choice liquor, *Garum.*
called garum, consisting of the guts of fish and the
other parts that would otherwise be considered refuse;
these are soaked in salt, so that garum is really liquor
from the putrefaction of these matters. Once this
used to be made from a fish that the Greeks called
garos; they shewed that by fumigation with its burning head the after-birth was brought away. Today
the most popular garum is made from the scomber [b]
in the fisheries of Carthago Spartaria [c]—it is called
garum of the allies—one thousand sesterces being

[b] Probably the mackerel.
[c] " Carthago where broom grows," New Carthage.

permutantibus congios fere binos. nec liquor ullus
paene praeter unguenta maiore in pretio esse coepit,
nobilitatis etiam gentibus. scombros quidem et
Mauretania Baeticaeque Carteia ex oceano intrantes
capiunt ad nihil aliud utiles. laudantur et Clazomenae garo Pompeique et Leptis, sicut muria Antipolis ac Thuri, iam vero et Delmatia.

95   XLIV. Vitium huius est allex atque imperfecta
nec colata faex. coepit tamen et privatim ex
inutili pisciculo minimoque confici. apuam nostri,
aphyen Graeci vocant, quoniam is pisciculus e pluvia
nascitur. Foroiulienses piscem ex quo faciunt lupum
appellant. transiit deinde in luxuriam, creveruntque
genera ad infinitum, sicuti garum ad colorem mulsi
veteris adeoque suavitatem dilutum [1] ut bibi possit.
aliud vero . . .[2] castimoniarum superstitioni etiam
sacrisque Iudaeis dicatum, quod fit e piscibus squama
carentibus. sic allex pervenit ad ostreas, echinos,
urticas maris, mullorum iocinera, innumerisque
generibus ad saporis gulae coepit sal tabescere.
96 haec obiter indicata sint desideriis vitae, et ipsa tamen
non nullius usus in medendo. namque et allece
scabies pecoris sanatur infusa per cutem incisam, et
contra canis morsus draconisve marini prodest, in
97 linteolis autem concerptis inponitur. Et garo ambusta recentia sanantur, si quis infundat ac non
nominet garum. contra canum quoque morsus

---

[1] suavitatem dilutum *Mayhoff*: dilutam suavitatem *codd.*
[2] ad *codd.*: est *Mayhoff*: *post* ad *lacunam indicat Detlefsen.*

---

[a] The *congius* was nearly six pints.
[b] As *allex* is feminine, and *aliud* neuter, it seems best to
suppose that there is a lacuna here, but Pliny may be thinking
of *garum*, to which he has just reverted.

exchanged for about two congii *a* of the fish. Scarcely any other liquid except unguents has come to be more highly valued, bringing fame even to the nations that make it. The scomber is caught also in Mauretania and at Carteia in Baetica; the scomber enters the Mediterranean from the Atlantic, but it is used only for making garum. Clazomenae too is famous for garum, and so are Pompeii and Leptis, just as Antipolis and Thurii are for muria, and today too also Delmatia.

XLIV. Allex is sediment of garum, the dregs, *Allex.* neither whole nor strained. It has, however, also begun to be made separately from a tiny fish, otherwise of no use. The Romans call it *apua*, the Greeks *aphye*, because this tiny fish is bred out of rain. The people of Forum Julii call *lupus* (*wolf*) the fish from which they make garum. Then allex became a luxury, and its various kinds have come to be innumerable; garum for instance has been blended to the colour of old honey wine, and to a taste so pleasant that it can be drunk. But another kind ⟨of garum⟩ *b* is devoted to superstitious sex-abstinence and Jewish rites, and is made from fish without scales. Thus allex has come to be made from oysters, sea urchins, sea anemones, and mullet's liver, and salt to be corrupted in numberless ways so as to suit all palates. These incidental remarks must suffice for the luxurious tastes of civilized man. Allex however itself is of some use in healing. For allex both cures itch in sheep, being poured into an incision in the skin, and is a good antidote for the bites of dog or sea draco; it is applied on pieces of lint. By garum too are fresh burns healed, if it is poured over them without mentioning garum. It is also good for dog-bites and

prodest maximeque crocodili et ulceribus quae serpunt aut sordidis. oris quoque et aurium ulceribus
aut doloribus mirifice prodest. muria quoque sive
illa salsugo spissat, mordet, extenuat, siccat, dysintericis utilis, etiam si nome intestina corripit, ischiadicis, coeliacis veteribus infunditur. fotu quoque
apud mediterraneos aquae marinae vicem pensat.

98 XLV. Salis natura per se ignea est et inimica
ignibus, fugiens eos, omnia erodens, corpora vero
adstringens, siccans, adligans, defuncta etiam a
putrescendi tabe [1] vindicans, ut durent ea per saecula,
in medendo vero mordens, adurens, repurgans, ex-
tenuans, dissolvens, stomacho tantum inutilis, praeterquam ad excitandam aviditatem. adversus serpentium morsus cum origano, melle, hysopo, contra
cerasten cum origano et cedria [2] aut pice aut melle.

99 auxiliatur contra scolopendras ex aceto potus, adversus scorpionum ictus cum quarta parte lini seminis
ex oleo vel aceto inlitus, adversus crabrones vero et
vespas similiaque ex aceto, ad heterocranias capitisque ulcera et pusulas papulasve et incipientes verrucas cum sebo vitulino, item [3] oculorum remediis et
ad excrescentes ibi carnes totiusque corporis pterygia,
sed in oculis peculiariter, ob id collyriis emplastrisque
additus—ad haec maxime probatur Tattaeus aut

100 Caunites—ex ictu vero suffusis cruore oculis suggilla-
tisque cum murrae pari pondere ac melle aut cum

---

[1] tabe *Ianus*: tabo *Detlefsen*: ta V: to R: ita E *vulg.*
[2] cedria *Hermolaus Barbarus*: cedro *codd.*
[3] *Post* item *velit* in *addere Mayhoff.*

---

[a] See § 92.   [b] Horned viper.

especially those of the crocodile, and for spreading or
foul ulcers. For ulcers too or pains in mouth or ears
it is wonderfully good. Muria too or the salsugo I
spoke of [a] is astringent, biting, reducing and drying,
useful for dysentery, even if there is ulceration of the
bowels. It is injected for sciatica and chronic
coeliac disease. Among inland peoples it also takes
the place of sea water for fomentations.

XLV. The nature of salt is of itself fiery, and yet *Use of salt*
it is hostile to fires, fleeing from them, corroding all *in medicine.*
things, but astringent to the body, drying it and
binding, preserving corpses also from corruption so
that they last for ages; in medicine however it is
mordent, caustic, cleansing, reducing, and resolvent,
injurious only to the stomach except in so far as it
stimulates the appetite. For the bites of serpents it
is used with origanum, honey, and hyssop, for the
cerastes [b] with origanum and cedar resin, or pitch, or
honey. It is helpful for bite of the scolopendra if
taken internally with vinegar, for scorpion stings if
applied in oil or vinegar with a fourth part of linseed,
but for hornets, wasps, and similar creatures, in
vinegar only, for migraine, ulcers on the head, blisters,
pimples, and incipient warts, with veal suet. It is
also used in eye remedies, for excrescences of flesh
there, and for *pterygia* [c] anywhere on the body, but
especially on the eyes, and so it is an ingredient of eye
salves and plasters; for these purposes Tattaean salt
or that of Caunus is the most approved. For eyes
bloodshot from a blow, however, and for bruised eyes,
it is used with an equal weight of myrrh and with
honey, or with hyssop in warm water, and the eyes

[c] Either (*a*) whitlows or (*b*) inflammatory swellings of the
eye.

hysopo ex aqua calida, utque foveantur salsugine.
ad haec Hispaniensis eligitur, contraque suffusiones
oculorum cum lacte in coticulis teritur, privatim sug-
gillationibus in linteolo involutus crebroque ex aqua
ferventi inpositus, ulceribus oris manantibus in linteolo
concerpto, gingivarum tumori infricatus et contra
101 scabritiem linguae fractus comminutusque. aiunt
dentes non erodi nec putrescere, si quis cotidie mane
ieiunus salem contineat sub lingua donec liquescat.
lepras idem et furunculos et lichenas et psoras emendat cum passa uva exempto eius ligno et sebo bubulo
atque origano ac fermento vel pane—maxime
Thebaicus ad haec et pruritus eligitur—tonsillis et
uvis cum melle prodest.[1] quicumque ad anginas,
hoc amplius cum oleo et aceto eodem tempore extra
102 faucibus inlitus cum pice liquida. emollit et alvum
vino mixto, innoxie[2] et taenearum genera pellit in vino
potus. aestus balnearum convalescentes ut tolerare
possint linguae subditus praestat. nervorum dolorem,
maxime circa umeros et renes, in saccis aqua ferventi
crebro candefactus levat, colum torminaque et cox-
arum dolores potus et in isdem saccis inpositus
candens, podagras cum farina ex melle et oleo tritus,
ibi maxime usurpanda observatione quae totis corporibus nihil esse utilius sale et sole dixit. itaque[3]
cornea videmus corpora piscatorum. sed hoc prae-
103 cipuum dicatur[4] in podagris. tollit et clavos pedum,
item perniones. ambustis ex oleo inponitur aut com-

---

[1] *Non post* prodest *sed* quicumque *comma Mayhoff.*
[2] innoxie dT *Mayhoff*: innoxio V *Detlefsen*: innoxia RE.
[3] itaque dTEr: utique *coni. Ianus.*
[4] dicatur *codd.*: iudicatur *Mayhoff.*

---

should be fomented with salsugo.ᵃ For these purposes Spanish salt is chosen. For cataract it is ground in a little stone mortar with milk; for bruises a specific is salt wrapped in linen, dipped frequently in boiling water, and applied; for running ulcers in the mouth it is applied in lint; it is rubbed on swollen gums, and for roughness of the tongue it is broken and ground up fine. They say that teeth neither rot nor decay if one daily while fasting in the morning keeps a piece of salt under the tongue until it melts. It also cures leprous sores, boils, lichen and psoriasis, used with stoned raisons, beef suet, origanum, and leaven or bread; for these purposes and for pruritus Theban salt is mostly chosen. For diseased tonsils and uvula salt with honey is beneficial. For quinsy any salt is good, but all the more when oil and vinegar are added, while at the same time salt and liquid pitch are applied externally to the throat. Mixed with wine salt also softens the belly, and taken in wine drives out harmlessly the various kinds of worms. Placed under the tongue salt enables convalescents to endure the heat of the bath. Pains of the sinews, especially in the region of the shoulders and kidneys, are relieved by salt in bags, kept hot by frequent dipping into boiling water; colitis, griping and sciatica by taking salt in drink and by hot applications in the same kind of bags; gout by salt pounded with flour, honey, and oil. Herein is especially applicable the saying that for the whole body nothing is more beneficial than salt and sun. Accordingly we see that the bodies of fishermen are horny, but the above remark should be applied especially to gout. It also removes corns on the feet and chilblains. It is applied to burns in oil or

manducatus pusulasque reprimit, ignibus vero sacris
ulceribusque quae serpant ex aceto aut hysopo,
carcinomatis cum uva taminia, phagedaenis ulcerum
tostus cum farina hordei, superinposito linteolo
madente vino. morbo regio laborantes, donec sudent
ad ignem, contra pruritus quos sentiunt ex oleo et
104 aceto infricatus iuvat, fatigatos ex oleo. multi et
hydropicos sale curavere fervoresque febrium cum
oleo perunxere et tussim veterem linctu eius dis-
cussere, clysteribus infudere ischiadicis, ulcerum
excrescentibus vel putrescentibus inposuere, crocodilorum morsibus ex aceto in linteolis ita ut battue-
rentur ante ulcera. bibitur et contra opium ex aceto
mulso, luxatis inponitur cum farina et melle, item
105 extuberationibus. dentium dolori cum aceto fotus
et inlitus cum resina prodest. ad omnia autem
spuma salis iucundior utiliorque. sed quicumque
sal acopis additur ad excalfactiones, item zmegmatis
ad extendendam [1] cutem levandamque. pecorum
quoque scabiem et boum inlitus tollit, daturque lin-
gendus et oculis iumentorum inspuitur. haec de sale
dicta sint.

106   XLVI. Non est differenda et nitri natura, non
multum a sale distans et eo diligentius dicenda, quia
palam est medicos qui de eo scripserunt ignorasse
naturam nec quemquam Theophrasto diligentius
tradidisse. exiguum fit apud Medos canescentibus

---

[1] extendendam E r *vulg.*: extenuendam VR: extenuandam
dT.

---

[a] Pliny seems to have confused the verbs βάπτω (Dioscorides) and τύπτω.

chewed. It checks blisters, but for erysipelas and for creeping ulcers vinegar or hyssop is added, for carcinomata taminian grapes, while for phagedaenic ulcers it is roasted with barley meal, a linen cloth being placed on top, soaked in wine. Sufferers from jaundice are helped by rubbing with salt, oil, and vinegar before a fire until they sweat; this relieves the itching caused by this disease. Oil should be used in cases of fatigue. Many have treated dropsy too with salt, rubbed with salt and oil hot feverish patients, stayed a chronic cough by licking it, injected salt enemas into sufferers from sciatica, applied it to swollen or festering ulcers, and treated crocodile bites by salt and vinegar in lint cloths, taking care first to flog *a* the sores with them. Salt is taken in oxymel for poisoning by poppy-juice, with flour and honey it is applied to dislocations, and also to tumours. Fomenting with salt and vinegar, or an application of salt and resin, is good for tooth-ache. But for all purposes foam of salt is more pleasant and more beneficial. Salt however of any kind is added to anodynes for a warming effect, also to detergents for stretching and smoothing the skin. An application of salt removes itch-scab in sheep and oxen; salt is also given to be licked, and it is spit into the eyes of draught animals. This must suffice for my account of salt.

XLVI. I must not put off describing the character *Soda.* of soda, which is very similar to salt; a more careful account must be given because it is plain that the physicians who have written about it were ignorant of its character, and that nobody has given a more careful description than Theophrastus. A little is formed in Media in valleys that are white through

siccitate convallibus, quod vocant halmyraga, minus
etiam in Thracia iuxta Philippos, sordidum terra quod
107 appellant agrium. nam quercu cremata numquam
multum factitatum est et iam pridem in totum
omissum. aquae vero nitrosae plurimis in locis repperiuntur, sed sine viribus densandi. optimum copio-
sumque in Clitis[1] Macedoniae, quod vocant Chales-
tricum, candidum purumque, proximum sali. lacus
est nitrosus exiliente e medio dulci fonticulo. ibi fit
nitrum circa canis ortum novenis diebus totidemque
108 cessat ac rursus innatat et deinde cessat. quo
apparet soli naturam esse quae gignat, quoniam
compertum est nec soles proficere quicquam, cum
cesset, nec imbres. mirum et illud, scatebra fonticuli
semper emicante lacum neque augeri neque effluere.
his autem diebus quibus gignitur si fuere imbres,
salsius nitrum faciunt, aquilones deterius, quia vali-
109 dius commovent limum. et hoc quidem nascitur,
in Aegypto autem conficitur multo abundantius,
sed deterius. nam fuscum lapidosumque est. fit
paene eodem modo quo sal, nisi quod salinis mare
infundunt, Nilum autem[2] nitrariis. hae † cedente †
Nilo[3] siccantur, † decedente † madent suco nitri XL

---

[1] in Clitis] *coni.* inclutis (aquis) *Mayhoff.*
[2] autem E: autem mo VRd: autumno *Mayhoff.*
[3] Nilo . . . decedente *om.* VR[1]dT: accedente Nilo rigantur, decedente *Mayhoff*: excedente Nilo siccantur, recedente
Detlefsen: cedente *codd.*: decedente (-tem E) Er: *uncos ego
posui.*

---

a I.e. "wild soda."
b Mayhoff's guess makes an adjective (*inclutis*) of "*in
Clitis*," meaning "famous."
c A *locus nondum sanatus*. From the next sentence it is
clear that the flow into the beds was controlled, so that it
appears that only the falling Nile was admitted. This would

drought; they call it *halmyrax*. It is also found in Thrace near Philippi, but in less quantities and contaminated with earth; it is called *agrium.*[a] But soda from burnt oak-wood was never made in large quantities, and the method has long been altogether abandoned. Alkaline water, however, is found in very many places, but the soda is not concentrated enough to solidify. At Clitae [b] in Macedonia is found in abundance the best, called soda of Chalestra, white and pure, very like salt. There is an alkaline lake there with a little spring of fresh water rising up in the centre. Soda forms in it about the rising of the Dog-star for nine days, ceases for nine days, comes to the top again and then ceases. This shows that it is the character of the soil that produces soda, since it has been discovered that, when it ceases, neither sunshine is of any help at all nor yet rain. Another wonderful thing about the lake is that although the spring is always bubbling up it neither gets larger nor overflows. But if, on those days on which soda forms, has been rain, it makes the soda more salty, while north winds on those days, by stirring up the mud too vigorously, makes it inferior. This soda is natural, but in Egypt it is made artificially, in much greater abundance but of inferior quality, for it is dark and stony. It is made in almost the same manner as is salt, except that they pour sea-water into the salt-beds but the Nile into the soda-beds. The latter † as the Nile rises become dry;[c] as it falls † they are moist with liquid soda for

require *accedente* and *decedente*. Mayhoff conjectured *accedente*, but read *rigantur* for *siccantur*, because he held that the rising Nile filled the beds. It is a pity that VRdT have a hiatus here, for the missing words might have thrown light on the difficulty.

diebus continuis, non ut in Macedonia statis.[1] si
etiam imbres adfuerunt, minus ex flumine addunt,
statimque ut densari coeptum est, rapitur, ne resolvatur in nitrariis. sic [2] quoque olei natura intervenit,
ad scabiem animalium utilis. ipsum autem conditum
110 in acervis durat. mirum in lacu Ascanio et quibusdam circa Chalcida fontibus summas aquas dulces
esse potarique, inferiores nitrosas. in nitro optimum
quod tenuissimum, et ideo spuma melior, ad aliqua
tamen sordidum, tamquam ad inficiendas purpuras
tincturasque omnes. magnus et vitro usus, qui
111 dicetur suo loco. nitrariae Aegypti circa Naucra-
tim et Memphin tantum solebant esse, circa Memphin
deteriores. nam et lapidescit ibi in acervis, multique
sunt cumuli ea de causa saxei. faciunt ex his vasa,
nec non et frequenter liquatum cum sulpure
coquentes. in corporibus [3] quoque quae [4] inveterari
volunt illo nitro utuntur. sunt ibi nitrariae in quibus
112 et rufum exit a colore terrae. spumam nitri, quae
maxime laudatur, antiqui negabant fieri nisi cum ros
cecidisset praegnantibus nitrariis, sed nondum pari-
entibus. itaque non fieri incitatis, etiamsi caderet.
113 alii acervorum fermento gigni existimavere. proxima
aetas medicorum aphronitrum tradidit in Asia colligi

---

[1] statis *codd.*: cessantis *coni. Mayhoff.*
[2] sic *codd.*: hic *vet. Dal., Mayhoff.*
[3] corporibus *coni. K. C. Bailey, Hermathena 1926 :* carnibus
Ianus, Detlefsen, Mayhoff: carbonibus *codd.*
[4] quae *Bailey*: quas *codd.*

---

[a] Or, with the reading *hic*, " here."

forty days on end, and not as in Macedonia during fixed periods. If rain also has fallen, they add less river water, and gather at once the soda that has begun to solidify, lest it should melt back into the soda-bed. Thus [a] too oily matter forms among the soda, useful for itch-scab on animals. Soda however, stored in heaps, lasts a long time. A wonder of Lake Ascanius and of certain springs around Chalcis is that the surface water is sweet and drinkable but underneath is alkaline. Of soda the best is the finest, and therefore froth of soda is superior, but for some purposes the impure is good, for example colouring purple cloths and all kinds of dyeing. Soda is of great use in the making of glass, as will be described in its proper place. [b] The soda-beds of Egypt used to be confined to the regions around Naucratis and Memphis, the beds around Memphis being inferior. For the soda becomes stone-like in heaps there, and many of the soda piles there are for the same reason quite rocky. From these they make vessels, and frequently by baking melted soda with sulphur. For the bodies too that they wish to embalm this is the soda they use. In this region are soda-beds from which red soda also is taken owing to the colour of the earth. Foam of soda, which is very highly prized, the ancients said was formed only when dew had fallen on beds teeming with soda but not yet bringing it forth; accordingly, even if dew fell, soda did not form on beds in agitated action. Others have thought that foam is produced by fermentation of the heaps. The last generation of physicians said that in Asia was gathered aphronitrum [c] oozing in

[b] XXXVI. § 193.
[c] A Greek word meaning " soda foam."

in speluncis mollibus [1] destillans—specus eos colligas [2] vocant—dein siccant sole. optimum putatur
Lydium; probatio, ut sit minime ponderosum et
maxime fricabile, colore paene purpureo. hoc in
pastillis adfertur, Aegyptium in vasis picatis,[3] ne
liquescat. vasa quoque ea sole inarescentia per-
114 ficiuntur. nitri probatio, ut sit tenuissimum et
quam maxime spongeosum fistulosumque. adulteratur in Aegypto calce, deprehenditur gustu.
sincerum enim statim resolvitur, adulteratum calce
pungit et asperum [4] reddit odorem vehementer.
uritur in testa opertum ne exultet, alias igni non
exilit nitrum, nihilque gignit aut alit, cum in salinis
herbae gignantur et in mari tot animalia, tantum algae.
115 sed maiorem esse acrimoniam nitri apparet non hoc
tantum argumento sed et illo quod nitrariae calcia-
menta protinus consumunt, alias salubres oculorum-
que claritati utiles. in nitrariis non lippiunt. ulcera
allata eo celerrime sanantur, ibi facta tarde. ciet et
sudores cum oleo perunctis corpusque emollit. in
pane salis vice utuntur Chalestraeo, ad raphanos
Aegyptio, teneriores eos facit, sed obsonia alba et
deteriora, olera viridiora. in medicina autem calfacit, extenuat, mordet, spissat, siccat, exulcerat,

[1] mollibus VRdTf: canalibus *Detlefsen*: molibus *Gelenius,
Mayhoff, qui etiam* nobilibus *vel* madidis *coni.*
[2] colligas (-gans E[1]) *codd.*, *Mayhoff*: Corycias *Detlefsen*:
alii alia.
[3] picatis d *vulg.*, *Mayhoff*: spissatum *Detlefsen*: spissatis
RE.
[4] asperum *cod.* a, *Detlefsen*: aspersum d *vulg. Mayhoff*:
aspersu VRf.

[a] Usually emended. But the word *mollis* may refer to a
cave with soft sides and floor, through which soda might ooze.
[b] This word is probably corrupt.

soft ^{a} caves—they are called *colligae* ^{b}—and then dried in the sun. The best is thought to be Lydian. The tests are that it should be the least heavy and the most friable, and of an almost purple colour. The last kind is imported in lozenges, but the Egyptian in vessels lined with pitch, lest it melt. These vessels too are finished off by being dried in the sun. The tests of soda are that it should be very fine and as spongy and full of holes as possible. In Egypt it is adulterated with lime, which is detected by the taste; for pure soda melts at once, but adulterated soda stings because of the lime, and gives out a strong, bitter ^{c} odour. It is burnt in an earthen jar with a lid, lest it should crackle out; otherwise soda does not crackle in fire; it produces nothing and nourishes nothing, whereas in salt-pits grow plants, and in the sea so many animals and so much sea-weed. ^{d} But that the pungency of soda is greater is shown not only by this evidence but also by the fact that soda-beds at once consume shoes, but are otherwise healthful and good for clearness of vision. In the soda-beds nobody has ophthalmia; sores brought there heal very quickly, but those that form there heal slowly. Soda and oil also make to sweat those who are rubbed with the mixture, which softens the flesh. They use Chalestran soda for bread instead of salt, Egyptian soda for radishes; it makes them more tender, but meats white and inferior and vegetables greener. In medicine soda warms, alleviates, stings, braces, dries, and clears away ^{e} ulcers, and is useful

^{c} With the reading *aspersum*: " when sprinkled it has a strong smell."
^{d} Or: " only sea-weeds."
^{e} See XXVII. § 22 and note on XXVII. § 105.

449

116 utile his quae evocanda sint aut discutienda et lenius
mordenda atque extenuanda, sicut in papulis pusulis-
que. quidam in hoc usu accensum vino austero
restingunt atque ita trito in balneis utuntur sine oleo.
sudores nimios inhibet cum iride arida adiecto oleo
viridi, extenuat et cicatrices oculorum et scabritias
genarum cum fico inlitum aut decoctum in passo ad
dimidias partes, item contra argema, oculorum ungues.

117 decoctum cum passo in[1] mali Punici calyce adiuvat
claritatem visus cum melle inunctum. prodest
dentium dolori ex vino, si cum pipere colluantur;
item cum porro decoctum nigrescentes dentes,
crematum dentrifricio, ad colorem reducit. capitis
animalia et lendes necat cum Samia terra inlitum ex
oleo. auribus purulentis vino liquatum infunditur,
sordes eiusdem partis erodit ex aceto, sonitus et tin-

118 nitus discutit siccum additum. vitiligines albas cum
creta Cimolia aequo pondere ex aceto in sole inlitum
emendat. furunculos admixtum resinae extrahit, aut[2]
cum uva alba passa nucleis eius simul tritis. testium
inflammationi occurrit, item eruptionibus pituitae
in toto corpore cum axungia, contraque canis morsus
addita et resina † inlitis †.[3] cum aceto inlinitur. sic
et serpentium morsibus, phagedaenis et ulceribus
quae serpunt aut putrescunt cum calce ex aceto.
hydropicis cum fico tusum datur inliniturque. discu-

---

[1] cum passo in *codd.*: in passo cum *Mayhoff.*
[2] extrahit aut *codd.*: extrahit *Mayhoff.*
[3] inlitis VVᵈR *Mayhoff*: initis E r *Detlefsen*: *uncos ego
addidi.*

---

[a] With Mayhoff's reading: "in raisin wine with pomegranate rind."
[b] In this part at any rate of Pliny the first words of each
clause seem to indicate the complaint. This fact should, I

for conditions where there must be withdrawal, dispersal, and gentle stinging and alleviation, as with pimples and blisters. Some for this purpose set it on fire and put it out with a dry wine, and use it so prepared and ground in the bath without oil. Excessive sweats are checked by soda with dried iris and the addition of green oil; it also improves scars on the eyes and roughness of the lids if applied with fig, or boiled down to one half in raisin wine, a preparation too which is used for white ulcers and inflamed swellings on the eyes. Boiled down with raisin wine in a pomegranate rind,[a] and applied with honey, it improves vision. Soda is good for toothache if a mouth-wash is made by adding pepper and wine. Boiled down too with leek, and burnt to make a dentifrice, it restores the colour of blackening teeth. Insects and nits on the head it kills if applied in oil with Samian earth. Dissolved in wine it is poured into purulent ears; wax in the same organ it eats away in vinegar; noises and singing it stops if added dry. Applied in sunshine with vinegar and an equal weight of Cimolian chalk it cures the white kinds of psoriasis. It brings to a head boils, either mixed with resin or with white raisins, the pips being ground up with them. With axle-grease it combats inflammation of the testicles, and also outbursts of phlegm on the whole body; it is applied with vinegar, resin being added, to dog-bites. This preparation is used for snake bites; for phagedaenic, creeping, or festering ulcers, with lime and vinegar; for dropsy it is pounded with figs and administered by the mouth and externally.[b] Griping pains too it

think, determine the punctuation. Editors differ widely in this.

119 tit et tormina, si decoctum bibatur pondere drachmae
cum ruta vel aneto vel cumino. reficit lassitudines cum
oleo et aceto perunctorum, et contra algores horrores-
que prodest manibus pedibusque confricatis cum
oleo. conprimit et pruritus suffusorum felle, maxime
cum aceto in sudore datum.[1] succurrit et venenis
fungorum ex posca potum aut, si buprestis hausta sit,
ex aqua, vomitionesque evocat. his qui sanguinem
120 tauri biberint cum lasere datur. in facie quoque
exulcerationes sanat cum melle et lacte bubulo.
ambustis tostum donec nigrescat tritumque inlinitur.
infunditur † urceis †[2] et renium dolori aut rigori
corporum nervorumve doloribus. paralysi in lingua
cum pane inponitur. suspiriosis in tisana sumitur.
121 tussim veterem sanat flore, mixto galbano resinae
terebinthinae, pari pondere omnium ita, ut fabae
magnitudo devoretur. coquitur dilutumque postea
cum pice liquida sorbendum in angina datur. flos
eius cum oleo cypreo et articulorum doloribus in sole
iucundus est. regium quoque morbum extenuat in
potione vini et inflationes discutit, sanguinis profluvium e naribus sistit ex ferventi aqua vapore naribus
122 rapto. porriginem alumine permixto tollit, alarum
virus ex aqua cottidiano fotu, ulcera ex pituita nata
cera permixtum, quo genere nervis quoque prodest.
coeliacis infunditur. perungui ante accessiones

---

[1] in sudore datum *Sillig*: instillatum *Mayhoff*: insudatum
codd.

[2] urceis *codd.*: ventris *Caesarius*: vesicae *Mayhoff*. *Warmington* umeris *coni.*

---

[a] The *urceis* of all the MSS. seems corrupt, and no proposed
emendation explains the cause of the corruption. Mayhoff's
vesicae is the word usually associated in Pliny with *renium*.

allays if there is taken a drachma by weight boiled
down with rue or dill or cummin. The pains of
fatigue are removed by rubbing all over with soda,
oil, and vinegar, while for chills and shivers it is of
advantage to rub the hands and feet thoroughly
with soda and oil. It also checks the itch of jaundice,
especially when administered with vinegar while the
patient is sweating. Taken in vinegar and water
soda is beneficial against the poisons of fungi; if a
buprestis has been swallowed it is taken in water;
it is also a good emetic. It is given in laser to those
who have drunk bull's blood. Ulcerations also on
the face it heals with honey and cow's milk. It is
applied to burns roasted until it turns black and
crushed to powder. It is injected for pain in the
. . . *a* and kidneys, or for rigors of the body, or for
pains of the sinews. For paralysis of the tongue it
is applied there with bread, and for asthma it is
taken in barley gruel. Chronic cough is cured by
flower of soda with galbanum mixed with terebrinth
resin, all equal in weight, but the piece to be swallowed must be of the size of a bean. Soda, boiled
and then combined with liquid pitch, is given to be
swallowed by patients with quinsy. Flower of soda
with oil of cyprus is also soothing if applied in the sun
for pains in the joints. Jaundice also it alleviates
taken in a draught of wine; this remedy relieves
flatulence. It checks epistaxis if inhaled in the steam
from boiling water. By soda mixed with alum is
removed scurf, rank smell of the armpits by daily
fomentation with soda and water, sores due to nose-
running by soda mixed with wax—a mixture also
good for the sinews—and it is injected for the coeliac
affection. Many have prescribed complete rubbing

frigidas nitro et oleo multi praecepere, sicut adversus lepras, lentigines; podagris in balneis uti.   solia nitri prosunt atrophis, opisthotonis, tetanis.   sal nitrum sulpuri concoctum in lapidem vertitur.

123   XLVII. Spongearum genera diximus in naturis aquatilium marinorum.   quidam ita distingunt: alias ex his mares tenui fistula spissioresque, persorbentes, quae et tinguntur in deliciis, aliquando et purpura; alias feminas maioribus fistulis ac perpetuis; maribus[1] alias duriores, quas appellant tragos, tenuissimis fistulis atque densissimis.   candidae cura fiunt: e mollissimis recentes per aestatem tinctae salis spuma ad lunam et pruinas sternuntur inversae, hoc est qua parte adhaesere, ut candorem bibant.   animal 124 esse docuimus, etiam cruore inhaerente.   aliqui narrant et auditu regi eas contrahique ad sonum, exprimentes abundantiam umoris, nec avelli petris posse, ideo abscidi ac saniem remittere.   quin et eas[2] quae ab aquilone sint genitae praeferunt ceteris, nec usquam diutius durare spiritum medici adfirmant. sic et prodesse corporibus, quia nostro suum misceant, et ideo magis recentes magisque umidas, sed minus

---

[1] maribus *codd.*: e maribus *Hermolaus Barbarus*: in maribus *Sillig.*

[2] *Ante* eas *lacunam indicat Mayhoff, qui fere* abscisas aliquamdiu vivere *excidisse putat.*

---

[a] Or: " the undernourished."

[b] For *nitrum* see Additional Note I, pp. 567–568.

[c] Book IX. § 148.

[d] The adjective *perpetuus* in this context is difficult.   It could mean "never closed," referring to sponges growing in the sea, or "connected with one another," used of the sponges of commerce.   See Additional Note J, p. 568.

[e] Or: *e* (or *in*) *maribus*: " of the males, the harder."

[f] A Greek word, τράγοι, " goats."   [g] See IX. § 149.

with soda and oil before the chills of fever come on, and so to use it for leprous sores and freckles; and they prescribe its use in the bath for gouty people. Soda baths are good for consumptives,[a] and for the victims of opisthotonus and other forms of tetanus. Salt and soda, when heated with sulphur, turn to stone.[b]

XLVII. Of the kinds of sponges I have spoken [c] *Sponges.* when describing the nature of marine creatures. Certain authorities classify them thus: some sponges, the males, have little holes, and are more compact and very absorbent; they are also dyed for the luxurious, sometimes even with purple; others, the females, have larger and uninterrupted [d] holes; others, harder [e] than the males, called *tragi,*[f] have very small holes that are very close together. Sponges are whitened artificially. Fresh sponges, of the softest kind, are soaked in foam of salt throughout the summer, and then laid open to the moon and hoar-frosts upside down, that is, with the side uppermost that adhered to the rocks, so that they may drink in whiteness. I have said [g] that sponges are animal, being even lined with a coating of blood. Some also declare that they are guided by a sense of hearing, and contract at a noise, sending out a great quantity of moisture; that they cannot be torn from the rocks, and therefore are cut off, bleeding sanies. Moreover, those [h] growing exposed to the north-east they prefer to others, and physicians declare that nowhere else does their breath last for a longer time. Such too, they say, are beneficial to the human body, because they mix their breath with

[h] The lacuna supposed by Mayhoff to be here he would fill up by words roughly meaning: " that cut off they live for a considerable time."

in calida aqua minusque unctas aut unctis corporibus
inpositas et spissas minus adhaerescere.[1] mollis-
125 simum genus earum penicilli. oculorum tumores sedant ex mulso inpositi, iidem abstergendae lippitudini,
utilissime ex aqua; tenuissimos esse mollissimosque
oportet. inponuntur et spongeae ipsae epiphoris ex
posca et aceto calido ad capitis dolores. de cetero
recentes discutiunt, mitigant, molliunt, veteres non
glutinant vulnera. usus earum ad abstergenda,
fovenda, operienda a fotu, dum aliud inponatur.
126 ulcera quoque umida et senilia inpositae siccant.
fracturae et vulnera spongeis utilissime foventur.
sanguis rapitur in secando, ut curatio perspici possit.
et ipsae vulnerum inflammationibus inponuntur nunc
siccae, nunc aceto adspersae nunc vino, nunc ex aqua
frigida; ex aqua vero caelesti inpositae secta recentia
127 non patiuntur intumescere. inponuntur et integris
partibus, sed fluctione occulta laborantibus quae discutienda sit, et his quae apostemata vocant melle
decocto perunctis, item articulis alias aceto salso
madidae, alias e posca; si ferveat impetus, ex

---

[1] adhaerescere E r *vulg.*: adhaerescente *aut* adhaerescentem
ceteri codd.: adhaerescentes *Mayhoff.*

ours; therefore fresh sponges are the more beneficial, as are also the moist, but less beneficial are those soaked in hot water, or those that are oily, or laid on oily bodies, while compact sponges are less adhesive. The softest kind of sponge is that used for bandage-rolls. Applied in honey wine these relieve swollen eyes. They are also good for wiping away the rheum of ophthalmia, which they do most efficiently with water. They should be very fine and very soft. Sponges themselves *a* are applied in vinegar and water for eye-fluxes, and in warm vinegar for headaches. For the rest, fresh sponges are dispersive, soothing, and emollient; old sponges do not close wounds. The uses of sponges are to be detergent, to foment, and after fomentation to cover until something else is applied. Applied also to wet ulcers of senile persons, sponges dry them, and they foment with the greatest benefit fractures and wounds. In surgery sponges quickly absorb the blood, so that treatment can easily be observed. Sponges themselves are applied to inflamed wounds, sometimes dry, at other times moistened with vinegar, or wine, or cold water; applied indeed in rain-water to fresh incisions they prevent their swelling. They are also laid on parts that are whole, but suffering from a hidden flux that has to be dispersed, and also on what are called *apostemata*,*b* after rubbing them with boiled honey; on joints also, sometimes moistened with salted vinegar, sometimes with vinegar and water; should the complaint be attended

---

a *Ipsae* can hardly mean " by themselves," as it apparently does in § 126, for *ex posca* seems to go with it. It may mark a contrast with the sponge ash of § 129.

b A Greek word, " abscesses."

aqua. eaedem[1] callo e salsa, at contra scorpionum
ictus ex aceto. in vulnerum curatione et sucidae
lanae vicem implent[2] ex eadem; differentia haec,
quod lanae emolliunt, spongeae coercent rapiuntque
128 vitia ulcerum. circumligantur et hydropicis siccae
vel ex aqua tepida poscave, utcumque blandiri opus
est operirive[3] aut siccare cutem. inponuntur et his
morbis quos vaporari oporteat, ferventi aqua perfusae
expressaeque inter duas tabulas. sic et stomacho
prosunt et in febri contra nimios ardores, sed splenicis
e posca, ignibus sacris ex aceto efficaciores quam
aliud; inponi oportet sic ut sanas quoque partes
129 spatiose operiant. sanguinis profluvium sistunt ex
aceto aut frigida, livorem ab ictu recentem ex aqua
salsa calida saepius mutata tollunt, testium tumorem
doloremque ex posca. ad canum morsus utiliter con-
cisae inponuntur ex aceto aut frigida aut melle,
abunde subinde umectandae. Africanae cinis cum
porri sectivi suco sanguinem reicientibus haustus,
aliis[4] ex frigida, prodest. idem cinis vel cum oleo vel
130 cum aceto fronti inlitus tertianas tollit. privatim
Africanae ex posca tumorem discutit, omnium autem
cinis cum pice crematarum sanguinem sistit vulnerum; aliqui raras tantum ad hoc cum pice urunt.

---

[1] eaedem *Mayhoff*: eadem *codd.*
[2] *Post* implent *add.* nunc ex vino et oleo nunc ex eadem
vulg. ante Ianum.
[3] operirive *plerique codd.*: operireve *cod. a Mayhoff.*
[4] haustus aliis *Mayhoff*: haustu salis *codd.*

---

[a] See Önnerfors *Pliniana*, pp. 167, 168 for *ve* after a short *-e*.
[b] This is a dubious reading, but *haustu salis* without *cum*
can scarcely be right.

with fever, water alone is to be used. With salt and
water sponges are also applied to callosities, but with
vinegar to scorpion stings. In the treatment of
wounds sponges with salt and water also act as a
substitute for greasy wool; the difference is that
wools soften, but sponges are astringent and absorb
quickly the diseased humours of ulcers. They are
also bound round dropsical parts, either dry or with
warm water or vinegar and water, whenever there
is need to soothe, or cover [a] the skin, or dry it. They
are applied also for such diseases as need a steamy
heat, steeped in boiling water, and pressed between
two boards. So applied they are also good for the
stomach, and for the excessive burnings of fever;
but for the spleen with vinegar and water, while for
erysipelas they are with vinegar more efficacious
than anything; they should be so placed that there
is ample covering for the healthy parts. With
vinegar or cold water they arrest haemorrhage, with
hot salt and water, often changed, they remove
fresh bruises caused by a blow, and with vinegar and
water they cure swollen and painful testicles. For
dog-bite are applied beneficially with vinegar, cold
water, or honey, cut-off pieces of sponge, which must
be thoroughly moistened every now and then. The
ash of the African sponge, swallowed with the juice
of cut-leek, is good for spitting of blood; for other [b]
complaints it should be taken in cold water. This
ash also, applied to the forehead with oil or vinegar,
cures tertian agues. African sponges are specific
with vinegar and water for reducing swellings, and
the ash of all sponges burnt with pitch arrest
haemorrhage from wounds; for this purpose some
burn with pitch only sponges of loose texture. For

et oculorum causa comburuntur in cruda olla figulini operis, plurimum proficiente eo cinere contra scabritias genarum excrescentesque carnes et quicquid opus sit ibi destringere, spissare, explere. utilius in eo usu lavare cinerem. praestant et strigilium vicem

131 linteorumque adfectis corporibus. et contra solem apte protegunt capita. medici inscitia ad duo nomina eas redegere, Africanas, quarum firmius sit robur, Rhodiacasque ad fovendum molliores. nunc autem mollissimae circa muros Antiphelli urbis reperiuntur. Trogus auctor est circa Lyciam penicillos mollissimos nasci in alto, unde ablatae sint spongeae, Polybius super aegrum suspensos quietiores facere noctes. nunc praevertemur ad marina animalia.

eye remedies sponges are burnt in an unbaked earthenware pot, this ash being very efficacious indeed for roughness or excrescences of the eyelids, and for any complaint in the region of the eyes that needs a remedy detergent, astringent, or expletive, but for this treatment it is better to rinse the ash. They also furnish a substitute for scrapers and towels when the body is diseased. Sponges protect also efficiently the head against the sun. In their ignorance physicians have reduced sponges to two classes: the African, which are firmer and harder, and the Rhodian, which are softer for fomentations. Today however *a* very soft sponges are found around the walls of Antiphellus.*b* Trogus informs us that around Lycia very soft tent-sponges grow out at sea, in places where sponges have been taken away; Polybius that hung over a sick man these give more peaceful nights. Now I shall turn my attention to the creatures of the sea.

a Warmington thinks that Pliny is translating the Greek νῦν δέ ("as things are").
b A city of Lycia.

BOOK XXXII

LIBER XXXII

1 I. Ventum est ad summa naturae exemplorumque
per rerum ordinem, et ipsum sua sponte occurrit inmensum potentiae occultae documentum, ut prorsus
neque aliud ultra quaeri debeat nec par ac simile.
possit inveniri, ipsa se vincente natura, et quidem
numerosis modis. quid enim violentius mari ventisve
et turbinibus ac procellis? quo maiore hominum
ingenio[1] in ulla sui parte adiuta est quam velis re-
misque? addatur his et reciproci aestus inenarrabilis
2 vis versumque totum mare in flumen. tamen omnia
haec pariterque eodem inpellentia unus ac parvus
admodum pisciculus, echenais appellatus, in se tenet.
ruant venti licet, saeviant procellae: imperat furori
viresque tantas compescit et cogit stare navigia,
quod non vincula ulla, non ancorae pondere inrevoca-
bili iactae.[2] infrenat impetus et domat mundi
rabiem nullo suo labore, non renitendo aut alio modo
3 quam adhaerendo. hoc tantulo[3] satis est, contra tot
impetus ut vetet ire navigia. sed[4] armatae classes
inponunt sibi turrium propugnacula, ut in mari quoque pugnetur velut e muris. heu vanitas humana,

---

[1] ingenio *codd.*: invento *coni. Mayhoff.*
[2] iactae *fere omnes codd.*: factae E.
[3] hoc tantulo *codd.*: hoc tantulum (-lū) *coni. Mayhoff.*
[4] sed *codd.*: ecce *coni. Mayhoff.*

---

[a] Or, with Mayhoff's conjecture, "invention."

BOOK XXXII

See *Index of Fishes* for identification of aquatic creatures.

I. THE course of my subject has brought me to the ^{*The sea and sea creatures.*} greatest of Nature's works, and I am actually met by such an unsought and overwhelming proof of hidden power that inquiry should really be pursued no further, and nothing equal or similar can be found, Nature surpassing herself, and that in numberless ways. For what is more violent than sea, winds, whirlwinds, and storms? By what greater skill [a] of man has Nature been aided in any part of herself than by sails and oars? Let there be added to these the indescribable force of tidal ebb and flow, the whole sea being turned into a river. All these, however, although acting in the same direction, are checked by a single specimen of the sucking fish, a very small fish. Gales may blow and storms may rage; this fish rules their fury, restrains their mighty strength, and brings vessels to a stop, a thing no cables can do, nor yet anchors of unmanageable weight that have been cast.[b] It checks their attacks and tames the madness of the Universe with no toil of its own, not by resistance, or in any way except by adhesion. This little creature suffices in the face of all these forces to prevent vessels from moving. But armoured fleets bear aloft on their decks a rampart of towers, so that fighting may take place even at sea as from the walls of a fortress.

[b] With the reading *factae*: " made of incalculable strength."

cum rostra illa aere ferroque ad ictus armata semi-
pedalis inhibere possit ac tenere devincta pisciculus!
fertur Actiaco Marte tenuisse praetoriam navem
Antoni properantis circumire et exhortari suos, donec
transiret in aliam, ideoque Caesariana classis impetu
maiore protinus venit. tenuit et nostra memoria Gai
4 principis ab Astura Antium renavigantis. ut res est,[1]
etiam auspicalis pisciculus, siquidem novissime tum
in urbem reversus ille imperator suis telis confossus
est, nec longa fuit illius morae admiratio, statim causa
intellecta, cum e tota classe quinqueremis sola non
proficeret, exilientibus protinus qui quaererent circa
navem. invenere adhaerentem gubernaculo osten-
deruntque Gaio indignanti hoc fuisse quod se revo-
caret quadringentorumque remigum obsequio contra
5 se intercederet. constabat peculiariter miratum,
quomodo adhaerens tenuisset nec idem polleret in
navigium receptus. qui tunc posteaque videre eum,
limaci magnae similem esse dicunt. nos plurium
opiniones posuimus in natura aquatilium, cum de eo
diceremus, nec dubitamus idem valere omnia ea[2]
genera, cum celebri et consecrato etiam exemplo
apud Cnidiam Venerem conchas quoque esse eius-
6 dem potentiae credi necesse sit. e nostris quidam
Latine moram appellavere eum, mirumque, e Graecis

---

[1] ut res est B, *Mayhoff.*
[2] ea B, *Mayhoff: om. ceteri codd.*

---

[a] See IX. § 79.
[b] That is: "delay." It has none of the powers ascribed to
it by Pliny.

How futile a creature is man, seeing that those rams, armed for striking with bronze and iron, can be checked and held fast by a little fish six inches long! It is said that at the battle of Actium the fish stopped the flagship of Antonius, who was hastening to go round and encourage his men, until he changed his ship for another one, and so the fleet of Caesar at once made a more violent attack. Within our memory the fish stayed the ship of the Emperor Gaius as he was sailing back from Astura to Antium. As it turned out, the little fish also proved ominous, because very soon after that Emperor's return to Rome on this occasion he was stabbed by his own men. This delay caused no long surprise, for the reason was immediately discovered; of the whole fleet the quinquereme alone making no progress, men at once dived and swam round the ship to trace the cause. They found this fish sticking to the rudder and showed it to Gaius, who was furious that it had been such a thing that was keeping him back and vetoing the obedience to himself of four hundred rowers. It was agreed that what astonished him in particular was how the fish had stopped him by sticking to the outside, yet when inside the ship it had not the same power. Those who saw the fish then or afterwards say that it is like a large slug. I have given *a* the views of the majority in my account of water creatures, where I discussed the fish, and I do not doubt all this kind of fish have the same power, since there is a famous and even divinely sanctioned example in the temple of the Cnidian Venus, where snails too, we are forced to believe, have the same potency. Of the Roman authorities some have given this fish the Latin name of *mora,b*

alii lubricos partus atque procidentes continere[1] ad
maturitatem adalligatum,[2] ut diximus, prodiderunt,
alii sale adservatum adalligatumque gravidis partus
solvere, ob id alio nomine odinolyten appellari.   quocumque modo ista se habent, quis ab hoc tenendi
navigia exemplo de ulla potentia naturae vique et
effectu in remediis sponte nascentium rerum dubitet?

7   II. Quid? non et sine hoc exemplo per se satis
esset ex eodem mari torpedo? etiam procul et e
longinquo, vel si hasta virgave attingatur, quamvis
praevalidos lacertos torpescere, quamlibet ad cursum
veloces alligari pedes? quod si necesse habemus
fateri hoc exemplo esse vim aliquam, quae odore
tantum et quadam aura corporis sui adficiat membra,
quid non de remediorum omnium momentis speran-
dum est?

8   III. Non sunt minus mira quae de lepore marino
traduntur. venenum est aliis in potu aut cibo datus,
aliis etiam visu, siquidem gravidae, si omnino ad-
spexerint feminam ex eo genere dumtaxat, statim
nausiant et redundatione stomachi vitium fatentur[3]
ac deinde abortum faciunt. remedio est mas ob id
induratus sale, ut in bracchialibus habeant. eadem
res in mari ne tactu quidem nocet. vescitur eo
unum tantum animalium, ut non intereat, mullus
piscis; tenerescit tantum et inertior[4] viliorque fit.

[1] continere B, *Mayhoff*: contineri *ceteri codd.*
[2] adalligatum *Mayhoff*: adalligato B: adalligato eo *plerique
codd., Detlefsen.*
[3] nausiam et redundationem stomachi vomitu fatentur
coni. Mayhoff.
[4] inertior B[1], *Ianus, Mayhoff*: ingratior *codd. vulg.,
Detlefsen.*

[a] See IX. § 79.
[b] That is: " deliverer from birth-pangs."

and a marvel is told by some Greeks, who have related, as I have said,[a] that worn as an amulet it arrests miscarriage, and by reducing procidence of the uterus allows the foetus to reach maturity ; others say that preserved in salt and worn as an amulet it delivers pregnant women, this being the reason why another name, *odinolytes*,[b] is given to it. However these things may be, would anybody after this instance of staying a ship's course entertain doubts about any power, force, and efficacy of nature, to be found in remedies from things that grow spontaneously ?

II. But surely, even without this example, evidence enough by itself could be found in the electric-ray, which also is a sea creature. Even at a distance, and that a long distance, or if it is touched with a spear or rod, to think that the strongest arms are numbed, feet as swift in racing as you like are paralysed ! But if this example forces us to confess that there is a force which by smell alone, and by what I may call the breath from the creature's body, so affects our limbs, what limits are there to our hopes based on the potency of all remedies ?

III. No less wonderful things are related of the sea-hare. To some it is poison if given in drink or food, to others if merely seen, since pregnant women, if they have but looked at one, the female, that is, of the species, at once feel nausea, show by regurgitation signs of a disordered stomach, and then miscarry. The remedy is a male specimen, specially hardened for this purpose with salt, to be worn in a bracelet. In the sea, however, it does not hurt, even by touch. There feeds on it without being killed one creature only, red mullet, which merely becomes flabby, more insipid, and coarser. Struck by it a human being

469

9 homines, quibus inpactus est, piscem olent ; hoc
primo argumento veneficium id deprehenditur.
cetero moriuntur totidem in diebus, quot vixerit
lepus, incertique temporis veneficium id esse auctor
est Licinius Macer. in India adfirmant non capi
viventem invicemque ibi hominem illi pro veneno
esse ac vel digito omnino in mari tactum mori, esse
autem multo ampliorem, sicuti reliqua animalia.

10 IV. Iuba in iis voluminibus, quae scripsit ad C. Caesarem Aug, f. de Arabia, tradit mitulos ternas
heminas capere, cetos sescentorum pedum longitudinis et trecentorum sexaginta latitudinis in flumen
Arabiae intrasse, pinguique eius mercatores negoti-
atos, et omnium piscium adipe camelos perungui in
eo situ, ut asilos ab iis fugent odore.

11 V. Mihi videntur mira et quae Ovidius prodidit
piscium ingenia i̶n̶ ̶e̶o̶ ̶v̶o̶l̶u̶m̶i̶n̶e̶,̶ ̶q̶u̶o̶d̶ ̶h̶a̶l̶i̶e̶u̶t̶i̶c̶o̶n̶ in-
scribitur : scarum inclusum nassis non fronte erum-
pere nec infestis viminibus caput inserere, sed aversum caudae ictibus crebris laxare fores atque ita
retrorsum repere, quem luctatum eius si forte alius
scarus extrinsecus videat, adprehensa mordicus cauda
adiuvare nisus erumpentis ; lupum rete circumdatum

12 harenas arare cauda atque ita condi dum transeat rete ;
murenam maculas adpetere ipsas consciam teretis ac
lubrici tergi, tum multiplici flexu laxare, donec eva-
dat ; polypum hamos adpetere bracchiisque com-

---

[a] Or, perhaps better: " In India they say that etc."
[b] I.e. " On fishing."

smells of fish ; this is the first symptom by which such
poisoning is detected. Furthermore, the victims die
in the same number of days as the hare has lived, and
Licinius Macer is authority for saying that this
poison has variable periods for its action. They say
that in India [a] the sea-hare is never caught alive ;
and that inversely man is there poisonous to the hare ;
that even a mere touch of a human finger in the sea
is fatal to it ; but that like all other animals the Indian
variety is far larger.

IV. In those volumes about Arabia which he
dedicated to Gaius Caesar, the son of Augustus,
Juba related that there are mussels there with shells
holding three heminae ; that a whale 600 feet long
and 360 feet broad entered a river of Arabia ; that
merchants did a trade with its blubber ; and that
camels in that district are rubbed all over with the
fat of any fish, so that gad-flies may be kept away
by the smell.

V. Wonderful too appear to me the characters of
fishes given by Ovid in his book entitled *Halieuticon* : [b]
how the scarus, caught in a weel, does not burst out
to the front, or thrust his head through the osiers
that imprison him, but turns round, widens the gaps
with repeated blows of his tail, and so creeps backwards. If by chance his struggles are seen by
another scarus outside, he seizing the other's tail with
his teeth helps the efforts to burst out. The basse,
he says, when surrounded by a net, ploughs a hole in
the sand with his tail, and so is buried until the net
passes over him. He says too that the murena,
knowing that his back is rounded and slippery, attacks
the meshes themselves, and then by involved
wriggling widens them until he escapes ; that the

471

plecti, non morsu, nec prius dimittere, quam escam
circumroserit, aut harundine levatum extra aquam.
scit et mugil esse in esca hamum insidiasque non
ignorat, aviditas tamen tanta est, ut cauda verber-
13 ando excutiat cibum. minus in providendo lupus
sollertiae habet, sed magnum robur in paenitendo.
nam ut[1] haesit in hamo tumultuoso discursu laxat
volnera, donec excidant insidiae. murenae amplius
devorant quam hamum, admovent dentibus lineas
atque ita erodunt. anthias[2] tradit idem infixo hamo
invertere se, quoniam sit in dorso cultellata spina,
eaque liniam praesecare.
14   Licinius Macer murenas feminini tantum sexus
esse tradit et concipere e serpentibus, ut diximus ob
id sibilo a piscatoribus tamquam a serpentibus evocari et capi.[3] . . . et pinguescere, iactato fusti non
interemi, easdem ferula protinus. animam in cauda
capitis ictum difficulter. novacula pisce qui attacti
sunt, ferrum olent. durissimum esse piscium constat
qui orbis vocetur; rotundus est, sine squamis totus-
que capite constat.
15   VI. Trebius Niger xiphian, id est gladium, rostro
mucronato esse, ab hoc naves perfossas mergi; in
oceano ad locum Mauretaniae, qui Cottae vocetur,
non procul Lixo flumine idem lolligines evolare ex
aqua tradit tanta multitudine, ut navigia demergant.

---

[1] ut *multi codd.*: si in B[1]: si ut B[2] *Sillig*: is, ut *Mayhoff.*
[2] anthias *Urlichs, Detlefsen, Mayhoff*: *varia codd.*
[3] *Hic Mayhoff lacunam esse coni.*

---

[a] See IX. § 76.

polypus attacks the hook, grips it with his tentacles, not teeth, and does not let it go before he has nibbled round the bait, or been lifted out of the water by the rod. The mugil too knows that in the bait is a hook, and is quite aware of the trap; his greed however is so great that by lashing with his tail he knocks off the food. The basse has less cunning insight, but great strength when he realizes his mistake. For when caught on the hook he dashes about wildly, widening the wounds until the snare is torn out. The murena swallows more than the hook, applies the line to his teeth, and so gnaws it through. Ovid also relates that the anthias, when the hook catches, turns over, since on his back is a spine with a knife-edge, with which he cuts through the line.

Licinius Macer relates that the murena is female only, and conceives out of serpents, as I have said,[a] and that therefore fishermen whistle in imitation of a serpent's call, and so catch the fish, and . . . grow fat; that a club hurled at them does not kill, but fennel-giant kills at once. It is certain that the seat of life is in their tail, for if this is struck they very quickly die, but it is difficult to kill them by blows on the head. Those touched by the razor-fish smell of iron. It is a well-known fact that the hardest fish is the *orbis*, which is round, without scales, and all head.[b]

VI. Trebius Niger tells us that the xiphias, that is the sword-fish, has a pointed beak, by which ships are pierced and sunk; in the open sea, off the place in Mauretania called Cottae, not far from the river Lixus, the same authority tells us that the lolligo flies out of the water in such numbers as to sink a

[b] The repetition of *constat* in different senses is very awkward; it is an instance of " unconscious repetition."

Lolligo quotiens cernatur extra aquam volitans,
tempestates mutari.[1]

16 VII. E manu vescuntur pisces in pluribus quidem
Caesaris villis, sed—quae veteres prodidere in stagnis,
non piscinis, admirati—in Heloro Siciliae castello non
procul Syracusis, item in Labrayndi Iovis fonte
anguillae et inaures additas gerunt, similiter in Chio
iuxta Senum delubrum, in Mesopotamiae quoque
fonte Chabura, de quo diximus, pisces.

17 VIII. Nam in Lycia Myris in fonte Apollinis, quem
Curium appellant, ter fistula vocati veniunt ad
augurium. diripere eos carnes abiectas laetum est
consultantibus, caudis abigere dirum. Hieropoli
Syriae in lacu Veneris aedituorum vocibus parent,
vocati veniunt exornati auro, adulantes scalpuntur,
ora hiantia manibus inserendis praebent. in Stabiano
Campaniae ad Herculis petram melanuri in mari
panem abiectum rapiunt, iidem ad nullum cibum, in
quo hamus sit, accedunt.

18 IX. Nec illa in novissimis mira, amaros esse pisces
ad Pelen insulam et ad Clazomenas, contra scopulum
Siciliae [2] ac Leptim Africae et Euboeam et Durra-
chium, rursus ita salsos, ut possint salsamenta existu-
mari, circa Cephallaniam et Ampelon, Paron et Deli
petras, in portu eiusdem insulae dulces. quam
19 differentiam pabulo constare non est dubium. Apion

---

[1] lolligo . . . mutari *post* demergant *transfert Mayhoff, qui*
nuntiari *pro* mutari *coni.*

[2] Siciliae *codd., Mayhoff:* Scyllae *Urlichs, Detlefsen.*

---

[a] The last sentence is transferred to this place from the end
of § 14 by Mayhoff (not in his text), who also reads *nuntiari,*
that is: " storms are indicated."

vessel. Whenever the lolligo, he says, is seen flying out of the water a change of weather occurs.[a]

VII. In several country seats indeed of the Emperor fish eat out of the hand, but—what our old writers have recorded with wonder as occurring in natural pools, not fish-ponds—at Helorus, a fortress of Sicily not far from Syracuse, and likewise in the spring of Jupiter of Labraynda, the eels even wear ear-rings, as do the fishes in Chios near the Shrine of the Old Men, and in the spring Chabura also in Mesopotamia, about which I have spoken.[b]

VIII. But at Myra in Lycia in the spring of Apollo called Curium, when summoned three times by the pipe the fishes come to give oracular responses. For the fish to snap at the meat thrown to them is a happy augury for enquirers, to cast it aside with their tails an augury of disaster. At Hieropolis in Syria the fish in the pond of Venus obey the voice of the temple ministers; they come at their call adorned with gold, fawning to be scratched, and offer gaping mouths to receive their hands. At Stabiae in Campania at the Rock of Hercules the melanuri in the sea seize the bread thrown to them, but they will not go near any food in which is a hook.

IX. Nor are these the last among the marvels we know of fishes: that they are bitter near the island of Pele and near Clazomenae, over against the rock of Sicily,[c] Leptis in Africa, Euboea, and Dyrrhachium; and again, so salt that they might be thought pickled, off Cephallania, Ampelos, Paros and the rocks of Delos; while in the harbour of Delos they are sweet. These differences depend without a doubt on the

---

[b] See XXXI. § 37.
[c] I.e. Scylla, which has been conjectured for Sicilia.

piscium maximum[1] esse tradit porcum, quem Lace-
daemoni orthagoriscum vocent; grunnire eum, cum
capiatur. esse vero illam naturae accidentiam—
quod magis miremur—etiam in locis quibusdam,
adposito occurrit exemplo, siquidem salsamenta
omnium generum in Italia Beneventi refici constat.

20   X. Pisces marinos in usu fuisse protinus a condita
Roma auctor est Cassius Hemina, cuius verba de ea
re subiciam : Numa constituit ut pisces, qui squamas
non essent, ni pollucerent, parsimonia commentus, ut
convivia publica et privata cenaeque ad pulvinaria
facilius compararentur, ni qui ad polluctum emerent
pretio minus parcerent eaque praemercarentur.

21   XI. Quantum apud nos Indicis margaritis pretium
est, de quis suo loco satis diximus, tantum apud Indos
curalio ; namque ista persuasione gentium constant.[2]
gignitur et in Rubro quidem mari, sed nigrius, item
in Persico—vocatur lace—laudatissimum in Gallico
sinu circa Stoechadas insulas et in Siculo circa Aeolias
ac Drepana. nascitur et apud Graviscas et ante
Neapolim Campaniae ; maximeque rubens, sed molle

22 et ideo vilissimum Erythris. forma est ei fruticis,
colos viridis. bacae eius candidae sub aqua ac molles,
exemptae confestim durantur et rubescunt qua corna

---

[1] maximum *codd.*: maxime mirum *Mayhoff, qui notam
addit*: " *an excidit (ante* Apion) *alterum exemplum piscis
aliquo loco non muti?* "
[2] constant *multi codd.*: constat BV: ita . . . constat *in
Appendice Mayhoff.*

---

    [a] With Mayhoff's reading: " most wonderful."
    [b] An historian who flourished about 140 B.C.
    [c] See IX. § 104 foll.
    [d] This phrase is generally taken with the preceding clause.
The punctuation is mine.

food. Apion tells us that the largest [a] of the fishes
is the pig-fish, which the Lacedaemonians call *orthagoriscus*, saying that it grunts when it is caught.
That this accident of nature, however (to increase our
wonder), is also met with in certain localities, is suggested by a ready example, seeing that salted foods
of every kind, as is well known, at Beneventum in
Italy have to be resalted.

X. That sea fish were commonly eaten immediately
after the foundation of Rome is told us by Cassius
Hemina,[b] whose very words on the subject I will
quote here. " Numa ordained that scaleless fish
should not be provided at sacrificial meals, being induced by reasons of economy, so that provision could
be more easily made for public and private banquets
and for feasts of the gods, to prevent caterers on
those sacred occasions from being extravagant and
buying up the market."

XI. Coral is as valuable among the Indians as
Indian pearls, about which I have spoken [c] in their
proper place, are among the Romans, for cost varies
with the demand of any particular people. Coral is
also found in the Red Sea, but this is of a darker
colour ; also in the Persian Gulf—this is called *lace*—
the most valued is in the Gallic Gulf around the
Stoechades Islands, in the Sicilian Gulf around the
Aeolian Islands, and around Drepana. Coral also
grows at Graviscae and before Naples in Campania ;
but that at Erythrae, which is very red indeed,[d] is soft
and therefore thought worthless.

In shape coral is like a shrub, and its colour is green.
Its berries are white under the water and soft ;
when taken out they immediately harden and grow
red, being like, in appearance and size, to those of

sativa specie atque magnitudine. aiunt tactu protinus lapidescere, si vivat; itaque occupari evellique retibus aut acri ferramento praecidi, qua de causa curalium vocitatum interpretantur, probatissimum quam maxime rubens et quam ramosissimum nec scabiosum aut lapideum aut rursus inane et concavum.

23 auctoritas bacarum eius non minus Indorum viris quoque pretiosa est quam feminis nostris uniones Indici. harispices eorum vatesque inprimis religiosum id gestamen amoliendis periculis arbitrantur. ita et decore et religione gaudent. prius quam hoc notesceret, Galli gladios, scuta, galeas, adornabant eo. nunc tanta paenuria est vendibili merce, ut per-

24 quam raro cernatur in suo orbe. surculi infantiae adalligati tutelam habere creduntur, contraque torminum ac vesicae et calculorum mala in pulverem igni redacti potique cum aqua auxiliantur, simili modo ex vino poti aut, si febris sit, ex aqua somnum adferunt—ignibus diu repugnat [1]—sed eodem medicamine saepius poto tradunt lienem quoque absumi. sanguinem reicientibus excreantibusve medetur cinis eorum; miscetur oculorum medicamentis, spissat enim ac refrigerat, ulcerum cava explet, cicatrices extenuat.

25 XII. Quod ad repugnantiam rerum attinet, quam Graeci antipathian vocant, nihil est usquam venenatius quam in mari pastinaca, utpote cum radio eius arbores necari dixerimus. hanc tamen persequitur

[1] " an ignibus diu repugnat *pertinet ad finem* § 22 *post* concavum? " *Mayhoff.*

[a] Greek κείρω, I cut. [b] See § 23.
[c] The reason for the proposed transposition is the sudden change from plural to singular (*creduntur, auxiliantur, adferunt, repugnat*).

cultivated cornel. It is said that at a touch it immediately petrifies, if it lives; and that therefore it is quickly seized and pulled away in nets or cut off by a sharp iron instrument. In this way they explain its name " coral." [a] The most valued coral is the reddest and most branchy, without being rough or stony, or again empty and hollow. Coral berries are no less valued by Indian men than are large Indian pearls by Roman women. Indian soothsayers and seers think that coral is a very powerful amulet [b] for warding off dangers. Accordingly they take pleasure in it both as a thing of beauty and as a thing of religious power. Before the Indian love of coral became known, the Gauls used to ornament with coral their swords, shields, and helmets. At the present day it has become so scarce because of the price it will fetch that it is very rarely to be seen in the countries where it grows. Branches of coral, worn as an amulet by babies, are believed to be protective, and reduced to powder by fire and taken with water are helpful in gripings, bladder trouble and stone; similarly, taken in wine, or, if fever is present, in water, coral is soporific. Coral resists fire for a long time,[c] but they say also that taken in drink repeatedly as medicine it consumes the spleen. The ash of coral branches is good treatment for bringing up or spitting of blood. It is a component of eye salves, for it is astringent and cooling, fills up the hollows of ulcers, and smooths out scars.

XII. As to the hostility between things, which the Greeks call *antipathia*, there is nowhere anything more venomous than the sting-ray in the sea, since we have said [d] that by its ray trees are killed. The

[d] See IX. § 155.

galeos, idem et alios quidem pisces, sed pastinacas
praecipue, sicut in terra mustela serpentes—tanta
est avidatas ipsius veneni—percussis vero ab ea
medentur et hic quidem, sed et mullus ac laser,

26   XIII. Spectabili naturae potentia, in iis quoque, quibus et in terris victus est, sicut fibris, quos castoras
vocant et castorea testes eorum. amputari hos ab
ipsis, cum capiantur, negat Sextius diligentissimus
medicinae, quin immo parvos esse substrictosque
et adhaerentes spinae, nec adimi sine vita animalis
posse; adulterari autem renibus eiusdem, qui sint
grandes cum veri testes parvi admodum reperiantur;

27 praeterea ne vesicas quidem esse, cum sint geminae,
quod nulli animalium; in iis folliculis inveniri[1]
liquorem et adservari sale; itaque inter probationes
falsi esse folliculos geminos ex uno nexu depen-
dentes, quod ipsum corrumpi fraude conicientium
cummin cum sanguine aut Hammoniacum, quoniam
Hammoniaci coloris esse debeant, circumdati liquore
veluti mellis cerosi, odore graves, gustu amaro et
acri, friabiles. efficacissimi e Ponto Galatiaque, mox

28 Africa. sternumenta olfactu movent. somnum con-
ciliant cum rosaceo et peucedano peruncto capite et
per se poti ex aqua, ob id phreneticis utiles; iidem

---

[1] " *locus adhuc corruptus videtur; exspectaveris potius* ne
vesicam quidem (*sc.* communem) esse, cum sint gemini folli-
culi . . . in iis inveniri *sqq. cfr. Diosc.*" *Mayhoff.*

---

[a] The plural (*efficacissimi, movent,* etc.) is due to *testes,* but
it seems more natural in English to use the singular, referring
to *castoreum.*

galeos however chases the sting-ray, and also indeed other fishes, but the sting-ray in particular, just as on land the weasel chases serpents, so great is its greed for the very poison itself. Those however stung by the sting-ray find good treatment in the *galeos*, as well as in red mullet and laser.

XIII. Equally remarkable is the might of Nature in those creatures also which are amphibious, such as the beaver, which they call *castor* and its testes *castoreum*. Sextius, a very careful inquirer into medical subjects, denies that the beaver himself bites off his own testes when it is being captured; he says that on the contrary these are small, tightly knit, attached to the spine, and not to be taken away without destroying the creature's life. Castoreum (beaver-oil) he says is however adulterated by beaver's kidneys, which are large, while the real testes are found to be very small. Moreover, they cannot even be the creature's bladders, for they are twin, and no animal has two bladders. In these pouches (he goes on) is found a liquid, which is preserved in salt. Accordingly one of the tests of fraud is whether two pouches hang down from one connection, while the liquid itself is adulterated by adding to it cummin and beaver blood or ammoniacum, because the testes ought to be of the colour of ammoniacum, coated with a liquid like waxy honey, with a strong smell, a bitter taste, and friable. The most efficacious castoreum comes *ᵃ* from Pontus and Galatia, the next best from Africa. Doctors cause sneezing by its smell. It is soporific if the head is rubbed all over with beaver oil, rose oil, and peucedanum, or if by itself it is taken in water, for which reason it is useful in brain fever. It also arouses, by

481

lethargicos odoris [1] suffitu excitant volvarumque exanimationes vel subditu, ac menses et secundas
29 cient II drachmis cum puleio ex aqua poti. medentur et vertigini, opisthotono, tremulis, spasticis, nervorum vitiis, ischiadicis, stomachicis, paralyticis, perunctis omnibus, vel triti ad crassitudinem mellis cum semine viticis ex aceto ac rosaceo. sic et contra comitiales sumpti, poti vero contra inflationes, tormina, venena. differentia tantum contra genera est
30 mixturae, quippe adversus scorpiones ex vino bibuntur, adversus phalangia et araneos ex mulso ita, ut vomitione reddantur aut ut contineantur cum ruta, adversus chalcidas cum myrtite, adversus cerasten et presteras cum panace aut ruta ex vino, adversus ceteras serpentes cum vino. dari binas drachmas
31 satis; eorum, quae adiciantur, singulas. auxiliantur privatim contra viscum ex aceto, adversus aconitum ex lacte aut aqua, adversus helleborum album ex aqua mulsa nitroque. medentur et dentibus infusi cum oleo triti in aurem, a cuius parte doleant, aurium dolori melius, si cum meconio. claritatem visus faciunt cum melle Attico inunctis. cohibent singultus ex aceto. urina quoque fibri resistit venenis et ob id in antidota additur. adservatur autem optume in sua vesica, ut aliqui existumant.
32 XIV. Geminus similiter victus in aquis terraque et testudinum effectusque par, honore habendo vel propter excellens in usu pretium figuraeque proprietatem. sunt ergo testudinum genera terrestres,

---

[1] odoris] " *an* odore? " *Mayhoff.*

[a] See Book XXIX. § 102.

the smell of fumigation, sufferers from coma and hysterical, fainting women, the latter also by a pessary; it is an emmenagogue and brings away the after-birth if two drachmae are taken in water with pennyroyal. It is also a remedy for vertigo, opisthotonus, palsied tremors, cramps, sinew pains, sciatica, stomach troubles, and paralysis; in all cases by rubbing all over, or ground to the consistency of honey with seed of vitex in vinegar and rose oil. In this form it is taken for epilepsy, but in drink for flatulence, griping and poisons. The only difference in its use for the various poisons lies in the ingredients with which it is mixed. For scorpion bites it is taken in wine; for the phalangium and other spiders in honey wine if it is to be vomited back or with rue if it is to be retained; for the chalcis *a* with myrtle wine; for the horned asp and prester with panaces or rue in wine; for the bites of other serpents with wine. Two drachmae are a sufficient dose, of the other ingredients one drachma. It is specific in vinegar for mistletoe poisoning, in milk or water for poisoning by aconite, for white hellebore in oxymel and soda. It also cures toothache if pounded with oil; it is poured into the ear on the side of the pain; for ear-ache it is better mixed with poppy juice. Added to Attic honey and used as an ointment it improves the vision. In vinegar it checks hiccoughs. Beaver urine, too, counteracts poisons, and therefore is added to antidotes. It is however best preserved, as some think, in the beaver's bladder.

XIV. Like the beaver the tortoise is amphibious, and of the same medical properties, distinguished by the high price given for its use, and by its peculiar shape. So there are various kinds: tortoises that live

marinae, lutariae et quae in dulci aqua vivunt. has
33 quidam e Graecis emydas appellant. Terrestrium
carnes suffitionibus propriae magicisque artibus
refutandis et contra venena salutares produntur.
plurimae in Africa. hae ibi amputato capite pedibusque pro antidoto dari dicuntur et e iure in cibo
sumptae strumas discutere, lienes tollere, item comitiales morbos. sanguis earum claritatem visus facit,
sistit[1] suffusiones oculorum. et contra serpentium
omnium et araneorum ac similium et ranarum
venena auxiliatur servato sanguine in farina pilulis
factis et, cum opus sit, in vino datis. felle testudinum cum Attico melle glaucomata inungui prodest,
34 scorpionum plagae instillari.[2] tegimenti cinis vino
et oleo subactus pedum rimas ulceraque sanat.
squamae e summa parte derasae et in potu datae
venerem cohibent. eo magis hoc mirum, quoniam
totius tegimenti farina accendere traditur libidinem.
urinam aliter earum quam in vesica dissectarum inveniri posse non arbitror et inter ea[3] esse hoc quoque,
quae portentose Magi demonstrant, adversus aspidum
ictus singularem, efficaciorem tamen, ut aiunt, cimicibus admixtis. ova durata inlinuntur strumis et
ulceribus frigore aut adustione factis. sorbentur in
35 stomachi doloribus. marinarum carnes admixtae
ranarum carnibus contra salamandras praeclare
auxiliantur, neque est testudine aliud salamandrae

---

[1] sistit *Brakman*: discutit *Mayhoff*: *in codd. lacuna.*
[2] instillari *codd.*: instillant *Mayhoff.*
[3] inter ea *codd.*: interanea *Detlefsen*: " *locus fortasse non-dum sanatus,*" *Mayhoff.*

---

[a] Brakman's *sistit* is perhaps the best supplement of the lacuna.
[b] Toads are included in *ranae.*

on land, in the sea, in muddy water, and in fresh
water. The last are called by some Greeks *emydes*.

The flesh of the land tortoise is reported to be
especially useful for fumigations, to keep off magical
tricks, and to counteract poisons. It is most common
in Africa. There the flesh of this tortoise, with its
head and feet cut off, is said to be given as an antidote,
and taken in its broth as food to disperse scrofulous
sores, to reduce the spleen, and to cure epilepsy.
The blood clarifies the vision and arrests [a] cataract.
For the poisons of all serpents, spiders and similar
creatures, and of frogs,[b] it is of service; the blood is
preserved in flour, made up into pills, and given in
wine when necessary. It is beneficial to use the gall
of tortoises with Attic honey as an eye-wash for
opaqueness of the lens, and to drop it [c] into the
wounds made by scorpions. The shell, reduced to
ash and kneaded with wine and oil, heals chaps and
sores on the feet. Shavings from the top of the shell
and given in drink are antaphrodisiac. This is all the
more surprising because the whole shell, reduced to
powder, is said to incite to lust. The urine of this
tortoise, I believe, is found only in the bladder of
dissected animals, and this is one of the substances
to which the Magi give supernatural virtues as being
specific for the bites of asps; a more efficacious one,
however, they say, if bugs are added. The eggs are
applied hard boiled to scrofulous sores, frost bites and
burns. They are swallowed for pains in the stomach.

The flesh of sea tortoises mixed with that of frogs
is an excellent remedy for salamander bites, and
nothing is more opposed to the salamander than the

[c] If a comma is placed at *prodest* the *instillari* of the MSS.
can perhaps be kept with *fel* as its understood subject.

adversius. sanguine alopeciarum inanitas et porrigo omniaque capitis ulcera curantur; inarescere eum oportet lenteque ablui. instillatur et dolori aurium cum lacte mulierum. adversus morbos comitiales manditur cum polline frumenti, miscetur autem san-

36 guinis heminis tribus aceti hemina. datur et suspiriosis, sed tum[1] hemina vini additur;[2] his et cum hordeacea farina, aceto quoque admixto, ut sit quod devoretur fabae magnitudine; et haec singula et matutina et vespera dantur, dein post aliquot dies bina vespera. comitialibus instillatur ore diducto;[3] iis, qui modice corripiantur spasmo, cum castoreo

37 clystere infunditur. quod si dentes ter annis[4] colluantur testudinum sanguine, immunes a dolore fiant. et anhelitus discutit quasque orthopnoeas vocant; ad has in polenta datur. fel testudinum claritatem oculorum facit, cicatrices extenuat, tonsillas sedat et anginas et omnia oris vitia, privatim nomas ibi, item testium. naribus inlitum comitiales erigit attollitque. idem cum vernatione anguium aceto admixto unice purulentis auribus prodest. quidam bubulum fel admiscent decoctarum[5] carnium testu-

38 dinis suco, addita aeque vernatione anguium; sed vino testudinem excocunt. oculorum utique vitia omnia fel inunctum cum melle emendat, suffusiones

---

[1] tum *Ianus, Mayhoff*: cum B *Sillig, Mayhoff*.

[2] additur B, *Sillig, Mayhoff*: addito VRdT, *Hard.*: *coni.* sed cum hemina vini. manditur his et *Mayhoff*.

[3] diducto B, *Gelenius*: diducis VR: deductis E, *vulg.*

[4] annis VRf, *Io. Müller*: minis B: *coni.* heminis *Mayhoff*.

[5] decoctarum *Mayhoff*: decoctarumve (decoctarumque) *codd.*

tortoise. Its blood is good treatment for the bare
patches of mange, for dandruff, and for all sores on
the head; it should be allowed to dry and then gently
washed off. With woman's milk it is poured by drops
into aching ears. For epilepsy it is taken with
wheaten flour, but three heminae of blood are diluted
with one hemina of vinegar. It is also given for
asthma, but with a hemina of wine added; for this
purpose also with barley flour, vinegar too being
added, so that the dose to be swallowed is the size of
a bean. One of these doses is given morning and
evening; then after a few days a double dose is given
in the evening. The mouths of epileptics are opened
and the blood poured by drops into them; to those
seized with a slight convulsion is given an enema of
the blood and beaver oil. If teeth are rinsed with
tortoise blood three times a year [a] they will become
immune to toothache. It is a remedy too for short-
ness of breath and for what is called orthopnoea;
when so used it is administered in pearl barley.
Tortoise gall gives clearness of vision, effaces scars,
relieves sore tonsils, quinsy, and all diseases of
the mouth, being specific for malignant sores there
and on the testicles. If the nostrils are smeared
with it, epileptics are roused and made to stand up.
The gall too with snakes' slough and vinegar is also a
sovereign remedy for pus in the ears. Some mix ox
gall with the broth of boiled tortoise-flesh, adding the
same amount of snakes' slough, but they boil the
tortoise in wine. An application of the gall with
honey cures especially all affections of the eyes;
cataract is also cured by the gall of sea tortoise with

[a] If we adopt Mayhoff's attractive emendation: "three
times with a hemina."

etiam marinae fel cum fluviatilis sanguine et lacte.
capillus mulierum inficitur felle. contra sala-
39 mandras vel sucum decoctae bibisse satis est. Tertium genus testudinum est in caeno et paludibus
viventium. latitudo his et in dorso pectori similis nec
convexo curvata calice, ingrata visu. ex hac quoque
tamen aliqua contingunt auxilia. tres namque in
succensa sarmenta coiectae dividentibus se tegu-
mentis rapiuntur, tum evolsae carnes earum cocuntur
in congio aquae sale modice addito; ita decoctarum
ad tertias partes sucus paralysim et articularios morbos sentientibus bibitur. detrahit idem fel pituitas
sanguinemque vitiatum. sistitur eo remedio alvus
40 aquae frigidae potu. E quarto genere testudinum,
quae sunt in amnibus, divolsarum pinguia cum aizoo
herba tunsa admixto unguento et semine lili, si ante
accessiones perunguantur aegri praeter caput, mox
convoluti calidam aquam bibant, quartanis liberare
dicuntur. hanc testudinem xv luna capi oportere,
ut plus pinguium reperiatur, verum aegrum xvi luna
perungui. ex eodem genere testudinum sanguis instillatus cerebro capitis dolores sedat, item strumas.
41 sunt qui testudinum sanguinem cultro aereo supin-
arum capitibus praecisis excipi novo fictili iubeant,
ignem sacrum cuiuscumque generis sanguine inlini,
item capitis ulcera manantia, verrucas. iidem pro-

ᵃ Evidently the Magi, but for some reason Pliny withholds
the name.

the blood of river tortoise and milk. Woman's hair is dyed by the gall. For salamander bites it is enough merely to drink the broth of a decoction.

A third kind of tortoise lives in mud and marshes. These have a level width, like that across the breast, over the back also; this is not rounded into a cup-like convexity—indeed an unpleasant sight. Yet from this creature also a few remedies are obtained. For three are together thrown on burning brush-wood, and when the shells separate they are at once taken off; the flesh is then torn away and boiled in a congius of water with a little salt added. The broth is boiled down to one third and taken for paralysis and diseases of the joints. The gall of this creature carries off phlegms and vitiated blood. This remedy taken in cold water acts astringently on the bowels.

There is a fourth kind of tortoise, which lives in rivers. The shells being torn off, the fats are beaten up with houseleek mixed with unguent and lily seed. If of a patient all the body except the head is rubbed with this preparation before the paroxysms come on, and he is then wrapped up and drinks hot water, he is cured, it is said, of quartan ague. This tortoise, they say, should be killed on the fifteenth of the moon, so that more fats may be obtained from it, but the patient should be rubbed on the sixteenth. The blood too of this kind of tortoise, poured in drops on the skull, relieves headache as well as scrofulous sores. There are some [a] who recommend tortoises to be laid on their backs, their heads chopped off with a bronze knife, and the blood caught in new earthenware; this blood is to be used as embrocation for all kinds of erysipelas, running sores on the head, and warts. The same authorities assure us that the dung of all

489

mittunt testudinum omnium fimo panos discuti; et, quod incredibile dictu sit, aliqui tradunt tardius ire navigia testudinis pedem dextrum vehentia.

42   XV. Hinc deinde in morbos digeremus aquatilia, non quia ignoremus gratiorem esse universitatem animalium maiorisque miraculi, sed hoc utilius est vitae, contributa habere remedia, cum aliud alii prosit, aliud alibi facilius inveniatur.

43   XVI. Venenatum mel diximus ubi nasceretur. auxilio est piscis aurata in cibo. vel si ex melle sincero fastidium cruditasve, quae fit gravissima, incidat, testudinem circumcisis pedibus, capite, cauda decoctam antidotum esse auctor est Pelops, scincum Apelles. quid esset scincus diximus, saepius vero

44 quantum veneficii in menstruis mulierum. contra ea omnia auxiliatur, ut diximus, mullus, item contra pastinacam et scorpiones terrestres marinosque et dracones, phalangia inlitus sumptusve in cibo, eiusdem recentis e capite cinis contra omnia venena, privatim contra fungos. mala medicamenta inferri negant posse aut certe nocere stella marina volpino sanguine inlita et adfixa limini superiori aut clavo aereo ianuae.

45   XVII. draconis marini scorpionumque ictus carnibus earum inpositis, item araneorum morsus sanantur. in summa contra omnia venena vel potu vel ictu vel morsu noxia sucus earum e iure decoctarum efficacissi-

<hr />

^a See XXI. § 74 foll.
^b See VIII. § 91 and XXVIII. § 119.
^c See VII. § 64 and XXVIII. § 82.

tortoises disperses superficial abscesses; and others tell us (an incredible remark) that vessels travel more slowly if the right foot of a tortoise is on board.

XV. From now on I will arrange water creatures according to diseases, not that I do not know that a complete account of each living thing is more attractive and more wonderful, but it is more useful to mankind to have remedies grouped into classes, since they vary with individuals, and are more easily found in one place than in another.

XVI. I have already said [a] where poisonous honey is found. A remedy is the gilthead fish taken in food. But if pure honey should cause nausea, or indigestion that becomes very acute, an antidote is, according to Pelops, the decoction of a tortoise with the feet, head, and tail cut off; according to Apelles, a similar decoction of a scincus; I have said what a scincus is.[b] Several times moreover I have said how poisonous is the menstrual fluid of women;[c] against all forms of it, as I have said, the red mullet is a help, as it is against the sting-ray, land- and sea-scorpions, the weever fish, and poisonous spiders. It may be applied locally or taken in food. A fresh red mullet's head, reduced to ash, is an antidote to all poisons, being specific against poisonous fungi. They say that noxious charms cannot enter, or at least cannot harm, homes where a star-fish, smeared with the blood of a fox, has been fastened to the upper lintel or to the door with a bronze nail.

XVII. By an application of tortoise flesh are healed the stings of weever fish, of scorpions, and also the bites of spiders. To sum up: the gravy of tortoise meat, that is, the broth obtained by boiling it down, is considered to be a most efficacious antidote for all

mus habetur. sunt et servatis piscibus medicinae, salsamentorumque cibus prodest a serpente percussis et contra bestiarum ictus mero subinde hausto ita, ut per se etiam [1] cibus vomitione reddatur, peculiariter
46 a chalcide, ceraste aut quas sepas vocant aut elope, dispsade percussis. contra scorpionem largius sumi, sed non evomi, salsamenta prodest ita, ut sitis toleretur; et inponere plagis eadem convenit. contra crocodilorum quidem morsus non aliud praesentius habetur. privatim contra presteris morsum sarda prodest. inponuntur salsamenta et contra canis
47 rabiosi; vel si non sint ferro ustae plagae corporaque clysteribus exinanita, hoc per se sufficit. et contra draconem marinum ex aceto inponuntur. idem et cybio effectus. draco quidem marinus ad spinae suae, qua ferit, venenum ipse inpositus vel cerebro toto [2] prodest.
48 XVIII. Ranarum marinarum ex vino et aceto decoctarum sucus contra venena bibitur, et contra ranae rubetae venenum et contra salamandras. fluviatilium [3] si carnes edantur iusve decoctarum sorbeatur, prosunt et contra leporem marinum et contra serpentes supra dictos, contra scorpiones ex vino.
49 Democritus quidem tradit, si quis extrahat ranae viventi linguam, nulla alia corporis parte adhaerente, ipsaque dimissa in aquam inponat supra cordis palpi-

---

[1] per se etiam B[2] *Sillig*: ad vesperam *multi codd.*: per satiem *Mayhoff.*

[2] toto *multi codd.*; toti B: poto *Mayhoff.*

[3] fluviatilium *Detlefsen*: fluviatilil/jiū B[2]: *ante ponunt* vel e *multi codd.*

---

[a] *Ad vesperam* would be "towards evening"; *ad satiem* "to a surfeit."

[b] *Poto*: "its brain taken in drink."

poisons, whether conveyed in drink, by sting, or by
bite. There are also remedies from preserved fish;
to eat salted fish is good for the bites of snakes and of
other venomous creatures, but now and then should
be drunk enough neat wine to bring back by vomiting
even the food whole; [a] the remedy is specially good
for those bitten by the chalcis lizard, horned viper,
what is called seps, elops, or dipsas. For scorpion
stings a bigger dose of salted fish is beneficial, but
not enough to cause the vomiting, or intolerable
thirst; it is also good to lay salted fish on the wounds.
Against the bites of crocodiles nothing else is considered to be a more sovereign remedy. The sarda
is specific against the bite of the prester. Salted fish
is also applied to the bite of a mad dog; even if the
wound has not been cauterised with a hot iron, and
the bowels emptied with a clyster, the fish by itself
is enough. Salted fish is also applied with vinegar to
the wound given by the weever fish. The tunny too
has the same property. The weever fish indeed, if
itself, or the whole [b] of its brain, if applied to the
poisoned wound caused by a blow of his own spine,
makes a good remedy.

XVIII. A decoction of sea frogs [c] boiled down in
wine and vinegar is drunk to counteract poisons, also
that of the bramble toad and salamander; if the
flesh of river frogs is eaten, or the broth drunk after
boiling them down, it counteracts the poison of the
sea-hare, of the snakes mentioned above, and of
scorpions if wine is used in the preparation. Democritus indeed tells us that if the tongue, with no other
flesh adhering, is extracted from a living frog, and
after the frog has been set free into water, placed

[c] Angler-fish.

tationem mulieri dormienti, quaecumque interroga-
verit, vera responsuram. addunt etiamnum alia
Magi, quae si vera sint, multo utiliores vitae existu-
mentur ranae quam leges; namque harundine
transfixis a [1] natura per os si surculus in menstruis
50 defigatur a marito, adulterorum taedium fieri. carnibus earum vel [2] in hamum additis praecipue purpuras
adlici certum est. iocur ranae geminum esse dicunt
abicique formicis oportere; eam partem, quam
adpetant, contra venena omnia esse pro antidoto.
sunt quae in vepribus tantum vivunt, ob id rubetarum
nomine, ut diximus, quas Graeci φρύνους vocant,
grandissimae cunctarum, geminis veluti cornibus,
plenae veneficiorum. mira de iis certatim tradunt
51 auctores: inlatis in populum silentium fieri; ossiculo,
quod sit in dextro latere, in aquam ferventem deiecto
refrigerari vas nec postea fervere nisi exempto, id
inveniri abiecta rana formicis carnibusque erosis,
singula in oleum [3] addi; esse in sinistro latere quo
52 deiecto fervere videatur, apocynon vocari, canum
impetus eo cohiberi, amorem concitari et iurgia
addito in potionem, venerem adalligato stimulari,

---

[1] transfixis a B: transfixa *multi codd.*: transfixa a *Ianus.*
[2] *Ante* vel *addit* nassis *Sillig cum vet. Dal.*
[3] oleum *fere omnes codd.*: solium *Hermolaus Barbarus*:
ollam *Ianus.*

---

[a] Something seems wrong with this sentence, which means,
if literally translated, that frogs are pierced with a reed, and
then the husband plants a shoot. There is no indication that
the shoot is the same as the reed. Perhaps there is a lacuna
after *os*; perhaps too the *transfixa* of most MSS. is correct,
although such a use of *transfigor* ("a reed having been
thrust") is rare.
[b] The addition of *nassis* is a better remedy than any other.
[c] See Book VIII. § 110. The word *rana* may be either
"frog" or "toad."

over the beating heart of a sleeping woman, she will
give true answers to all questions.

The Magi add also other details, and if there is any
truth in them, frogs should be considered more
beneficial than laws to the life of mankind. They
say that if frogs are pierced [a] with a reed from the
genitals through the mouth, and if the husband plants
a shoot in his wife's menstrual discharge she conceives
an aversion to adulterous lovers. It is certain that
frogs' flesh placed ⟨in weels⟩ [b] or on a hook makes excellent bait for the purple-fish. It is said that the liver
of a frog is double, and should be thrown in the way
of ants; that the part the ants attack is an antidote
for all poisons. Some frogs there are that live only
in brambles, and so they are called bramble-toads, as
I have said,[c] and by the Greeks φρῦνοι. These are
the largest of all frogs, have as it were a pair of horns,
and are full of poison. Our authorities vie with one
another in relating marvellous stories about the
toad: that when brought into a meeting of the people
silence reigns; that if the little bone found in its
right side is let fall into boiling water, the vessel
cools, and does not afterwards boil unless the bone is
taken out; that it is found when a frog has been
thrown to ants and the flesh gnawed away; that one
at a time these bones are put into oil; [d] that there is
in a frog's left side a bone called " dog's bane,"
which dropped ⟨into oil⟩ gives the appearance of
boiling; by it the attacks of dogs are repelled, and
if it is put in drink love and quarrels [e] brought about;
that worn as an amulet it acts as an aphrodisiac; that

---

[d] With the reading *solium*, " tub "; with *ollam*, " pot."

[e] Is there a zeugma here, " love aroused and quarrels
settled." Perhaps read *conciliari*.

rursus e dextro latere refrigerari ferventia; hoc et
quartanas sanari adalligato in pellicula agnina recenti
aliasque febres, amorem inhiberi, ex isdem his ranis
lien contra venena, quae fiant ex ipsis, auxiliatur,
iocur vero etiam efficacius.

53 XIX. Est colubra in aqua vivens. huius adipem
et fel habentes qui crocodilos venentur mire adiuvari
dicunt, nihil contra belua audente, efficacius etiamnum, si herba potamogiton[a] misceatur. cancri fluviatiles triti potique ex aqua recentes seu cinere adser-
vato contra venena omnia prosunt, privatim contra
scorpionum ictus cum lacte asinino, si non sit, caprino
aut quocumque; addi et vinum oportet. necant eos
54 triti cum ocimo admoti. eadem vis contra venena-
torum omnium morsus, privatim scytalen[b] et angues
et contra leporem marinum ac ranam rubetam. cinis
eorum servatus prodest pavore potus periclitantibus
ex canis rabiosi morsu. quidam adiciunt gentianam
et dant in vino, et si iam pavor occupaverit, pastillos
55 vino subactos devorandos ita praecipiunt. decem
vero cancris cum ocimi manipulo adligatis omnes, qui
ibi sint, scorpiones ad eum locum coituros Magi
dicunt, et cum ocimo ipsos cineremve eorum percussis inponunt. minus in omnibus his marini prosunt. Thrasyllus auctor est nihil aeque adversari
serpentibus quam cancros; sues percussas[1] hoc
pabulo sibi mederi; cum sol sit in cancro, serpentes
56 torqueri. ictibus scorpionum carnes et fluviatilium

[1] percussas] percussos B.

---

[a] Pondweed; see *Index of Plants* in Vol. VII.
[b] A snake of equal thickness throughout. The word means
a cylinder.

the bone again on the right side cools boiling liquids;
that worn in fresh lamb's skin as an amulet this bone
also cures quartan and other fevers, but love is
restrained. The spleen of these frogs is also a
remedy for the poisons that come from them, while
their liver is even more efficacious.

XIX. There is a snake, a colubra, that lives in the
water. It is said that, if they have its fat or gall on
their persons, crocodile hunters are helped wonderfully, as the brute dares not attack it at all; it is still
more efficacious when combined with the plant potamogiton.[a] Fresh river-crabs pounded and taken in
water, or their ash preserved, are good for all poisons,
being specific for scorpion stings, if taken with asses'
milk, or failing that with goat's or any other milk;
wine too should be added. Pounded with basil and
applied to scorpions, river-crabs kill them. Their
property avails also against the bites of all venomous
creatures, being specific against the scytale,[b] snakes,
sea-hare, and bramble toad. Their ash preserved is
good for those threatened with hydrophobia from the
bite of a mad dog. Some add gentian and administer
in wine, and if hydrophobia has already set in, prescribe lozenges made with the ash and wine to be
swallowed. The Magi indeed assert that if ten
crabs with a handful of basil are tied together, all
the scorpions of the district will collect to the spot,
and to those wounded by scorpions they apply with
basil either crabs themselves or else their ash. For
all these purposes sea crabs are less efficacious.
Thrasyllus avows that no antidote for snake bite is
as good as crabs; that pigs, when bitten, cure themselves by taking crabs as food; and that when the
sun is in Cancer snakes are in torture. The stings

coclearum resistunt crudae vel coctae. quidam ob
id salsas quoque adservant. inponunt et plagis
ipsis. coracini pisces Nilo quidem peculiares sunt,
sed nos haec omnibus terris demonstramus. carnes
eorum adversus scorpiones valent inpositae. inter
venena piscium sunt porci marini spinae in dorso,
cruciatu magno laesorum. remedio est limus ex liquore [1] piscium eorum corporis.

57 XX. Canis rabidi morsibus potum expavescentibus
faciem perungunt adipe vituli marini, efficacius, si
medulla hyaenae et oleum e lentisco et cera misceatur.[2] murenae morsus ipsarum capitis cinere
58 sanantur. et pastinaca contra suum ictum remedio
est cinere suo ex aceto inlito vel alterius. cibi causa
extrahi debet ex dorso eius quidquid croco simile est
caputque totum; et haec [3] autem et omnia testacea
modice collui [4] cibis, quia saporis gratia perit. e
lepore marino veneficium restingunt poti hippocampi.
contra dorycnium echini maxime prosunt, et iis, qui
sucum carpathii biberint, praecipue e iure sumpti.
et cancri marini decocti ius contra dorycnium efficax
habetur, peculiariter vero contra leporis marini
venena.

59 XXI. Et ostrea adversantur isdem, nec potest
videri satis dictum esse de iis, cum palma mensarum

---

[1] liquore *coni. Mayhoff* (reliquiis *in textu*): reliquo *aut* liquo
codd.
[2] misceatur *codd.*: misceantur *vet. Dal., Mayhoff.*
[3] haec *Ianus*: hanc *codd., Mayhoff.*
[4] collui in *codd.*: colluunt *coni. Mayhoff, qui dativi* (cibis)
multa exempla dat.

---

[a] Thorn-apple. See *Index of Plants* in Vol. VII.
[b] A narcotic plant.

of scorpions are counteracted also by the flesh of
river snails, raw or cooked. Some too keep them for
this purpose preserved in salt. They also apply
them to the wounds themselves. Though the fish
called coracini are peculiar to the Nile, I am giving
this information for the benefit of all lands. Application of their flesh is good for scorpion stings.
Among poisonous parts of fishes are the prickles on
the back of the sea-pig, a wound from which causes
severe torture. A remedy is the slime from the
liquid part of the body of these fishes.

XX. When the bite of a mad dog causes a dread
of drink they rub the face with the fat of a seal, with
more effect if there are mixed with it the marrow of
a hyaena, mastic oil, and wax. The bites of the
murry are healed by the head of the murry itself,
reduced to ash. For the wound of the sting-ray a
remedy is the ash, of the same ray itself or of any
other specimen, applied locally in vinegar. When
the fish is used as food there should be taken from
its back whatever is like saffron, and the whole head
removed, while the ray, and all shell fish, when used
as food, should not be over-washed, as to do so spoils
the flavour. The poison of the sea-hare is counteracted by the sea-horse taken in drink. Sea-urchins
are very good as an antidote to dorycnium,[a] as they
are also for those who have drunk juice of carpathium,[b]
especially if they are taken in their broth. Effective
against dorycnium is also considered a decoction of
sea-crab, and indeed specific for the poison of the sea-
hare.

XXI. The same poisons are counteracted also by
oysters. About these it cannot appear that enough
has been said, seeing that they have long been con-

diu iam tribuatur illis. gaudent dulcibus aquis et
ubi plurumi influant[1] amnes; ideo pelagia parva et
rara sunt. gignuntur tamen et in petrosis carenti-
busque aquarum dulcium adventu, sicut circa Grynium et Myrinam. grandescunt sideris quidem
ratione maxime, ut in natura aquatilium diximus, sed
privatim circa initia aestatis multo lacte praegnatia
60 atque ubi sol penetret in vada. haec videtur causa,
quare minora in alto reperiantur; opacitas cohibet
incrementum, et tristitia minus adpetunt cibos.
variantur coloribus, rufa Hispaniae, fusca Illyrico,
nigra et carne et testa Cerceis, praecipua vero habentur in quacumque gente spissa nec saliva sua lubrica,
crassitudine potius spectanda quam latitudine, neque
in lutosis capta neque in harenosis, sed solido vado,
spondylo brevi atque non carnoso, nec fibris laciniosa
61 ac tota in alvo. addunt peritiores notam ambiente
purpureo crine fibras, eoque argumento generosa
interpretantur calliblephara ea[2] appellantes. gaudent et peregrinatione transferrique in ignotas aquas.
sic Brundisina in Averno compasta et suum retinere
sucum et a Lucrino adoptare creduntur.
62   Haec sint dicta de corpore; dicemus et de nationi-
bus, ne fraudentur gloria sua litora, sed dicemus

---

[1] influant *Mayhoff*: influunt *codd.*: *cf.* penetret *infra.*
[2] calliblephara ea *Ianus*: calliblepharata d: *varia ceteri
codd.*

---

[a] See IX. § 96.

sidered the prize delicacy of our tables. Oysters love fresh water, and where there is an inflow from many rivers; wherefore deep-sea oysters are small and far between. They also breed, however, in rocky districts and places where no fresh water in comes, such as around Grynium and Myrina. Their growth corresponds very closely to the increase of the moon, as I said [a] when dealing with water-creatures, but they grow most about the beginning of summer, and where sunshine makes its way into shallows, for then they swell with copious, milky, juice. This appears to be the reason why oysters found in deep water are rather small; darkness hinders their growth, and their gloom robs them of appetite.

Oysters vary in colour; red in Spain they are tawny in Illyricum, and black, both flesh and shell, in Circeii. In every country, however, those are most prized that are compact, not greasy with their own slime, remarkable for thickness rather than breadth, taken from water neither muddy nor sandy, but from that with a hard bottom, those whose meat is short and not fleshy, those without fringed edges, and lying wholly in the hollow of the shell.

Experts add a mark of distinction: if a purple line encircle the beard, they consider such oysters to be of a nobler type, and call them " beautifully eye-browed." Oysters like to travel and be moved into strange waters. And so oysters of Brundisium that have fed in Lake Avernus are believed to retain their own flavour as well as acquire that of the oysters of Lake Lucrinus.

So much for their bodies. I will now speak of the countries that breed oysters, lest the shores should be cheated of their proper fame; but I shall do so

aliena lingua quaeque peritissima huius censurae in
nostro aevo fuit.[1] sunt ergo Muciani verba, quae
subiciam: Cyzicena maiora Lucrinis, dulciora Britt-
annicis, suaviora Medullis, acriora Ephesis, pleniora
Iliciensibus, sicciora Coryphantenis, teneriora Histri-
cis, candidiora Cerceiensibus. sed his neque dulciora

63 neque teneriora ulla esse compertum est. in Indico
mari Alexandri rerum auctores pedalia inveniri prodidere, nec non inter nos nepotis[2] cuiusdam nomen-
clatura tridacna appellavit, tantae amplitudinis
intellegi cupiens, ut ter mordenda essent.

64 Dos eorum medica hoc in loco tota dicetur; stomachum unice reficiunt, fastidiis medentur, addidtque
luxuria frigus obrutis nive, summa montium et maris
ima miscens. emolliunt alvum leniter. eademque
cocta cum mulso tenesmo, qui sine exulceratione sit,
liberant. vesicarum ulcera quoque repurgant. cocta
in conchis suis, uti clusa invenerint, mire destilla-

65 tionibus prosunt. testae ostreorum cinis uvam sedat
et tonsillas admixto melle, eodem modo parotidas,
panos mammarumque duritias, capitum ulcera ex
aqua cutemque mulierum extendit; inspergitur et
ambustis. et dentifricio placet. pruritibus quoque
et eruptionibus pituitae ex aceto medetur. purpurae

---

[1] fuit *codd.*: fuerit *vel* fit *coni. Mayhoff.*
[2] nepotis] *Fröhner* Nepotis *coni.*

---

[a] A tax-free colony on the coast of Spain.
[b] There is a difference of opinion as to where the quotation
ends. Some stop here, some at *Cerceiensibus*, Jan at *essent.*
[c] With Fröhner's emendation " one Nepos."
[d] From τρίς " thrice " and δάκνω " I bite."

in the words of another, one who was the greatest connoisseur of such matters in our time. These then are the words of Mucianus, which I will quote:—

Oysters of Cyzicus are larger than those of Lake Lucrinus, fresher than the British, sweeter than those of Medullae, sharper than the Ephesian, fuller than those of Ilici,^a less slimy than those of Coryphas, softer than those of Histria, whiter than those of Circeii.

It is agreed, however, that none are fresher or softer than the last.^b The writers of Alexander's expedition tell us that in the Indian sea are found oysters a foot long, and among ourselves a spend-thrift ^c has invented the nickname *tridacna*,^d wishing it to be used of oysters so large that they require three bites.

I shall give all their medical virtues at this point. Oysters are specific for settling the stomach, they restore lost appetite, and luxury has added coolness by burying them in snow, thus wedding the tops of the mountains to the bottom of the sea. They are a gentle laxative. They also, if boiled with honey wine, cure tenesmus if there is no ulceration. They also clean an ulcerated bladder. Boiled, unopened as gathered, in their shells, they are wonderfully good for streaming colds. Reduced to ash and mixed with honey oyster shells relieve troubles of the uvula and tonsils, similarly parotid swellings, superficial abscesses and indurations of the breasts. Applied with water the ash cures sores on the head and smooths the skin of women. It is sprinkled on burns and is popular as a dentifrice. Applied also with vinegar it cures itch and eruptions of phlegm. The purple-fish too is a good antidote to poisons.

quoque contra venena prosunt. crudae si tundantur, strumas sanant et perniones pedum.

66 XXII. Et algam maris theriacen esse Nicander tradit. plura eius genera, ut diximus, longo folio et rubente, latiore alia vel crispo. laudatissima quae in Creta insula iuxta terram in petris nascitur, tinguendis etiam lanis, ita colorem alligans, ut elui postea non possit. e vino iubet eam dari.

67 XXIII. Alopecias replet hippocampi cinis nitro et adipe suillo mixtus aut sincerus ex aceto, praeparat autem saepiarum crustae farina medicamentis cutem ; replet et muris marini cinis cum oleo, item echini cum carnibus suis cremati, fel scorpionis marini, ranarum quoque trium, si vivae in olla concrementur, cinis cum melle, melius cum pice liquida. capillum denigrant sanguisugae, quae in vino nigro diebus xxxx com-
68 putuere. alii in aceti sextariis duobus sanguisugarum sextarium in vase plumbeo putrescere iubent totidem diebus, mox inlini in sole. Sornatius tantam vim hanc tradit, ut, nisi oleum ore contineant qui tinguent, dentes quoque *suco* [1] earum denigrari dicat. Capitis ulceribus muricum vel purpurarum testae cinis cum melle utiliter inlinitur, conchyliorum vel, si non uratur, farina ex aqua, doloribus castoreum cum peucedano et rosaceo.

---

[1] *Post* quoque *add.* suco *Mayhoff.*

---

[a] See *Theriaca* 845.
[b] Book XXVI. § 103.
[c] Or, " close to dry land."

Beaten up raw, oysters cure scrofulous sores and chilblains on the feet.

XXII. Seaweed too is said by Nicander[a] to be an antidote. There are many kinds of it, as I have said:[b] one with a long, red leaf, another with a broader leaf, and a third with a curly one. The most prized is the one growing near the ground[c] in the island of Crete among the rocks, for this dyes even wool with a colour so fixed that it cannot be washed out afterwards. Nicander recommends it to be given in wine.

XXIII. Hair lost through mange is restored by ashes of the sea-horse, either mixed with soda and pig's lard, or else by itself in vinegar; the skin however must be prepared for medicaments by the rind of the sepia cuttle-fish ground to powder. It is restored also by the ash of the sea-mouse with oil, by that of the sea-urchin burnt with its flesh, by the gall of the sea-scorpion, also by the ash of three frogs with honey, better with liquid pitch, but the frogs must be burnt together alive in a jar. Leeches blacken the hair if they have rotted for forty days in a red wine. Others recommend that for the same number of days a sextarius of leeches be allowed to rot in a leaden vessel containing two sextarii of vinegar, and that then they should be applied in the sun. Sornatius tells us that they have such power that unless those who are going to dye keep oil in the mouth, the extract from the leeches blackens the teeth as well. To sores on the head are applied with honey beneficially shells of murex or purple-fish, reduced to ash; those of any shell-fish, ground to powder if not burned, and applied in water, are also beneficial. For headache use beaver-oil with peucedanum and rose-oil.

69 XXIV. Omnium piscium fluviatilium marinorumque adipes liquefacti sole admixto melle oculorum claritati plurimum conferunt, item castoreum cum melle. callionymi fel cicatrices sanat et carnes oculorum supervacuas consumit. nulli hoc piscium copiosius, ut existumavit Menander quoque in comoediis. idem piscis et uranoscopos vocatur ab oculo,

70 quem in capite habet. et coracini fel excitat visum, et marini scorpionis rufi cum oleo vetere aut melle Attico incipientes suffusiones discutit; inungui ter oportet intermissis diebus. eadem ratio albugines oculorum tollit. mullorum cibo aciem oculorum hebetari tradunt. lepus marinus ipse quidem venenatus est, sed cinis eius in palpebris pilos inutiles evolsos cohibet. ad hunc usum utilissimi minimi, item pectunculi salsi triti cum cedria, ranae, quas diopetas et calamitas vocant; earum sanguis cum

71 lacrima vitis evolso pilo palpebris inlinatur. tumorem oculorum ruboremque saepiae cortex cum lacte mulieris inlitus sedat et per se scabritias emendat; invertunt ita genas et medicamentum auferunt post paulum rosaceoque inungunt et pane inposito mitigant. eodem cortice et nyctalopes curantur, in farinam trito ex aceto inlito. extrahit et squamas

72 eius cinis. cicatrices oculorum cum melle sanat, pterygia cum sale et cadmia singulis drachmis, emendat et albugines iumentorum. aiunt et ossiculo eius genas, si terantur, sanari. echini ex aceto

---

a In Aelian XIII. 4; Meineke IV. p. 79.
b I.e. " stargazer."
c I.e. " fallen from Jupiter."   d The " green-frog."

XXIV. Of all fish, river or sea, the fats, melted in the sun and mixed with honey, are very good for clearness of vision, and so is beaver oil and honey. The gall of the star-gazer heals scars, and removes superfluous flesh about the eyes. No other fish has a greater abundance of gall; this opinion, Menander[a] too expresses in his comedies. This fish is also called uranoscopos,[b] from the eye which it has in its head. The gall of the coracinus too improves vision, and that of the red sea-scorpion with old oil and Attic honey disperses incipient cataract; it should be applied as ointment three times, once every other day. The same treatment removes albugo from the eyes. A diet of mullet is said to dull the eye-sight. Though the sea-hare itself is poisonous, yet reduced to ash it prevents from growing again superfluous hair on the eyelids that has been plucked out. For this purpose the most useful specimens are the smallest; also small scallops, salted and pounded with cedar rezin, frogs called *diopetae*[c] or *calamitae*;[d] their blood, with vine tear-gum, should be rubbed on the lids after plucking out the hair. Swellings and redness of the eyes are soothed by an application of sepia bone with woman's milk, and by itself it is good for roughness of the lids. In this cure they turn up the lids, taking off the ointment after a little time, treat the part with rose-oil and soothe with a bread-poultice. The bone is also good treatment for night-blindness, if ground to powder and applied in vinegar. Reduced to ash it brings away scales; with honey it heals scars on the eyes; with salt and cadmia, a drachma of each, it heals inflammatory swellings, and also albugo in cattle. They say that eyelids, if rubbed by its small bone, are healed. Urchins in vinegar remove night

epinyctidas tollunt. eundem comburi cum viperinis pellibus ranisque et cinerem aspergi potionibus
73 iubent Magi, claritatem visus promittentes. ichthyocolla appellatur piscis, cui glutinosum est corium. idem nomen glutino eius; hoc epinyctidas tollit. quidam ex ventre, non e corio, fieri dicunt ichthyocollam, ut glutinum taurinum. laudatur Pontica, candida et carens venis squamisque et quae celerrime liquescit. madescere autem debet concisa in aqua aut aceto nocte ac die, mox tundi marinis lapidibus, ut facilius liquescat. utilem eam et capitis doloribus
74 adfirmant et tetanis. ranae dexter oculus dextri, sinister laevi, suspensi e collo nativi coloris panno lippitudines sanant; quod si per coitum lunae eruantur, albuginem quoque, adalligati, similiter in putamine ovi. reliquae carnes inpositae suggillationem rapiunt. cancri etiam oculos adalligatos collo mederi
75 lippitudini dicunt. est parva rana in harundinetis et herbis maxime vivens, muta ac sine voce, viridis, si forte hauriatur, ventres boum distendens. huius corporis umorem derasum specillis claritatem oculis inunctis narrant adferre. et ipsas carnes doloribus oculorum superponunt. ranas xv coiectas in fictile novum iuncis configunt quidam sucoque earum, qui ita effluxerit, admiscent vitis albae lacrimam atque ita palpebras emendant, inutilibus pilis exemptis acu

---

^a The fish is our sturgeon, and its glue is isinglass.

rashes. The Magi recommend the same to be burnt with vipers' skins and frogs, and the ash to be sprinkled into drinks; they assure us that clearer vision will result. Ichthyocolla *a* is the name of a fish that has a sticky skin; the same name is given to the glue of the fish; this disperses night rashes. Some say that ichthyocolla is made from the belly and not from the skin, just as is bull glue. Pontic ichthyocolla is popular, being white, free from veins and scales, and melting very quickly. It ought, however, to be cut up and soaked in water or vinegar for a night and a day, and then to be pounded by sea-pebbles, to make it melt more readily. They assure us that it is useful both for headache and for all tetanus. The right eye of a frog hung round the neck in a piece of undyed cloth cures ophthalmia in the right eye; the left eye similarly tied cures ophthalmia in the left. But if the frog's eyes are gouged out when the moon is in conjunction, and worn similarly by the patient, enclosed in an egg-shell, it will also cure albugo. The rest of the flesh, if applied, quickly takes away bruises. An amulet of crabs' eyes also, worn on the neck, is said to cure ophthalmia. There is a small frog, found living especially in reed-beds and grasses, deaf, without a croak, and green, which, if it by chance is swallowed, swells up the bellies of oxen. They say that the fluid of its body, scraped off with a spatula and applied to the eyes, improves vision. The flesh by itself is placed over painful eyes. Some put together into a new earthen jar fifteen frogs, piercing them with rushes; to the fluid that thus exudes they add the gum of the white vine, and so treat eyelids; superfluous hairs are plucked out, and the mixture dropped with a needle

instillantes hunc sucum in vestigia evolsorum.
76 Meges psilotrum palpebrarum faciebat in aceto
enecans putrescentes et ad hoc utebatur multis variis-
que per aquationes autumni nascentibus. idem praestare sanguisugarum cinis ex aceto inlitus putatur—
comburi eas oportet in novo vaso—idem thynni iocur
siccatum pondere X IIII cum oleo cedrino perunctis
pilis novem mensibus.

77   XXV. Auribus utilissimum batiae piscis fel recens,
sed et inveteratum vino,[1] item bacchi, quem quidam
mizyenem [2] vocant, item callionymi cum rosaceo in-
fusam vel castoreum cum papaveris suco. vocant et
in mari peduculos eosque tritos instillari ex aceto
auribus iubent. et per se [3] et conchylio infecta lana
magnopere prodest; quidam aceto et nitro made-
78 faciunt. sunt qui praecipue contra omnia aurium
vitia laudent gari excellentis cyathum, mellis dimidio
amplius, aceti cyathum in calice novo leni pruna decoquere subinde spuma pinnis detersa et, postquam
desierit spumare, tepidum infundere. si tumeant
aures, coriandri suco prius mitigandas iidem praecipiunt. ranarum adips instillatus statim dolores tollit.
cancrorum fluviatilium sucus cum farina hordeacea
aurium volneribus efficacissime prodest. parotides
muricum testae cinere cum melle vel conchyliorum
ex mulso curantur.

---

[1] vino *codd.*: nitro *Mayhoff, qui XXXI,* 111 (117) *confert.*
[2] mizyenem B, *Detlefsen, Mayhoff: varia codd.*
[3] ex per se *codd.*: operire *coni. Mayhoff ex Marcello.*

into the holes made by the plucked-out hairs. Meges used to make a depilatory for the eyelids by killing frogs in vinegar and letting them putrefy; for this purpose he used the many spotted frogs that breed in the autumn rains. The same effect is thought to be produced by leeches reduced to ash and applied in vinegar; they must be burnt in a new vessel. The same effects too by the dried liver of a tunny, in doses of four denarii added to cedar oil and applied to the hairs for nine months.

XXV. Most beneficial to the ears is the fresh gall of the skate, but also when preserved in wine, the gall of grey mullet, which some call mizyene, and also that of the star-gazer with rose-oil poured into the ears, or beaver oil poured into the ears with poppy juice. There is a creature called the sea-louse, and they recommend sea-lice to be crushed and dropped into the ears in vinegar. Wool, both by itself and dyed with the purple fish, is very good for ear troubles; some moisten it with vinegar and soda. Some there are who recommend as a sovereign remedy for all ear troubles a cyathus of first-grade garum, half as much again honey, with a cyathus of vinegar, to be boiled down in a new cup over a slow fire, every now and then wiping away the froth with feathers, and when the mixture has ceased to froth, to pour it into the ears when tepid. Should the ears be swollen, the same authorities prescribe that the swellings should be first reduced with juice of coriander. Frog fat dropped into the ears immediately takes away pains. The juice of river crabs with barley flour is most beneficial for wounds of the ears. The ash of murex shell with honey, or that of other shell-fish in honey wine, is good treatment for parotid swellings.

79   XXVI. Dentium dolcres sedantur ossibus draconis
marini scariphatis gingivis, cerebro caniculae in oleo
decocto adservatoque, ut ex eo dentes semel anno
colluantur. pastinacae quoque radio scariphari gingivas in dolore utilissimum contritus. is et cum helleboro albo inlitus dentes sine vexatione extrahit.
salsamentorum etiam ⟨in⟩ [1] fictili vase combustorum
80 cinis addita farina marmoris inter remedia est. et
cybia vetera eluta in novo vase, dein trita prosunt
doloribus. aeque prodesse dicuntur omnium salsamentorum spinae combustae tritaeque et inlitae.
decocuntur et ranae singulae in aceti heminis, ut
dentes ita colluantur contineaturque in ore sucus.
si fastidium obstaret, suspendebat pedibus posteriori-
bus eas Sallustius Dionysius, ut ex ore virus deflueret
in acetum fervens, idque e pluribus ranis ; fortioribus
stomachis ex iure mandendas dabat. maxillaresque
ita sanari praecipue dentes putant, mobiles vero
81 supra dicto aceto stabiliri. ad hoc quidam ranarum
corpora binarum praecisis pedibus in vini hemina
macerant et ita collui dentium labantes iubent.
aliqui totas adalligant maxillis. alii denas in sextariis
tribus aceti decoxere ad tertias partes, ut mobiles
dentium stabilirent. nec non XLVI [2] ranarum corda
in olei veteris sextario sub aereo testo discoxere, ut
infunderent per aurem dolentis maxillae. alii iocur
ranae decoctum et tritum cum melle inposuere denti-

---

[1] in *post* etiam *add.* Mayhoff.
[2] XLVI B, *Sillig* : XXXVI *ceteri codd.*

XXVI. Toothache is relieved by scraping the gums with the bones of the weever fish, or by the brain of a dog-fish boiled down in oil and kept, so that the teeth may be washed with it once every year. To scrape the gums too with the ray of the sting-ray is very beneficial for toothache. This ray if pounded and applied with white hellebore brings out teeth without any distress. Salted fish also, reduced to ash in an earthen vessel and mixed with powdered marble, is another remedy. Old slices of tunny rinsed in a new vessel and then beaten up, are good for toothaches. Equally good are said to be the backbones of any salted fish, burnt, pounded, and applied. A single frog is boiled down in one hemina of vinegar, so that the teeth may be rinsed with the juice, which should be held in the mouth. Should the nasty taste be an objection, Sallustius Dionysius used to hang frogs by their hind legs so that the fluid from their mouths might drop into boiling vinegar, and that from several frogs. For stronger stomachs he prescribed the frogs themselves, to be eaten with their broth. It is thought that double teeth yield best to this treatment, when loose indeed the vinegar spoken of above is thought to make them firm. For this purpose some cut off the feet of two frogs and soak the bodies in a hemina of wine, and recommend loose teeth to be rinsed with the liquid. Some tie whole frogs on the jaws as an amulet; others have boiled down ten frogs in three sextarii of vinegar to one third the volume, in order to strengthen loose teeth. Furthermore they have boiled the hearts of 46 frogs under a copper vessel in one sextarius of old oil, to be poured into the ear on the side of the aching jaw. Others have boiled the liver of a frog, beaten

513

82 bus. omnia supra scripta ex marina efficaciora. si
cariosi et faetidi sint, cetum in furno arefieri per
noctem praecipiunt, postea tantundem salis addi
atque ita fricari. enhydris vocatur a Graecis colubra
in aqua vivens. huius quattuor dentibus superioribus
in dolore superiorum gingivas scariphant, inferiorum
inferioribus; aliqui canino tantum earum contenti
sunt. utuntur et cancrorum cinere, nam muricum
cinis dentifricium est.

83 XXVII. Lichenas et lepras tollit adips vituli
marini, menarum cinis cum mellis obolis ternis, iocur
pastinacae in oleo coctum, hippocampi aut delphini [1]
cinis ex aqua inlitus. exulcerationem sequi debet
curatio, quae perducit ad cicatricem. quidam delphini in fictili torrent, donec pinguitudo similis oleo
84 fluat; hac [2] perungunt. muricum vel conchyliorum
testae cinis maculas in facie mulierum purgat cum
melle inlitus cutemque erugat et extendit septenis
diebus inlitus ita, ut octavo candido ovorum foveantur. muricum generis sunt quae vocant Graeci
coluthia, alii coryphia, turbinata aeque, sed minora,
multo efficaciora, etiam oris halitum custodientia.
ichthyocolla erugat cutem extenditque in aqua
decocta horis quattuor, dein contusa et subacta
85 ad liquorem usque mellis. ita praeparata in vase
novo conditur et in usu quattuor drachmis eius

---

[1] delphini *Mayhoff*: delphinū B²dT: delphini iecur *vulg.*
[2] hac *Mayhoff*: ac *fere omnes codd.*

---

a Apparently *pinguitudom* is to be understood with
delphini.

it up with honey, and placed it on the teeth. All the above prescriptions are more efficacious if the sea frog is used. If the teeth are decayed and foul, they recommend whale's flesh to be dried for a night in a furnace, and then the same amount of salt to be added and the whole to be used as a dentifrice. The enhydris is a snake so-called by the Greeks and living in water. With four upper teeth of this creature they scrape the upper gums, when there is aching of the upper teeth, and with four lower teeth the lower gums when there is aching in the lower teeth. Some are content to use the canine tooth only of these creatures. They also use the ash of crabs, but the ash of the murex makes a dentifrice.

XXVII. Lichens and leprous sores are removed by the fat of the seal, the ash of *menae* with three oboli of honey, the liver of the sting-ray boiled in oil, or the ash of the sea-horse or dolphin applied with water. Ulceration should be followed by treatment, which results in a scar. Some roast dolphin fat [a] in an earthen jar until it flows like oil; this they use as ointment. The shell of murex or other shell-fish reduced to ash clears spots from the faces of women, remove wrinkles, and fill out the skin, if applied with honey for seven days, but on the eighth day there should be fomentation with white of egg. To the class murex belong the shell-fish called by the Greeks *coluthia*, by others *coryphia*, equally conical but smaller and much more efficacious, and they also keep the breath sweet. Fish-glue removes wrinkles and fills out the skin; prepared by boiling down in water for four hours and then kneading until liquid like honey. After being thus prepared it is stored away in a new vessel, and when used four drachmae

binae sulpuris et anchusae totidem, octo spumae
argenteae adduntur aspersaque aqua teruntur una.
sic inlita facies post quattuor horas abluitur. medetur et lentigini ceterisque vitiis ex ossibus saepiarum
cinis. idem et carnes excrescentes tollit et umida
ulcera. psoras tollit rana decocta in heminis quinque
aquae marinae; excoqui debet, donec sit lentitudo
86 mellis. Fit in mari alcyoneum appellatum, e nidis,
ut aliqui existumant, alcyonum et ceycum, ut alii,
sordibus spumarum crassescentibus, alii e limo vel
quadam maris languine. quattuor eius genera:
cinereum, spissum, odoris asperi, alterum molle,
lenius odore et fere algae, tertium [1] candidioris vermiculi, quartum pumicosius, spongeae putri simile.
87 paene purpureum quod optimum; hoc et Milesium
vocatur. quo candidius autem, hoc minus probabile
est. vis eorum ut exulcerent, purgent. usus tostis [2]
et sine oleo. mire lepras, lichenas, lentigines tollunt
cum lupino et sulpuris duobus obolis. alcyoneo
utuntur et ad oculorum cicatrices. Andreas ad lepras
cancri cinere cum oleo usus est, Attalus thynni adipe
recenti.
88 XXVIII. Oris ulcera menarum muria et capitum
cinis cum melle sanat. strumas pungi piscis eius, qui
rana marina appellatur, ossiculo e cauda ita, ut non
volneret, prodest. faciendum id cotidie, donec per-

[1] *Post* tertium *velit* formā *supplere Mayhoff.*
[2] tostis] *an* lotis? *Mayhoff.*

---

[a] *Exulcerare* may mean " to clear away ulcers."
[b] Mayhoff suggests " washed."

of it, two of sulphur, two of alkanet, eight of litharge, are mixed, sprinkled with water, and pounded together. Applied to the face this mixture is washed off after four hours. Freckles too and the other facial affections are treated by the calcined bones of cuttle-fish; they also remove excrescences of flesh and running sores. Itch-scab is removed by the decoction of a frog in five heminae of sea-water; the boiling should continue until the consistency is that of honey. In the sea is found a substance called alcyoneum, some think out of the nests of the alcyon and the ceyx, others out of clotted sea-foam, others from the slime of the sea or from what might be called its down. There are four kinds of it: the first is ash-coloured, compact, and of a pungent smell; the second is milder in smell, which is almost that of sea-weed; the third is in shape like a whitish grub; the fourth is rather like pumice, resembling rotten sponge. The best is almost purple, and is also called Milesian. The whiter alcyoneum is the less valuable it is. The property of alcyoneum is to ulcerate [a] and to cleanse. When used it is parched,[b] and applied without oil. With lupins and two oboli of sulphur it removes wonderfully well leprous sores, lichens, and freckles. It is also used for scars on the eyes. Andreas used for leprous sores crabs reduced to ash and applied with oil, Attalus the fresh fat of the tunny.

XXVIII. Ulcers in the mouth are healed by the brine of menae, and by their heads reduced to ash and applied with honey. For scrofulous sores it is good to prick them, but not causing a wound, with the little bone from the tail of the fish called the sea-frog. This should be done daily, until the cure is

curentur. eadem vis est pastinacae radio et lepori marino inposito ita, ut celeriter removeatur, echini testis contusis et ex aceto inlitis, item scolopendrae marinae e melle, cancro fluviatili contrito vel combusto ex melle. mirifice prosunt et saepiae ossa cum

89 axungia vetere contusa et inlita. sic et ad parotidas utuntur, et sauri piscis marini iocineribus, quin et testis cadi salsamentarii tusis cum axungia vetere, muricum cinere ex oleo ad parotidas strumasque. rigor cervicis mollitur et marinis, qui pediculi vocantur, drachma pota, castoreo poto cum pipere ex mulso mixto, ranis decoctis ex oleo et sale, ut sorbeatur sucus. sic et opisthotono medentur et tetano,

90 spasticis vero pipere adiecto. Anginas menarum salsarum e capitibus cinis ex melle inlitus abolet, ranarum decoctarum aceto sucus; hic et contra tonsillas prodest. cancri fluviatiles triti singuli in hemina aquae anginis medentur gargarizati, aut e vino et calida aqua poti. uvae medetur garum coclearibus subditum. vocem siluri recentes salsive in cibo sumpti adiuvant.

91 XXIX. Vomitiones mulli inveterati tritique in potione concitant. Suspiriosis castorea cum Hammoniaci exigua portione ex aceto mulso ieiunis utilissima potu. eadem potio spasmos stomachi sedat ex

complete. The same property is possessed by the sting of the sting-ray and by the sea-hare, but the application must be quickly removed, with the shells of the urchin crushed and applied in vinegar, by the sea-scolopendra too applied in honey, and by river-crabs, crushed or burnt and applied in honey. Wonderfully good too are the bones of cuttle-fish crushed with old axle-grease and applied. The same prescription is used for parotid swellings as well, as is the liver of the horse-mackerel, and even the crushed pieces of a jar in which fish have been salted, applied with old axle-grease; the ash of the murex is applied with oil for parotid swellings and scrofulous sores.

A stiff neck is softened by what are called sea-lice, the dose being a drachma taken in drink, by beaver oil mixed with pepper and taken in honey-wine, and by frogs boiled down in oil and salt for the liquor to be swallowed. This prescription is treatment for opisthotonus and tetanus. For spasms, however, pepper is added. Quinsy is cured by an application in honey of the heads of salted menae, and by the liquor of frogs boiled down in vinegar, which last is also good for diseased tonsils. River crabs pounded one by one in a hemina of water make a healing gargle for quinsy, or they may be taken in wine and warm water. Garum, placed beneath the uvula with a spoon, is good treatment for it. Fresh or salted silurus taken as food improve the voice.

XXIX. Red mullet, preserved, crushed and taken in drink, is an emetic. For asthma is very beneficial beaver oil taken fasting in oxymel with a small quantity of sal ammoniac. This draught also calms stomach spasms when taken in warm oxymel. A

92 aceto mulso caldo. Tussim sanare dicuntur piscium
modo e iure decoctae in patinis ranae. suspensae
autem pedibus, cum destillaverit in patinas saliva
earum, exinterari iubentur abiectisque interaneis
condiri. est rana parva arborem scandens atque ex
ea vociferans; in huius os si quis expuat ipsamque
dimittat, tussi liberari narratur. praecipiunt et
cocleae crudae carnem tritam bibere ex aqua calda
in tussi cruenta.

93 XXX. Iocineris doloribus . . . scorpio marinus in
vino necatus, ut inde bibatur, conchae longae carnes
ex mulso potae cum aquae pari modo aut, si febres
sint, ex aqua mulsa. Lateris dolores leniunt hippocampi tosti sumpti tetheaque similis ostreo in cibo
sumpta, ischiadicorum muria siluri clystere infusa.
dantur autem conchae ternis obolis dilutis in vini
sextariis duobus per dies xv.

94 XXXI. Alvum emollit silurus e iure et torpedo in
cibo et olus marinum simile sativo—stomacho inimicum alvum facillime purgat, sed propter acrimoniam
cum pingui carne coquitur—et omnium piscium ius.
idem et urinas ciet, e vino maxime. optimum e
scorpionibus et iulide et saxatilibus nec virus resi-
pientibus nec pinguibus. coci debent cum aneto,
95 apio, coriandro, porro, additis oleo, sale. purgant et
cybia vetera, privatimque cruditates, pituitas, bilem
trahunt.

---

a In taste? *Tethea* is a sea-squirt.

cough is said to be cured by frogs boiled down in a pan as are fish in their own liquor. A prescription is: the frogs to be hung up by the feet, their saliva allowed to drip into a pan, and then, after being gutted, they are preserved after the entrails have been cast aside. There is a small frog that climbs trees and croaks loudly out of them. If a person with a cough spits into the mouth of one of these and lets it go, he is said to be cured of the complaint. For a cough with spitting of blood is prescribed the raw flesh of a snail beaten up and taken in warm water.

XXX. For liver pains are good: . . . a sea scorpion drowned in wine, so that the liquor may be drunk, or the flesh of the long mussel taken in honey wine with an equal quantity of water, or if there is fever in hydromel. Pains in the side are relieved by eating the flesh of the sea-horse roasted, or the tethea, which resembles [a] the oyster, taken in the food; sciatica is relieved by the brine of the silurus, injected as an enema. Mussels too are given for fifteen days in doses of three oboli soaked in two sextarii of wine.

XXXI. The bowels are relaxed by the silurus, taken with its broth, by the torpedo, taken in food, by the sea-cabbage, which is like the cultivated kind —it is bad for the stomach but readily purges the bowels, and owing to its pungency is boiled with fat meat—and by the liquor of any boiled fish; the last is also diuretic, especially when taken in wine. The best is from the sea-scorpion, the wrasse, and the rock-fish, which are neither of a rank taste nor fatty. They should be boiled with dill, parsley, coriander, leeks, and with oil added and salt. Purgative too is stale tunny sliced, and it is specific for bringing away undigested food, phlegm and bile.

Purgant et myaces, quorum natura tota in hoc loco dicetur. acervantur muricum modo vivuntque in algosis, gratissimi autumno et ubi multa dulcis aqua miscetur mari, ob id in Aegypto laudatissimi. procedente hieme amaritudinem trahunt coloremque
96 rubrum. horum ius traditur alvum et vesicas exinanire, interanea destringere, omnia adaperire, renes purgare, sanguinem adipemque minuere. itaque utilissimi sunt hydropicis, mulierum purgationibus, morbo regio, articulario, inflationibus, item obesis, fellis pituitae,[1] pulmonis, iocineris, lienis vitiis, rheumatismis. fauces tantum vexant vocemque
97 obtundunt. ulcera, quae serpant aut sint purganda, sanant, item carcinomata cremati ut murices; et morsus canum hominumque cum melle, lepras, lentigines. cinis eorum lotus emendat caligines, scabritias, albugines, gingivarum et dentium vitia, eruptiones pituitae; et contra dorycnium aut opo-
98 carpathum antidoti vicem optinent. degenerant in duas species: mitulos, qui salem virusque resipiunt, myiscas quae rotunditate differunt, minores aliquanto atque hirtae, tenuioribus testis, carne dulciores. mituli quoque ut murices cinere causticam vim habent et ad lepras, lentigines, maculas. lavantur[2] quoque plumbi modo ad genarum crassitudines et oculorum albugines caliginesque atque in aliis partibus sordida ulcera capitisque pusulas. carnes eorum ad canis morsus inponuntur.
99 At pelorides emolliunt alvum, item castorea in

[1] pituitae *multi codd.*: pituitaeque B: pituitae quoque *Mayhoff*: *an* felli?
[2] lavantur] lavatur *coni. Mayhoff*.

---

[a] With Mayhoff's conjecture (probably correct) " the ash is washed."

The myax also is purgative, and in this place shall be set forth all its characteristics. These animals form clusters, as does the murex, and live where sea-weed lies thick, for which reason they are most delicious in autumn, and from regions where much fresh water mingles with salt, for which reason it is in Egypt that they are most esteemed. As the winter advances, they contract a bitter taste, and a red colour. Their liquor is said to be a thorough purge of belly and bladder, cleanses the intestines, is a universal aperient, purges the kidneys, and reduces blood and fat. Hence these shell-fish are very beneficial for dropsy, menstruation, jaundice, diseases of the joints, flatulence, obesity also, bilious phlegm, affections of lungs, liver, and spleen, and for catarrhs, Their only drawback is that they harm the throat and obstruct the voice. Ulcers that are creeping or need cleansing they heal, and also, if burnt as is the murex, malignant growths. With honey added they heal the bites of dogs and men, leprous sores, and freckles. Their ash, washed, is good for dim vision, roughness and white ulcers of the eyes, affections of the gums and teeth and outbursts of phlegm. Against dorycnium and opocarpathum they serve as an antidote. There are two inferior kinds : the mitulus, with a salty, strong taste; the myisca, different in its roundness, rather smaller and hairy, with thinner shell and sweeter flesh. The mitulus too like the murex has a caustic ash good for leprous sores, freckles, and spots. They are washed [a] also as is lead for thick eye-lids, white ulcers, dim vision, dirty ulcers in other parts. and pustules on the head. Their flesh makes an application for dog bites.

But clams relax the bowels, as does beaver oil in

aqua mulsa drachmis binis. qui vehementius volunt
uti, addunt cucumeris sativi radicis siccae drachmam
et aphronitri duas tethea. torminibus et inflationibus
occurrunt. inveniuntur haec in foliis maris sugentia,
fungorum verius generis quam piscium. eadem et
100 tenesmum dissolvunt reniumque vitia. nascitur et in
mari apsinthium, quod aliqui seriphum vocant, circa
Taposirim maxime Aegypti, exilius terrestri. alvum
solvit et noxiis animalibus intestina liberat—solvunt
101 et saepiae—; in cibo datur cum oleo et sale et farina
coctum. menae salsae cum felle taurino inlitae
umbilico alvum solvunt. piscium ius in patina coctorum cum lactucis tenesmum discutit. cancri fluviatiles triti et ex aqua poti alvum sistunt, urinam cient
in vino albo. ademptis bracchiis calculos pellunt
tribus obolis cum murra et iride singulis earum drachmis, ileos et inflationes castorea cum dauci semine et
petroselino quantum ternis digitis sumatur, ex mulsi
calidi cyathis IIII, tormina vero cum aneto ex vino
mixto. erythini in cibo sumpti sistunt alvum.
dysentericis medentur ranae cum scilla decoctae ita,
ut pastilli fiant, vel cor earum cum melle tritum, ut
tradit Niceratus, morbo regio salsamentum cum
pipere ita, ut reliqua carne abstineatur.
102   XXXII. Lieni medetur solea piscis inpositus, item
torpedo, item rhombus; vivus dein remittitur in mare.
scorpio marinus necatus in vino vesicae vitia et cal-

hydromel, the dose being two drachmae. Those who wish to use a more drastic laxative add a drachma of dried root of cultivated cucumber and two drachmae of saltpetre. Tethea cures griping and flatulence. It is found as a parasite on sea plants, more a kind of fungus rather than a fish. They also cure tenesmus and affections of the kidneys. There also grows in the sea apsinthium, which some call seriphum, found chiefly around Taposiris in Egypt, and is more slender than the land variety. It relaxes the bowels and brings away harmful creatures from the intestines. The cuttle-fish too is laxative. The apsinthium is given in food, being boiled with oil, salt, and flour. Salted menae applied to the navel with bull's gall relax the bowels. The liquor of fish boiled in a pan with lettuce cures tenesmus. River crabs beaten up and taken in water are constipating but diuretic in a white wine. If their legs are taken off they bring away stone, the dose being three oboli with a drachma each of myrrh and iris; iliac colic and flatulence are cured by beaver oil with daucus seed and of rock parsley as much as can be picked up in three fingers, taken in four cyathi of warm honey-wine; while for griping it should be taken with a mixture of dill and wine. The erythinus taken in food is constipating. Dysentery can be treated by frogs boiled with squills to make lozenges, or by their heart beaten up with honey, as Niceratus prescribes, jaundice by salted fish with pepper, but the patient must abstain from all other meat.

XXXII. Splenic trouble is treated by the application of the fish sole, of the torpedo, or of the turbot, but the fish is then put back living into the sea. Bladder troubles and stone are cured by the

culos sanat, lapis, qui invenitur in scorpionis marini
cauda, pondere oboli potus, enhydridis iecur, blen-
diorum cinis cum ruta. inveniuntur et in bacchi piscis
capite ceu lapilli; hi poti ex aqua calculosis praeclare
medentur. aiunt et urticam marinam in vino potam
prodesse, item pulmonem marinum decoctum in aqua.

103 ova saepiae urinam movent reniumque pituitas extrahunt. rupta, convolsa cancri fluviatiles triti in asinino
maxime lacte sanant, echini vero cum spinis suis contusi et e vino poti calculos—modus singulis hemina;
bibitur, donec prosit—et alias in cibis ad hoc proficiunt. purgatur vesica et pectinum cibo. ex iis
mares alii δόνακας vocant, alii αὐλούς, feminas ὄνυχας.
urinam mares movent. dulciores feminae sunt et
unicolores. [saepiae quoque ova urinam movent et
renes purgant].[1]

104 XXXIII. Enterocelicis lepus marinus inlinitur
tritus cum melle. iecur aquaticae colubrae, item
hydri tritum potumque calculosis prodest. ischiadicos liberant salsamenta e siluro infusa clysterio,
evacuata prius alvo, sedis attritus cinis e capite mugi-
lum et mullorum; comburuntur autem in fictili vase,

105 inlini cum melle debent. item capitis menarum cinis
et ad rhagadas et ad condylomata utilis, sicut pelamy-

---

[1] *Uncos addunt Hard., Mayhoff.*

sea scorpion killed in wine, by the stone which is found
in the tail of the sea-scorpion, the dose being an
obolus, taken in drink, by the liver of the enhydris,
and by the ash of the blenny with rue.   There are
found too in the head of the fish bacchus as it were
pebbles; these taken in water are excellent treatment for stone.   It is said that the sea-nettle taken
in wine is also good for it, and likewise the pulmo
marinus boiled down in water.   The eggs of the
cuttle-fish are diuretic and bring away phlegms from
the kidneys.   Ruptures and sprains are healed by
river-crabs beaten up in milk, by preference asses',
stone however by sea-urchins, spines and all, crushed
in wine and taken in doses of a hemina to each urchin,
this amount being drunk until benefit is apparent;
urchins are also beneficial generally for stone when
taken as food.   The bladder is cleansed by a diet of
scallops.   The male scallops are called by some
σόνακες (reeds), by others αὐλοί (pipes); the female
they call ὄνυχες (nails).   The males are diuretic; the
females are sweeter and of a uniform colour.   [The
eggs of the cuttle-fish also are diuretic and cleanse
the kidneys].

XXXIII. For intestinal hernia is applied sea-hare
beaten up with honey.   The liver of the water-
coluber, likewise that of the water-snake, beaten up
and taken in drink, is good for stone.   Sciatica is
cured by the brine of pickled silurus, injected as an
enema, after previous thorough cleansing of the
bowels; chafing of the seat by the head of grey or
red mullet reduced to ash.   The fish are burnt in an
earthen vessel and should be applied with honey.
The heads too of menae, reduced to ash, are useful
for chaps and condylomata, just as the heads of salted

527

dum salsarum capitum cinis vel cybiorum cum melle. torpedo adposita procidentis interanei morbum ibi coercet. cancrorum fluviatilium cinis ex oleo et cera rimas in eadem parte emendat, idem et marini cancri pollent.

106   XXXIV. Panos salsamenta coracini [1] discutiunt, sciaenae interanea et squamae combustae, scorpio in vino decoctus ita, ut foveantur ex illo. at echinorum testae contusae et ex aqua inlitae incipientibus panis resistunt, muricum vel purpurarum cinis utroque modo, sive discutere opus sit incipientes sive concoctos emittere. quidam its componunt medicamentum: cerae et turis drachmas xx, spumae argenti xxxx,
107 cineris muricum x, olei veteris heminam. prosunt per se salsamenta cocta, cancri fluviatiles triti; [2] verendorum pusulas cinis e capite menarum, item carnes decoctae et inpositae, similiter percae salsae e capite cinis melle addito, pelamydum capitis cinis aut
108 squatinae cutis combustae. haec est qua diximus lignum poliri, quoniam et a mari fabriles usus exeunt. prosunt et zmarides inlitae, item muricum vel purpurarum testae cinis cum melle, efficacius crematarum cum carnibus suis. carbunculos verendorum privatim salsamenta cocta cum melle restingunt. testem, si descenderit, coclearum spuma inlini volunt.

---

[1] coracini *Hermolaus Barbarus*: coracina (*fortasse adiectivum*) *multi codd.*: coracinosa B: coracinorū *Mayhoff.*
[2] *Hic vult addere* ad *vel* contra *Mayhoff.*

---

[a] To govern *pusulas* Mayhoff adds *ad.* It is easy however to understand e.g. *emendat.*

pelamids, or sliced tunny, reduced to ash and applied with honey. An application of the torpedo to the intestinal region reduces a morbid procidence there. The ash of river-crabs in oil and wax heals cracks in that part; sea-crabs too have the same healing property.

XXXIV. The pickle of the coracinus disperses superficial abscesses, as do the burnt intestines and scales of the sciaena, or the sea-scorpion boiled down in wine for fomentation with that decoction. But the shells of sea-urchins crushed and applied with water are a remedy for these abscesses when incipient; the murex or purple-fish reduced to ash is beneficial for either purpose, whether it is necessary to disperse incipient abscesses or to mature them and make them discharge. Some make up the following prescription: wax and frankincense twenty drachmae, litharge forty drachmae, ash of the murex ten drachmae, old oil one hemina. By themselves are beneficial boiled salted-fish, and pounded river-crabs. For [a] pustules on the pudenda, ash of the head of menae, likewise their flesh boiled down and applied, similarly the ash of the head of salted perch, with honey added, ash of pelamids' heads, or the skin of burnt squatina. This skin is the one used, as I have said,[b] to polish wood, for from the sea too come useful things for our craftsmen. Zmarides also are beneficial when applied, likewise with honey the shells of the murex or purple-fish reduced to ash, more effectively if burnt with their flesh. Boiled salted fish are specific for reducing carbuncles on the pudenda. It is recommended, if a testicle hangs down, that the froth of snails be applied.

[b] See IX. § 40.

109   XXXV. Urinae incontinentiam hippocampi tosti et
in cibo saepius sumpti emendant, ophidion pisciculus
congro similis cum lili radice, pisciculi minuti ex
ventre eius, qui devoraverit, exempti cremati ita, ut
cinis eorum bibatur ex aqua. iubent et cocleas
Africanas cum sua carne comburi cineremque ex
vino Signino dari.

110   XXXVI. Podagris articulariisque morbis utile est
oleum, in quo decocta sint ranarum intestina, et
rubetae cinis cum adipe vetere. quidam et hordei
cinerem adiciunt trium rerum aequo pondere.
iubent et lepore marino recenti podagram fricari,
fibrinis quoque pellibus calceari, maxime Pontici fibri,
item vituli marini, cuius et adips prodest isdem, nec
non et bryon, de quo diximus, lactucae simile, rugo-
111 sioribus foliis, sine caule. natura ei styptica, inposi-
tumque lenit impetus podagrae. item alga, de qua
et ipsa dictum est. observatur in ea, ne arida inponatur. perniones emendat pulmo marinus, cancri
marini cinis ex oleo, item fluviatiles triti ustique,
cinere [1] et ex oleo subacto,[2] siluri adips. in articulis
morborum impetus sedant ranae subinde recentes
inpositae; quidam dissectas iubent inponi. corpus
auget ius mitulorum et concharum.

112   XXXVII. Comitiales, ut diximus, coagulum vituli
marini bibunt cum lacte equino asinaeve aut cum

[1] cinere *codd.*: in cinerem *coni. Sillig.*
[2] subacto *Mayhoff*: subacti *codd.*

---

[a] Green laver. See *Index of Plants* in Vol. VII.
[b] See XXVII. § 56.
[c] See § 66 of this book.   [d] See VIII. § 111.

XXXV. Incontinence of urine is remedied by the sea-horse, roasted and taken often as food, by the ophidion, a little fish like the conger, with lily-root added, and by the tiny fish in the belly of the fish that has swallowed them, taken out and burnt for their ash to be taken in water. They also recommend African snails to be burnt with their flesh, and the ash to be given in Signian wine.

XXXVI. For gouty pains and for diseases of the joints oil is useful in which the intestines of frogs have been boiled down, and also the ash of bramble-toads mixed with stale grease. There are some who add to these also barley ash, taking equal weights of three ingredients. They recommend too a gouty foot to be rubbed with a fresh sea-hare, and the patient also to be shod with beaver skin, by preference that of the Pontic beaver, or else with seal skin, seal fat also being good for gout. Good also is bryon,[a] about which I have spoken,[b] a plant like the lettuce, but with more wrinkled leaves and without a stem. Its nature is styptic, and applied to the painful part it soothes the paroxysms of gout. Sea-weed too is good, about which by itself also I have spoken.[c] Care is taken with sea-weed, not to apply it dry. An application of pulmo marinus is a cure for chilblains, and so is the ash of a sea-crab in oil, river-crabs too pounded and burnt, the ash also being kneaded with oil, and the fat of the silurus. In diseases of the joints paroxysms are soothed by applying fresh frogs every now and then; some recommend them to be cut up before being applied. Flesh is put on by the liquid of mussels and of shell-fish generally.

XXXVII. Epilepsy, as I have said,[d] is treated by doses of seals' rennet with mares' or asses' milk, or

Punici suco, quidam ex aceto mulso. nec non aliqui
per se pilulas devorant. castoreum in aceti mulsi
cyathis tribus ieiunis datur, iis vero, qui saepius corripiantur, clystere infusum mirifice prodest. castorei
drachmae duae esse debebunt, mellis et olei sextarius
et aquae tantundem. ad praesens vero correptis
olfactu subvenit cum aceto. datur et mustelae
marinae iocur, item muris, vel testudinum sanguis.

113   XXXVIII. Febrium circuitus tollit iocur delphini
gustatum ante accessiones. hippocampi necantur in
rosaceo, ut perunguantur aegri frigidis febribus, et
ipsi adalligantur aegris. item ex asello pisce lapilli,
qui plena luna inveniuntur in capite, alligantur in
linteolo. phagri fluviatilis longissimus dens capillo
adalligatus ita, ut quinque diebus eum, qui adalligaverit, non cernat aeger, ranae in trivio decoctae
oleo abiectis carnibus perunctos liberant quartanis.

114 sunt qui strangulatas in oleo ipsas clam adalligent
oleoque eo perunguant. cor earum adalligatum frigora febrium minuit et oleum, in quo intestina decocta sint. maxime autem quartanis liberant ablatis
unguibus ranae atque[1] adalligatae et rubeta, si
iocur eius vel cor adalligetur in panno leucophaeo.
cancri fluviatiles triti in oleo et aqua perunctis ante

---

[1] atque *codd.*: aeque *Mayhoff.*

with pomegranate juice; some prescribe it in oxymel. Some too swallow the rennet by itself, made up into pills. Beaver oil in three cyathi of oxymel is given on an empty stomach; those however frequently attacked are benefited wonderfully by a clyster; of the beaver oil there should be two drachmae, of honey and oil a sextarius, and the same quantity of water. If indeed persons have a momentary seizure it is beneficial to give the patients beaver oil and vinegar to smell. There is also given the liver of the sea-weasel, or of the sea-mouse, or the blood of tortoises.

XXXVIII. Recurrent fevers are cured by a dolphin's liver, taken before the paroxysms. Sea-horses are killed in rose-oil, to make ointment for those sick of chill fevers, and sea-horses themselves are worn as an amulet by the patients. The little stones also that at a full moon are found in the head of the fish asellus, are tied on the patient in a linen cloth. Quartans are cured by the longest tooth of the river fish phagrus, tied with a hair on the patient as an amulet, but the patient must not discern the person who attached it for five days; also by rubbing with the grease of frogs boiled in oil at a place where three roads meet, the flesh being first thrown away. Some drown frogs in oil, attach secretly as an amulet, and rub the patient thoroughly with the oil. The heart of frogs attached as an amulet, and the oil in which their entrails have been boiled, relieve the chills of fevers. The best cure for quartans, however, is a frog, worn as an amulet with its claws taken off, or a bramble-toad, if its liver or heart is worn as an amulet in a piece of ash-coloured cloth. River-crabs, pounded in oil and water and thoroughly

accessiones in febribus prosunt; aliqui et piper
115 addunt. alii decoctos ad quartas in vino e balineo
egressis bibere suadent in quartanis, aliqui vero sinistrum oculum devorare. Magi oculis eorum ante
solis ortum adalligatis aegro ita, ut caecos dimittant in
116 aquam, tertianas abigi promittunt. eosdem oculos
cum carnibus lusciniae in pelle cervina inligatos
praestare vigiliam somno fugato tradunt. in lethargum vergentibus coagulo ballaenae aut vituli marini
ad olfactum utuntur. alii sanguinem testudinis
lethargicis inlinunt. tertianis mederi dicitur et
spondylus percae adalligatus, quartanis cocleae
fluviatiles in cibo recentes; quidam ob id adservant
sale, ut dent tritas in potu.
117   XXXIX. Strombi in aceto putrefacti lethargicos
excitant odore. prosunt et cardiacis. cachectis,
quorum corpus macie conficitur, tethea utilia sunt
cum ruta ac melle. hydropicis medetur adips
delphini liquatus et cum vino potus. gravitati
saporis occurritur tactis naribus unguento aut odori-
bus vel quoquo modo opturatis. strombi quoque
carnes tritae et in mulsi tribus heminis pari modo
aquae aut, si febres sint, ex aqua mulsa datae pro-
118 ficiunt, item sucus cancrorum fluviatilium cum melle,
rana quoque aquatica in vino vetere et farre decocta
ac pro cibo sumpta ita, ut bibatur ex eodem vase, vel

---

a Or: turtle.

rubbed over the patient before the paroxysms, are beneficial in fevers; some add pepper also. Others prescribe them for quartans boiled down to a quarter in wine, to be taken after leaving the bath; some, however, the left eye to be swallowed. The Magi assure us that tertian fevers are driven away by crabs' eyes, attached as an amulet before sunrise to the patient, but the blinded crabs must be set free into water. The Magi also teach that crabs' eyes, tied on with the flesh of a nightingale in deer skin, drive away sleep and cause watchfulness. For those sinking into lethargus they prescribe that the patient smell the rennet of the whale or that of the seal. Others use as embrocation for lethargus the blood of a tortoise.[a] It is also said that tertians are treated successfully by the vertebra of a perch worn as an amulet; quartans by fresh river snails taken as food. Some preserve them in salt for this purpose, to administer them, beaten up, in a draught.

XXXIX. Strombi rotted in vinegar rouse by the smell the victims of lethargus. They are also good for those with stomach complaints. Those in a decline, with a body seriously wasting away, find beneficial tethea with rue and honey. Dropsy is treated with melted dolphin fat taken with wine. The nauseating taste is neutralised by touching the nostrils with unguent or scents, or plugging them in any suitable way. The flesh of the strombus also, pounded and given in three heminae of honey wine and an equal measure of water, or should there be fever, in hydromel, benefit the dropsical; likewise the juice of river crabs with honey; water frogs too are boiled down in old wine and emmer wheat, and then taken as food but out of the same vessel as

testudo decisis pedibus, capite, cauda et intestinis
exemptis, reliqua carne ita condita, ut citra fastidium
sumi possit. cancri fluviatiles ex iure sumpti et
phthisicis prodesse traduntur.

119 XL. Adusta sanantur cancri marini vel fluviatilis
cinere ex oleo; ichthyocolla, ranarum cinere ea, quae
ferventi aqua combusta sint; haec curatio etiam pilos
restituit.[1] cancrorum fluviatilium cinere putant
utendum cum cera et adipe ursino. prodest et
fibrinarum pellium cinis. ignes sacros restingunt
ranarum viventium ventres inpositi, pedibus post-
erioribus pronas adalligari iubent, ut crebriore an-
helitu prosint. utuntur et silurorum salsamenti
capitum cinere ex aceto. pruritum scabiemque non
hominum modo, sed et quadripedum efficacissime
sedat iecur pastinacae decoctum in oleo.

120 XLI. Nervos vel praecisos purpurarum callum, quo
se operiunt, tusum glutinat. tetanicos coagulum
vituli adiuvat ex vino potum oboli pondere, item
ichthyocolla, tremulos castoreum, si ex oleo perunguantur. mullos in cibo inutiles [2] nervis invenio.

121 XLII. Sanguinem fieri piscium cibo putant, sisti
polypo tuso inlito, de quo et haec traduntur: muriam
ipsum ex sese emittere et ideo non debere addi in
coquendo, secari harundine, ferro enim infici vitium-
que trahere natura dissidente. ad sanguinem sisten-

---

[1] Hic codd. cum habent: item Mayhoff. Fortasse cum ex aut
cancrorum aut cum cera ortum.
[2] utiles coni. Warmington.

---

[a] Or: turtle.
[b] In a Book dealing with fish remedies vituli cannot mean
an ordinary " calf."
[c] I so translate because of ex.

cooked; a tortoise [a] with feet, head, tail, and entrails taken out, the remaining flesh being so seasoned that it can be taken without nausea. River crabs taken in their juice are also reported to be beneficial to consumptives.

XL. Burns are healed by the ash in oil of a sea crab or river crab; by fish glue, or by the ash of frogs, the scalds caused by boiling water; this treatment also restores the lost hair. They think that the ash of river crabs should be used with wax and bear's grease. Beneficial also is the ash of beaver pelts. Erysipelas disappears under the application of the bellies of live frogs; they recommend the frogs to be tied on upside down by their hind legs, so that their rapid breathing may be of benefit. They also use the ash in vinegar of the heads of salted siluri. Pruritus and itch-scab in quadrupeds as well as in man are relieved with great efficacy by the liver of the sting-ray boiled down in oil.

XLI. The hard operculum, with which the purple-fish shuts its body from view, when beaten up, unites cut sinews even when severed. Patients with tetanus are relieved by an obolus by weight of seal's [b] rennet taken in wine; also by fish glue. The palsied obtain benefit from beaver oil, if they are thoroughly rubbed with it and olive oil.[c] I find that red mullet as a food is injurious to the sinews.

XLII. They think that to eat fish causes bleeding, but that haemorrhage is stopped by crushing and applying the polypus, about which are current the following reports. It of itself gives out of itself brine, and therefore none should be added in cooking; it should be cut with a reed, for iron spoils it and leaves a taint, as the natures of the two quarrel. To stop bleeding

dum et ranarum inlinunt cinerem vel sanguinem
122 arefactum. quidam ex ea rana, quam Graeci cala-
miten vocant, quoniam inter harundines fruticesque
vivat, minima omnium et viridissima, sanguinem
cineremve fieri[1] iubent, aliqui et nascentium ranarum
in aqua, quibus adhuc cauda est, in calice novo combustarum cinerem, si per nares fluat, inferciendum.[2]
123 diversus hirudinum, quas sanguisugas vocant, ad
extrahendum sanguinem usus est. quippe eadem
ratio earum, quae cucurbitarum medicinalium, ad
corpora levanda sanguine, spiramenta laxanda iudicatur, sed vitium, quod admissae semel desiderium
faciunt circa eadem tempora anni semper eiusdem
medicinae. multi podagris quoque admittendas
censuere. decidunt satiatae et pondere ipso sanguinis detractae aut sale adspersae; aliquando tamen
relinquunt adfixa capita, quae causa volnera insanabilia facit et multos interemit, sicut Messalinum e
consularibus patriciis, cum ad genu[3] admisisset, in
veneni[4] virus remedio verso. maxime rufae ita
124 formidantur; ergo sugentes[5] forficibus praecidunt,
ac velut siphonibus defluit sanguis, paulatimque
morientium capita se contrahunt, nec relinquuntur.
natura earum adversatur cimicibus, suffitu necat eos.
fibrinarum pellium cum pice liquida combustarum
cinis narium profluvia sistit suco porri mollitus.
125　XLIII. Extrahit corpori tela inhaerentia saepiarum

---

[1] Inlini coni. Warmington.
[2] inferciendum Ianus: imperficiendum codd.
[3] genu B[2] E: genum B[1]RdT: genam coni. Mayhoff.
[4] in veneni Ianus: invenit B: inveniunt multi codd.
[5] sugentes Sillig: (sugere?) ursas B[1]: oras VRdT: sugere
orsas Mayhoff ex multis lectionibus et coniecturis.

---

[a] The Greek κάλαμος means a reed.

they also apply the ash of frogs or their dried blood. Some recommend the blood or ash to come from the frog called by the Greeks calamites,[a] because it lives among reeds and shrubs, the smallest and greenest of all frogs; some that the ash of frogs at their birth in water, while still tadpoles with a tail, and calcined in a new earthen vessel, should be stuffed into the nostrils of those with epistaxis. Opposite is the use of leeches, called sanguisugae,[b] which are employed to extract blood. For these are supposed to have the same purpose as that of cupping-glasses, to relieve the body of blood and to open the pores of the skin; but an objection is that once applied they create a craving for the same treatment every year at about the same time. Many have been of opinion that leeches should be applied also for gout. When gorged leeches fall off, detached by the mere weight of blood or by a sprinkle of salt; sometimes however they leave their heads stuck fast in the flesh, thus causing incurable wounds that have often proved fatal. An instance is Messalinus, a patrician of consular rank, who applied leeches to his knee,[c] and the remedy turned to a virulent poison. It is especially red leeches that are so dreaded; so they cut them off with scissors while they are sucking, and the blood runs down as it were through tubes; as they die their heads little by little contract, and are not left in the bite. The nature of leeches is adverse to that of bugs, which are killed if fumigated with leeches. Beaver skins, burnt with liquid pitch and softened with leek juice, arrest discharges from the nostrils.

XLIII. Weapons sticking in the flesh are drawn

---

[b] I.e. " blood-suckers."
[c] With Mayhoff's suggestion, " cheek."

testae cinis, item purpurarum testae ex aqua, salsamentorum carnes, cancri fluviatiles triti, siluri
fluviatilis, qui et alibi quam in Nilo nascitur,
carnes inpositae, recentis sive salsi. eiusdem cinis
extrahit, adips et cinis spinae eius vicem spodii
praebet.

126   XLIV. Ulcera, quae serpunt, et quae in iis ex-
crescunt capitis menarum cinis vel siluri coercet, carcinomata percarum capita salsarum, efficacius si
cinere earum misceantur [1] sal et cunila capitata
oleoque subigantur. cancri marini cinis usti cum
plumbo carcinomata compescit. ad hoc et fluviatilis
sufficit cum melle lineaque lanugine; aliqui malunt
alumen melque miscere [2] cineri. phagedaenae siluro
inveterato et cum sandaraca trito, cacoëthe et nomae
et putrescentia cybio vetere sanantur; vermes innati
127 ranarum felle tolluntur. fistulae aperiuntur siccan-
turque salsamentis cum linteolo inmissis, intraque
alterum diem callum omnem auferunt et putrescentia ulcerum quaeque serpant emplastri modo
subacta et inlita. et allex purgat ulcera in linteolis
concerptis, item echinorum testae cinis. carbunculos
coracinorum salsamenta inlita discutiunt, item mullorum salsamenti cinis—quidam capite tantum utuntur cum melle—vel coracinorum carnes. muricum
cinis cum oleo tumores tollit, cicatrices fel scorpionis
marini.
128   XLV. Verrucas tollit glani iocur inlitum, capitis

---

[1] misceantur *coni. Mayhoff*: misceatur *codd.*
[2] miscere *multi codd.*: misceri B. *Sillig, Mayhoff.*

---

[a] See List of Diseases.
[b] See *Index of Plants* in Vol. VII.
[c] *Allex* (variously spelt) is fish pickle.

out by the ash in water of the shell of the cuttle-fish, also of the shell of the purple-fish, by the flesh of salted fish, by river-crabs beaten up, by an application of the flesh of the river silurus (which is found in other rivers besides the Nile), whether fresh or preserved in salt. The ash of the same fish draws out sharp bodies; its fat and the ash of its back-bone take the place of spodium.

XLIV. Creeping ulcers and the excrescences that form in them are checked by ash of menae or of the silurus, carcinomata [a] by heads of salted perch, with more effect if with their ash are mixed salt and headed cunila,[b] and the whole kneaded with oil. The ash of a sea crab that has been burnt with lead checks carcinomata. For this purpose river crab too suffices with honey and fine lint. Some prefer to mix alum and honey with the ash. Phagedaenic ulcers are healed by silurus kept till stale and beaten up with sandarach; malignant ulcers, corrosive ulcers, and festering sores by old tunny sliced; the maggots that breed in them are removed by frogs' gall. Fistulas are opened and dried up by salted fish inserted with lint; within two days such fish remove all callus, festering sores, and creeping ulcers, if kneaded up as for a plaster and applied. Allex [c] also applied in strips of lint cleans sores; likewise the shell of sea-urchins, reduced to ash. Carbuncles are dispersed if treated with salted coracinus, likewise with the ash of salted red mullet—some use the head only with honey—or with the flesh of coracinus. Ash of murex with oil removes swellings, and the gall of the sea scorpion scars.

XLV. Warts are removed by an application of the liver of the glanus, of menae ash beaten up with

menarum cinis cum alio tritus—ad thymia crudis
utuntur—fel scorpionis marini rufi, zmarides tritae
inlitae, allex defervefacta. unguium scabritiam cinis
e capite menarum extenuat.

129   XLVI. Mulieribus lactis copiam facit glauciscus e
iure sumptus et zmarides cum tisana sumptae vel cum
feniculo decoctae. mammas ipsas muricum vel purpurae testarum cinis cum melle efficaciter sanat, item
cancri inliti fluviatiles vel marini. pilos in mamma
muricum carnes inpositae tollunt. squatinae inlitae
crescere mammas non patiuntur. delphini adipe
linamenta tincta [1] accensa excitant volva strangulata

130 oppressas, item strombi in aceto putrefacti. percarum vel menarum capitis cinis sale admixto et cunila
oleoque volvae medetur, suffitione quoque secundas
detrahit. item vituli marini adips instillatur igni
naribus intermortuarum volvae vitio, coagulo eiusdem
in vellere inposito. pulmo marinus alligatus purgat
egregie profluvia, echini viventes tusi et in vino dulci
poti sistunt et cancri fluviatiles triti in vino potique.

131 item siluri suffitu, praecipue Africi, faciliores partus
facere dicuntur, cancri ex aqua poti profluvia sistere,
ex hysopo purgare. et si partus strangulet,[2] similiter
poti auxiliantur. eosdem recentes vel aridos bibunt

---

[1] tincta *add. Brakman,* inlita *Mayhoff, post C. F. W. Müller.*
[2] strangulet VR: stranguletur d (?).

---

[a] The Greek θύμιον, a large wart.
[b] Brakman's *tincta* is perhaps better than Mayhoff's *inlita,*
as *illino* in Pliny is regularly used of applying medicaments
to the human body.
[c] Or: " ailment of the womb."

garlic—for thymion [a] warts they use the materials raw—by the gall of the red sea scorpion, by zmarides beaten up and applied, and by allex thoroughly boiled. Rough nails are smoothed by the ash of menae heads.

XLVI. Milk in women is made plentiful by glauciscus taken with its liquor, by zmarides taken with barley water or boiled down with fennel. The breasts themselves are treated efficaciously by shells of murex or purple fish reduced to ash and combined with honey; by crabs too, river or sea, applied locally. The flesh of the murex if applied removes hair growing on the breasts. Squatinae applied prevent their swelling. Lint, smeared [b] with dolphin's fat and then set alight, arouse women suffering from hysterical suffocations; likewise strombi rotted in vinegar. The ash of the heads of perch or menae, mixed with salt, cunila, and oil, is healing to the uterus; by fumigation also it brings away the after-birth. The fat of the seal melted in the fire is inserted into the nostrils of women swooning from hysterical suffocation,[c] or else seal's rennet used as a pessary in a piece of fleece. The pulmo marinus, tied on,[d] is an excellent promoter of menstruation, which is checked by living sea urchins pounded up and taken in a sweet wine or by river crabs beaten up and so taken. Siluri also, especially the African, are said to make easier the birth of children, crabs taken in water to arrest menstruation, taken in hyssop to promote it. If birth causes choking,[e] the same medicament taken in drink is a help. Crabs, fresh

[d] Here apparently as an amulet, although that is usually *adalligare*.

[e] With the reading *stranguletur*: " if the child chokes."

ad partus continendos. Hippocrates et ad purgationes mortuosque partus utitur illis, cum quinis, lapathi radice rutaeque, et fuligine trita et in mulso
132 data potui. iidem in iure cocti cum lapatho et apio menstruas purgationes expediunt lactisque ubertatem faciunt, iidem in febri, quae sit cum capitis doloribus et oculorum palpitatione, mulieribus in vino austero dati prodesse dicuntur. castoreum ex mulso potum purgationibus prodest contraque volvam olfactum cum aceto et pice aut subditum pastillis. ad
133 secundas etiam uti eodem prodest cum panace in quattuor cyathis vini et a frigore laborantibus ternis obolis. sed si castoreum fibrumve supergrediatur gravida, abortum facere dicitur et periclitari partu, si superferatur. mirum est et quod de torpedine invenio, si capiatur cum luna in libra sit, triduoque adservetur sub diu, faciles partus facere postea, quotiens inferatur. adiuvare et pastinacae radius adalligatus umbilico existumatur, si viventi ablatus
134 sit, ipsa in mare dimissa. invenio apud quosdam ostraceum vocari quod aliqui onychen vocent; hoc suffitum volvae poenis mire resistere; odorem esse castorei, meliusque cum eo ustum proficere; vetera quoque ulcera et cacoëthe eiusdem cinere sanari. nam carbunculos et carcinomata in muliebri parte praesentissimo remedio sanari tradunt cancro femina

---

a See *Women's Diseases*, Littré VIII, p. 220.   In the Greek it is five crabs, etc., to be taken thrice fasting.

b A little shell.

c A nail or claw.

or dried, are taken in drink to prevent miscarriage. Hippocrates [a] uses them to promote menstruation and to withdraw a dead foetus; five crabs, root of lapathum and of rue, with some soot, are beaten up, and given to drink in honey wine. Crabs, boiled in their liquor with lapathum and celery, hasten on the monthly flow and produce a plentiful supply of milk; in fever accompanied by pains in the head and palpitation of the eyes, are said to be good for women when given in a dry wine. Beaver oil taken in honey wine is good for menstruation, as also for troubles of the uterus if given to smell with vinegar and pitch, or made into tablets for a pessary. To bring away the afterbirth it is also useful to use beaver oil with panaces in four cyathi of wine, and three-obol doses for those suffering from chill. If, however, a pregnant woman steps over beaver oil or a beaver, it is said to cause a miscarriage, and a dangerous confinement if it is carried over her. What I find about the torpedo is also wonderful: that, if it is caught when the moon is in Libra and kept for three days in the open, it makes parturition easy every time afterwards that it is brought into the room. It is thought to be helpful too if the sting of the sting-ray is worn as an amulet on the navel, but it must be taken from a living fish, which itself must be cast into the sea. I find in some writers that there is a substance called ostraceum,[b] called by some onyx [c]; that this by fumigation wonderfully counteracts severe pains of the uterus; that it has the smell of beaver oil, and is more efficacious if burnt with it; that the ash also of the same substance cures chronic or malignant ulcers. But carbuncles and cancerous sores on a woman's privates have, they say, a sovereign remedy in a female crab

545

cum salis flore contuso post plenam lunam et ex aqua
inlito.

135 XLVII. Psilotrum est thynni sanguis, fel, iocur,
sive recentia sive servata, iocur etiam tritum mixto-
que cedrio plumbea pyxide adservatum. ita pueros
mangonicavit Salpe obstetrix. eadem vis pulmoni [1]
marino [2] leporis marini sanguini [3] et felli [4] vel si in
oleo lepus hic necetur. . . .[5] cancri, scolopendrae
marinae cinis cum oleo, urtica marina trita ex aceto
scillite, torpedinis cerebrum cum alumine inlitum XVI

136 luna. ranae parvae, quam in oculorum curatione
descripsimus, sanies efficacissimum psilotrum est, si
recens inlinatur, et ipsa arefacta ac tusa, mox decocta
tribus heminis ad tertias vel in oleo decocta aereis
vasis. eadem mensura alii ex XV ranis conficiunt
psilotrum, sicut in oculis diximus. sanguisugae quoque tostae in vase fictili et ex aceto inlitae eundem
contra pilos habent effectum. [Hic suffitus urentium
eas necat cimices]. inuncto castoreo quoque cum
melle pro psilotro usi pluribus diebus reperiuntur.
in omni autem psilotro evellendi prius sunt pili.

137 XLVIII. Infantium   gingivis   dentitionibusque

---

[1] pulmoni *codd.*: pulmonis *vulg., Mayhoff.*
[2] marino VRd: marini Bb. *vulg. Mayhoff.*
[3] sanguini *multi codd.*: sanguine E, *vulg., Mayhoff.*
[4] felli dT *Hard.*: felle *Mayhoff cum multis codd.*
[5] *Hic lacunam indicat Mayhoff.*

---

[a] The best kind of salt.
[b] Mayhoff suggests that the words *item adhibetur* or the like
have fallen out here. The ending *-etur* may have caused the
omission of one verb.

crushed up with flower of salt [a] after a full moon and applied in water.

XLVII. Superfluous hair is removed by blood, gall, and liver of the tunny, whether fresh or preserved, by the liver too when beaten up, mixed with cedar oil, and stored in a leaden box. In this way slave boys were prepared for market by Salpe the midwife. The same property is found in the pulmo marinus, in the blood and gall of the sea hare, or this hare itself killed in oil.[b] There is also used the ash of the crab or of the sea scolopendra with oil, the sea anemone beaten up in squill vinegar, or the brain of the torpedo applied with alum on the sixteenth day of the moon. The blood-like matter (sanies) given out by the small frog, that we have spoken of [c] in the treatment of the eyes, is a most efficacious depilatory if applied fresh; and so is the frog itself, dried and pounded up, and then boiled down to one third in three heminae, or boiled down in oil in brazen vessels. Others make a depilatory out of fifteen frogs treated with the same proportions of liquid, as we mentioned when treating of the eyes.[d] Leeches also, roasted in an earthen vessel and applied with vinegar, have the same effect in extracting hair. The fumes that come from those burning the leeches kill bugs.[e] There are also found those who have used for several days as a depilatory rubbing with beaver oil and honey. Before using however any depilatory the hairs must first be pulled out.

XLVIII. The gums and the teething of infants are

[c] See § 74 of this Book.
[d] See § 75 of this Book; *eadem mensura* could be taken with the preceding sentence.
[e] This sentence is bracketed by Mayhoff.

547

plurimum confert delphini cum melle dentium cinis et si ipso dente gingivae tangantur. adalligatus idem pavores repentinos tollit. idem effectus et caniculae dentis. ulcera vero, quae in auribus aut ulla parte corporis fiant, cancrorum fluviatilium sucus cum 138 farina hordeacea sanat. et ad reliquos morbos triti in oleo perunctis prosunt. siriasim infantium spongea frigida cerebro umefacto rana inversa adalligata efficacissime sanat. aridam inveniri adfirmant.

XLIX. Mullus in vino necatus vel piscis rubellio vel anguillae duae, item uva marina in vino putrefacta iis, qui inde biberint, taedium vini adfert.

139 L. Venerem inhibet echeneis, hippopotamii frontis e sinistra parte pellis in agnina adalligata, fel torpedinis vivae genitalibus inlitum. concitant coclearum fluviatilium carnes sale adservatae et in potu ex vino datae, erythini in cibo sumpti, iocur ranae diopetis vel calamitis in pellicula gruis adalligatum vel dens crocodili maxillaris adnexus bracchio vel hippocampus vel nervi rubetae dextro lacerto adalligati. amorem finit in pecoris recenti corio rubeta adalligata.

140 LI. Equorum scabiem ranae decoctae in aqua ex-

---

helped very much by a dolphin's teeth reduced to ash and added to honey, and also if the gums are touched with a tooth itself. As an amulet a dolphin's tooth removes a child's sudden terrors. The same also is the effect of a tooth of the canicula. The sores however that form in the ears or on any part of the body are cured by the juice of river crabs with barley meal. The other diseases too are relieved if the patients are thoroughly rubbed with river crabs pounded in oil. For siriasis [a] in babies a very efficacious cure is a frog tied as an amulet back to front on the infant's skull [b] moistened with a cold sponge. The sponge is said to be found dry afterwards.

XLIX. Red mullet killed in wine, or the fish rubellio, or two eels, also a sea grape rotted in wine, brings a distaste for wine to those who have drunk of the liquor.

L. Antaphrodisiac are the echeneis, hide from the left side of the forehead of a hippopotamus attached as an amulet in lamb skin, or the gall of the torpedo, while it is still alive, applied to the genitals. Aphrodisiac is the flesh of river snails preserved in salt and given to drink in wine, erythini taken as food, the liver of the frog diopetes or calamites, attached as an amulet in a little piece of crane's skin, or the maxillary tooth of a crocodile tied to the forearm, or the hippocampus, or the sinews of a bramble toad worn as an amulet on the right upper arm. Love is killed by a bramble toad worn as an amulet in a fresh piece of sheep's skin.

LI. Itch scab in horses is relieved by frogs boiled

[b] The Bohn translation suggests that *crebro*, " from time to time " is the correct reading. It is not mentioned by Mayhoff.

tenuant, donec inlini possint. aiunt[1] ita curatos[2] non repeti postea. Salpe negat canes latrare, quibus in offa rana viva data sit.

LII. Inter aquatilia dici debet et calamochnus, Latine adarca appellata. nascitur circa harundines tenues e spuma aquae dulcis ac marinae, ubi se miscent. vim habet causticam, ideo acopis utilis et contra perfrictionum vitia. tollit et mulierum lenti-

141 gines in facie. et calami simul dici debent: phrag-mitis radix recens tusa luxatis medetur et spinae doloribus ex aceto inlita, Cyprii vero, qui et donax vocatur, cortex alopeciis medetur ustus et ulceribus veteratis,[3] folia extrahendis quae infixa sint corpori et igni sacro. paniculae flos aures si intravit, exsurdat. sepiae atramento tanta vis est, ut in lu-cernam[4] addito Aethiopas videri ablato priore lumine Anaxilaus tradat. rubeta excocta aqua potui data suum morbis medetur vel cuiuscumque ranae cinis. pulmone marino si confricetur lignum, ardere videtur adeo, ut baculum ita praeluceat.

142 LIII. Peractis aquatilium dotibus non alienum videtur indicare per tot maria, tam vasta et tot milibus passuum terrae infusa extraque circumdata mensura, paene ipsius mundi quae intellegatur, animalia cen-

---

[1] aiunt et *coni. Mayhoff.*
[2] curatos sic *coni. Mayhoff.*
[3] inveteratis *coni. Mayhoff*: veratis; folia <utilia> *coni. Warmington.*
[4] lucernam *Mayhoff*: lucerna *codd.*

---

[a] Probably e.g. at strangers. The Bohn translators have: "lose the power of barking." Perhaps when they see the frog.

down in water until they can be used as ointment. It is said that a horse so treated is never attacked again afterwards. Salpe says that dogs do not bark [a] if a live frog has been put into their mess.

LII. Among water creatures ought also to be mentioned calamochnus, the Latin name of which is adarca. It collects around thin reeds from the foam forming where fresh and sea water mingle. It has a caustic property, and is therefore useful for tonic pills and to cure cold shiverings. It also removes freckles on the face of women. At the same time reeds should be spoken of. The root of phragmites, pounded fresh, cures dislocations, and applied with vinegar pains in the spine; the Cyprian reed indeed, also called donax, has a bark which when calcined cures mange and chronic ulcers, and its leaves extract things embedded in the flesh, and help erysipelas. The flower of the reed panicula causes complete deafness if it has entered the ears. The ink of the cuttle fish has so great power that Anaxilaus reports that poured into a lamp the former light utterly vanishes, and people appear as black as Ethiopians. A bramble toad thoroughly boiled in water and given to drink cures pigs' diseases, as does the ash of any frog or toad. If wood is thoroughly rubbed with pulmo marinus it seems to be on fire, so much so that a walking-stick, so treated, throws a light forward.

LIII. Now that I have completed my account of the natural qualities of aquatic plants and animals, it seems to me not foreign to my purpose to point out that, throughout all the seas which are so numerous and spacious and come flooding into the landmass over so many miles and surround it outside to

tum quadraginta quattuor omnino generum esse
eaque nominatim complecti, quod in terrestribus
143 volucribusque fieri non quit. neque enim omnes
Indiae Aethiopiaeque aut Scythiae desertorumve
novimus feras aut volucres, cum hominum ipsorum
multo plurimae sint differentiae, quas invenire potui-
mus. accedat his Taprobane insulaeque aliae atque
aliae [1] oceani fabulose narratae. profecto conveniet
non posse omnia genera in contemplationem universam vocari. at, Hercules, in tanto mari oceanoque
quae nascuntur certa sunt, notioraque, quod miremur,
quae profundo natura mersit.

144   Ut a beluis ordiamur, arbores, physeteres, ballaenae, pistrices, Tritones, Nereides, elephanti, homines
qui marini vocantur, rotae, orcae, arietes, musculi et
alii piscium forma [arietes],[2] delphini celebresque
Homero vituli, luxuriae vero testudines et medicis
fibri—quorum generis lutras nusquam mari accepi-
145 mus mergi, tantum marina dicentes—iam caniculae,
drinones, cornutae, gladii, serrae, communesque
terrae, mari, amni hippopotami, crocodili, et amni
tantum ac mari thynni, thynnides, siluri, coracini,
percae.

Peculiares autem maris acipenser, aurata, asellus,

[1] aliae atque *Mayhoff*: aliaeq B: *omm. rell.*
[2] arietes *seclud. Warmington*: quadripedes *Birt*: terrestres
coni. Mayhoff.

---

[a] *Od.*, IV, 436.
[b] In fact otters do sometimes enter the sea at estuaries,
while beavers do not.
[c] In sections 145–153 there are many variants in the
names of fish. We note a few only. See Index of Fishes.

an extent which might be thought of as almost
equal to that of the world itself—there are one
hundred and forty-four species in all; and that they
can be included each under its own name, a thing
which, in the case of creatures of the land and those
which fly, cannot be done. For in fact we do not
know all the wild animals and flying creatures of
India and Ethiopia and Syria; while even of mankind
itself the varieties which we have been able to discover are the greatest in number by far. Add to this
Ceylon and various other islands of the ocean about
which fabulous tales are told. Surely it will be
agreed that not all the species can be brought under
one general view for our consideration. On the
other hand, upon my solemn word, in the sea, vast
though it is, and in the ocean, the number of animals
produced is known; and—we may well wonder at
this—we are better acquainted with the things which
nature has sunk down in the deep.

To begin with large beasts, there are " sea-trees,"
blower-whales, other whales, saw-fish, Tritons,
Nereids, walruses (?) so-called " men of the sea,"
" wheels," grampuses, " sea-rams," whalebone whales,
and others having the shape of fishes, dolphins, and
seals well known to Homer,[a] tortoises on the other
hand well known to luxury, beavers to medical
people (of the class of beavers we have never found
record, speaking as we are of marine animals, that
otters anywhere frequent the sea [b]); also sharks,
" drinones," horned rays (?), sword-fish, saw-fish;
hippopotamuses and crocodiles common to land, sea,
and river; and, common to river and sea only, tunnies, other tunnies, " siluri," " coracini," and perches.

Belonging [c] to the sea only are sturgeon, gilt-head,

553

acharne, aphye, alopex, anguilla, araneus, boca, batia, bacchus, batrachus, belonae, quos aculeatos vocamus-balanus, corvus, citharus, rhomborum generis pessi-
146 mus, chalcis, cobio, callarias, asellorum generis, ni minor esset, colias[1] sive Parianus sive Sexitanus a patria Baetica, lacertorum minimi, † ab iis mon-creses †[2] cybium—ita vocatur concisa pelamys, quae post XL dies a Ponto in Maeotim revertitur—cordyla—et haec pelamys pusilla ; cum in Pontum a Maeotide exit, hoc nomen habet—cantharus, callionymus sive uranoscopos, cinaedi, soli piscium lutei, cnide, quam
147 nos urticam vocamus, cancrorum genera, chemae striatae, chemae leves, chemae peloridum generis, varietate distantes et rotunditate, chemae glycymar-ides, quae sunt maiores quam pelorides, coluthia sive coryphia, concharum genera, inter quae et margariti-ferae, cochloe,[3] quorum generis pentadactyli, item helices (ab aliis[4] actinophoroe dicuntur), quibus radii ; . . . cantant—extra haec sunt rotundae in
148 oleario usu cocleae—cucumis, cynops, cammarus, cynosdexia, draco—quidam aliud volunt esse dracun-culum ; est autem gerriculae amplae similis, aculeos

---

[1] colias *Hermolaus Barbarus*; coliae *Birt*: collia B: colla *multi codd.*
[2] moncreses B: nostrates *Mayhoff*: *varia rell. codd.*
[3] conchoe *coni. Mayhoff.*
[4] helices ab aliis *Ianus*: h. ab his B: halicembalis *vel sim. rell.*

---

[a] Not of the island Paros, but of the city Parium on the Propontis.
[b] Of the town Sex in Spain.
[c] The Latin text is here corrupt.
[d] This is puzzling. What are *radii* in the case of shell-bearing molluscs? " The spokes on whose shells are used for

" asellus," " acharne," small fry, thresher-shark, eel, weever-fish, bogue, skate, grey mullet, angler-fish, garfish ?—fish which we call thorny, sea-acorn, " sea-crow," " cithari " the worst esteemed of the turbot kind, shad (?), goby, " callarias " of the " aselli " kind were it not smaller, Spanish mackerel also known as the Parian [a] and as Sexitan [b] from its native land Baetica, the smallest of the mackerels, . . .,[c] " cybium " (this is the name given, when it has been sliced, to the young tunny which returns from the Black Sea into Lake Maeotis after forty days), " cordyla " (this too is a very small young tunny ; it has this name when it goes out from Lake Maeotis into the Black Sea), black bream, the " callionymus " or " uranoscopus," " cinaedi "-wrasse—the only fishes which are yellow, sea-anemone, which we call nettle, species of crab, furrowed clams, smooth clams, clams of the kind " peloris," differing in variety of roundness of their shells, " glycymarides "-clams, which are larger than " pelorides," " coluthia " or " coryphia," species of bivalves amongst which are also the pearl-bearers, " cochloe " (to the class of these belong the " five-fingered," also " helices " called by others " actinophorae "), whose rays give a singing sound [d] (outside these [e] there are round shells used in dealing with oil), sea-cucumber, " cynops," shrimps,[f] " dog's right-hand," weever-fish ; (certain people want the " little weever " to be regarded as a different animal ; in fact it is like a large " gerricula,"

musical purposes "—Bostock and Riley. Perhaps the gastropod mollusc " pelican's foot " is meant.

[e] *haec*, neuter plural, is another problem. Mayhoff may be right in suggesting a lacuna after *radii*.

[f] Or prawns.

in branchiis habet ad caudam spectantes; sic ut
scorpio laedit, dum manu tollitur—erythinus, echenais, echinus, elephanti locustarum generis nigri, pedibus quaternis bisulcis—praeterea bracchia iis [1] II
binis articulis singulisque forcipibus denticulatis—
fabri sive zaei,[2] glauciscus, glanis, gonger, girres,
149 galeos, garos, hippos, hippuros, hirundo, halipleumon,
hippocampos, hepar, ictinus, iulis, lacertorum genera,
lolligo volitans, locustae, lucerna, lelepris,[3] lamirus,[4]
lepus, leones, quorum bracchia cancris similia sunt—
reliqua pars locustae—mullus, merula inter saxatiles
laudata, mugil, melanurus, mena, maeotes, murena,
mys, mitulus, myiscus, murex, oculata, ophidion,
ostreae, otia, orcynus—hic est pelamydum generis
maximus neque ipse redit in Maeotim, similis tritomi,
150 vetustate melior—orbis, orthagoriscus, phager, phycis
saxatilium quaedam, pelamys—earum generis
maxima apolectum vocatur, durius tritomo—porcus,
phthir, passer, pastinaca, polyporum genera, pectines—maximi et in his nigerrimi aestate laudatissimi, hi autem Mytilenis, Tyndaride, Salonis,
Altini, Chia in insula, Alexandriae in Aegypto—pectunculi, purpurae, pegrides, pina, pinoteres, rhine,
quem squatum vocamus, rhombus, scarus, principalis
151 hodie, solea, sargus, squilla, sarda—ita vocatur

---

[1] iis *add. Mayhoff.*
[2] zaei *Mayhoff:* ʀaes *codd.* (zais B).
[3] lelepris *Janus coll. Hesych.: varia codd.*
[4] lamirus] larinus *Sillig coll. Hesych.*

and has on its gills prickles which look towards the tail; and when it is lifted in the hand, it inflicts a wound like a scorpion), "erythrinus," sucking-fish, sea-urchin, black "elephants" of the lobster kind, having four forked legs (they also have two arms, each with double joints and a single pair of pincers having a toothed edge), "fabri" or "zaei," "glauciscus," cat-fish, conger eel, "girres," dogfish, "garos," runner-crab (?) "horsetail," flying-fish, jellyfish, sea-horse, "hepar," flying gurnard (?), rainbow-wrasse (?), species of mackerel, fluttering squid, crawfishes, "lantern-fish," "lelepris," "lamirus," sea-hare, "lion"-lobsters, whose arms are like crabs' and the rest is like the crawfish, red mullet, a wrasse highly praised amongst rock-fish, grey mullet, "black-tail," "mena," "maeotes," murry, "mys"-mussel, mussel, bearded mussel (?), purple-mollusc, "eyed" fish, eel (?), species of bivalves, sea-ear, large tunny (this is the largest of the pelamys kind and it never comes back to Lake Maeotis; it is like the "tritomum" and is best in its old age), globe-fish, "orthagoriscus", "phager," "phycis" one of the rock-fish, "pelamys"-tunny, of which kind the largest is called "choice piece," tougher than the "tritomus," "pig"-fish, sea-louse, plaice (?), sting-ray, species of octopus, scallops (the very large ones, and, among these, those which are very black in summer time, being the most highly esteemed; moreover, these are found at Mytilene, Tyndaris, Salonae, Altinum, the island of Chios, and Alexandria in Egypt), small scallops, purple-molluscs, "pegrides" (?), pinna, hermit crab (*or pinna-guard crab*), angel-fish which we call "squatus," turbot, parrot-wrasse, which is of first rank to-day, sole, sargue, prawn (*or shrimp*), "sarda"

pelamys longa ex oceano veniens—scomber, salpa,
sorus, scorpaena, scorpio, salax, sciaena, sciadeus,
scolopendra, smyrus, sepia, strombus, solen sive aulos
sive donax sive onyx sive dactylus, spondyli, smarides,
stellae, spongeae, turdus, inter saxatiles nobilis,
thynnis, thranis, quem alii xiphian vocant, thrissa,
torpedo, tethea, tritomum pelamydum generis magni,
152 ex quo terna cybia fiunt, veneria, uva, xiphias.

LIV. His adiciemus ab Ovidio posita animalia,
quae apud neminem alium reperiuntur, sed fortassis in
Ponto nascentia, ubi id volumen supremis suis temporibus inchoavit: bovem, cercyrum in scopulis viventem,
orphum rubentemque erythinum, iulum, pictas mor-
myras aureique coloris chrysophryn, praeterea per-
cam, tragum et placentem cauda melanurum, epodas
153 lati generis. praeter haec insignia piscium tradit:
channen ex se ipsam concipere, glaucum aestate numquam apparere, pompilum, qui semper comitetur
navium cursus, chromin,[1] qui nidificet in aquis. helo-
pem dicit esse nostris incognitum undis, ex quo
apparet falli eos, qui eundem acipenserem existi-
maverint. helopi palmam saporis inter pisces multi
dedere.

154 Sunt praeterea a nullo auctore nominati. sudis
Latine appellatur, Graece sphyraena, rostro similis

---

[1] *varia codd. Mayhoff sequimur.*

---

[a] *Hal.* 94, 102, 104, 110–113, 126.
[b] *Hal.* 96, 101, 108, 117, 121.

(this is the name given to an elongated pelamys-tunny which comes from the Ocean), mackerel, saupe, " sorus," two kinds of sculpin, two kinds of maigre, scolopendra-worm, " smyrus," cuttle-fish, spiral molluscs, razor-shells variously called " solen," " aulos ", " donax," " onyx," and " dactylus "; thorny oysters, picarels, starfishes, sponges, " turdus "-wrasse, famous amongst rock-fish, tunny, " thranis," which others call sword-fish, " thrissa," electric ray, sea-squirt, " tritomum " (" three-cut ") belonging to a large kind of tunny, from each of which three " cybia " can be cut, " veneria," cuttle-egg (?), sword-fish. LIV. We will add to these some animals, mentioned by Ovid,[a] which are found in no other writer, but which are perhaps native to the Black Sea, where he began that unfinished book in the last days of his life: horned ray, " cercyrus " which lives amongst rocks, " orphus," and red " erythinus," " iulus," tinted sea-breams and gilt-head of golden colour; and, besides these, perch, " tragus," " black-tail " with pretty tail, " epodes " of the flat kind. Besides these remarkable kinds of fishes he records: that the sea-perch conceives of herself, that the " glaucus " never appears in summer; and he mentions the pilot-fish as always accompanying ships on their course, and the " chromis " which makes its nest in the waves. He says that the " helops " is " unknown to our waters ";[b] from which it is clear that those who have believed that acipenser (*sturgeon*) is the same are in error. Many people have given the first prize for taste to the helops among all fish.

Moreover, there are some fish named by no author. There is one barracuda called " sudis " in Latin, " sphyraena " in Greek, in its muzzle resembling its

nomini, magnitudine inter amplissimos; rarus is et non degenerat. appellantur et pernae concharum generis, circa Pontias insulas frequentissimae. stant velut suillum crus e longo in harena defixae hiantesque, qua [1] latitudo est, pedali non minus spatio cibum venantur; dentes circuitu marginum habent pectinatim spissatos; intus spondyli grandis caro est. et hyaenam piscem vidi in Aenaria insula captum.— Exeunt praeter haec et purgamenta aliqua relatu indigna et algis potius adnumeranda quam animalibus.

quae *coni. Warmington.*

name ("stake"); it is in size amongst the largest; it is uncommon, and does not degenerate by inter-breeding. There are also shells (*pinnas*) of a kind for which the name " perna " is given; they are abundant round the Pontiae islands. They stand like pigs' hams fixed bolt upright in the sand; and, gaping not less than a foot wide where there is broad enough space,[a] they lie in wait for food. They have, all round the edges of the shells, teeth set thick like those of a comb; inside is a large fleshy muscle. I once saw also a " hyaena "-fish (*puntazzo*) which was taken in the island Aenaria.

Besides all these creatures, certain off-scourings also come out of the sea; they are not worth a description and are to be counted amongst sea-weeds and not amongst living creatures.

---

[a] Or, if we read *quae*, " according to their expansiveness."

ADDITIONAL NOTES

Additional Note A.

Mensa.

When used in reference to food *mensa* may have various meanings :—

(1) Dining-table.
(2) Small table, which when of many shelves was called *repositorium*. See Petronius *Satyr.* 34 : suam cuique mensam assignari.
(3) Course.
(4) Square slice of bread (*quadra*), used as a plate. See *Aeneid* VII 115 : patulis nec parcere quadris; " Heus, etiam mensas consumimus," inquit Iulus.
(5) A round plate, *lanx* or *discus*. See Pliny XXXIII § 140 : iam vero et mensas repositoriis imponimus ad sustinenda opsonia.

In Pliny XXVIII we have :

§ 24 nam si mensa adsit.   Meaning (1).
§ 26 aquis sub mensam profusis.   Meaning (1).
§ 26 mensam vel repositorium tolli.   Either (2) or (5).
§ 26 mensa linquenda non sit, nondum enim plures quam convivae numerabantur.   The first seems to be (1) but *plures* to be (2).   See, however, Wolters *ad loc.*
§ 27 utique per mensas.   This is (2) on the usual interpretation, but (3) on that of Wolters.
§ 27 in mensa utique id reponi.   This might be either (1) or (2).

Additional Note B.

The Hyaena.

The Romans were rather puzzled, and perhaps a little frightened, by the hyaena and its strange habits.   Pliny has

563

a short chapter (VIII §§ 105, 106) in which he refers to many popular beliefs about the animal: that it is bi-sexual, becoming male and female in alternate years; that it can imitate human speech, a belief arising perhaps from its laughing cry; that it imitates a person being sick, so as to attract dogs; that it digs up graves in search of corpses; and that it is an animal possessing magic powers.

Pliny seems to have obtained most if not all his information from books on magic, for perhaps none of the seventy-nine " remedies " in chapter XXVII of the twenty-eighth book can be considered rational. Neither Serenus nor Sextus Placitus mentions the animal, but Scribonius Largus makes use of hyaena's gall in an eye-salve (XXXVIII), and has much to say about a recipe for hydrophobia which he obtained *pro magno munere* from a *medicus* called Zopyrus (CLXXI and CLXXII). It turned out to be a piece of hyaena skin wrapped up in cloth. Scribonius took great pains to prepare the amulet and keep it ready, but confesses that he had not yet had a chance to put it to the test. Many of the hyaena remedies were probably fraudulent imitations, although hyaenas must have formed part of the wild-beast shows of which the Romans were so fond.

Additional Note C.

Sympathy and Antipathy.

" The Greeks have applied the terms ' sympathy ' and ' antipathy ' to the principle of Nature that water puts out fire . . . the magnetic stone draws iron to itself while another kind repels it . . . the diamond, unbreakable by any other force, is broken by goat's blood." So says Pliny (XX §§ 1, 2). At the beginning of Book XXIV he gives a longer list, from which examples are : oak and olive; oak and walnut; cabbage and vine; cabbage and cyclamen or marjoram; all being contraries. The affinities include : pitch and oil, both being fatty; gum and vinegar, which washes gum out; ink and water, which combine readily.

In the working out of this theory there must inevitably be, to modern minds, some inconsistency and much sheer fancy. The theory itself is fanciful, and more akin to the " Love and Hate " of Empedocles than to the *convenientia* of the Stoics,

ADDITIONAL NOTES

although parallels or analogies might be found in the scientific concepts of today. There was a tendency in Greek speculation to take an attractive idea, work it to death, and ignore or brush aside objections to it. Pliny says (XXIV § 4) of sympathy and antipathy: "Hence medicine was born." But it is not always clear whether a remedy is a cure because of antipathy to the disease or because of sympathy with it. The neutralization of disease suggests the former; the "doctrine of signatures" the latter. When, however, Pliny says (XXVIII § 147) that the power of sympathy under the influence of *religio* is great enough to render harmless the drinking of bull's blood by the priestess of Earth at Aegira, the reasoning is hard to follow. Various explanations could be given, but most modern minds would have been more satisfied if Pliny had said that the power of *religio* is so great that it can turn antipathy into sympathy.

Dr. W. T. Fernie, *Animal Simples*, pp. 63–65, says that bull's blood was once a favourite beverage! He also refers to Grote's suggestion that imperfect prussic acid, which may be obtained from blood, may have been called "ox-blood." There was a story that Themistocles committed suicide by drinking bull's blood, and the belief in its poisonous nature long persisted.

There is an article on "sympathy," *Der Heilmagnetismus bei Plinius*, by Th. Steinwender, in *Zeitschrift für die Oesterreichischen Gymnasien*, LXIX 1–20.

Additional Note D.

Pliny says (XXVIII. 108) that there are two kinds of crocodile, the second being smaller, living on land only, and eating scented plants so that in its bowels is formed a much-prized substance called *crocodilea*.

Actually Egypt has today but one crocodile, the *Crocodilus niloticus*, which has, however, two musk glands, one under the throat and the other in its cloaca.

We can only guess why Pliny says that the scent was taken from small crocodiles living on land. Pliny seems to have misunderstood his authorities; perhaps the perfumers kept baby crocodiles in semi-domestication.

565

ADDITIONAL NOTES

Additional Note E.

P. Fournier, writing in the *Revue de Philologie* for 1952 and 1953, has a few *Notulae Plinianae* which did not come to my attention in time to be mentioned in vol. VII. He thinks that *populus* should often be replaced by *opulus*, and *ornus* by *cornus*. For purely botanical reasons, he suggests the following emendations:

> In XXV. § 125, *in ulvis* for *in silvis*.
> In XXVI. § 56, *paleali* for *pallioli*.
> In XXVI. § 95, *tensior* for *tenuior*.
> In XXVII. § 104, *seridis* for *iridis*.

Additional Note F.
Pliny Book XXX.

In XXX. § 24, taking the best attested readings, we have: *is quoque vermiculus . . . mire prodest. nam urucae brassicae eius contactu cadunt et e malva cimices infunduntur auribus.* This gives: " The grub also . . . is wonderfully good (sc. for the teeth). For (*or* But) cabbage caterpillars fall at its touch, and bugs from the mallow are poured into ears." This is rather a *non sequitur.* Mayhoff emends: *urucae e brassicae foliis.* That is: " But at the touch of the caterpillar from the leaves of cabbages teeth fall out, and bugs, etc."

Professor Warmington would read: " *mire prodest, nam eius contactu cadunt ; urucae brassicae et e malva cimices,* etc."— a simple transposition: " is wonderfully good, for at its touch teeth fall out; cabbage caterpillars and bugs from mallow, etc."

Additional Note G.
Pliny Book XXX.

In XXX. 64 the best MSS. have: *in dolore si quis aquam per pedes fluentes* (or *fluentis*) *haurire sustineat.* Mayhoff has: *in dolore si quis aquam ter pedes eluens haurire sustineat.* The order of the words suggests that *ter* goes with *eluens,* but the sense that it goes with *haurire.*

Professor Warmington would keep *per* and change *fluentes* to *fluentem.* " If anyone when in pain can bring himself to swallow the water that swirls about his feet."

ADDITIONAL NOTES

Additional Note H.
Pliny Book XXXI, § 38.

The MSS. read: certior subtilitas, inter pares meliorem esse quae calefiat refrigereturque celerius, quin et haustam vasis ne manus pendeant depositisque in humum tepescere adfirmant.

The second sentence is very difficult, and one is reminded of Mayhoff's warning in the Appendix to Vol. IV. (p. 497): verum in talibus rebus, quae omni ratione careant, rectius est desperare quam nullo testimoniorum adiumento e solis litterarum vestigiis inanem coniecturam facere. Although it cannot be said that *omnis ratio* is wanting, yet the *ratio* is very obscure, and is perhaps irrecoverable.

The subject of the passage is the wholesomeness or " lightness " of water. It has just been said that the lightness cannot be determined by a pair of scales or steelyard. A more delicate test is the increase in heat when the water is placed in pots on the ground. The problem is: was Pliny's intention to say, " don't weigh " or " don't warm by touching "? Either alternative would require considerable emendation. Mayhoff adopts from a Dalechamp variant *manu* for *manus*, and adds *portatis* after *vasis* in order to balance *impositisque*, " in pots carried without weighing by hand and placed etc."; Detlefsen, aiming at much the same sense, reads *manus suspendant*, and leaves the *-que* difficult to explain. The other interpretation would require a radical change of *pendeant* to *tangant* or *tepeant*, and perhaps other changes as well. The difficulty of *que* might be overcome by reading *impositam*, and if the avoidance of warming by touch is the point of the *ne*-clause, *ansatis*, " with handles," a Plinian word, would be better than Mayhoff's *portatis*.

On the whole it is best to confess that the sentence is a puzzle hitherto unsolved, and that two meanings are possible, with a preference for the one that implies weighing.

Additional Note I.
Pliny XXXI. Ch. 46.

Nitrum, from the Arabic *natron*, was probably a mixture of sodium carbonate, calcium carbonate, and various chlorides. It was often obtained from pools N.W. of Cairo,

ADDITIONAL NOTES

From the account of Pliny we can conclude with certainty that *nitrum* was to a great extent soda, but not entirely so. We are told, for instance, that it could be used instead of salt in making bread, that it turned green vegetables greener, that with dill, cummin, or rue it relieved gripes, that it dissolved in the mouth, and that sometimes, but not always, it crackled in fire.

Soda scum (*spuma nitri, aphronitrum*) was said to ooze from the sides of certain caves in Asia and also to come from Egypt. It was probably carbonates and nitrates of soda and potash, coloured by copper and iron oxides. See the Loeb *Pliny*, vol. II, p. LII.

Additional Note J.

Pliny discusses sponges in IX. Ch. 69, and XXXI. Ch. 47.

In the former he says that sponges have four or five *fistulae*, going all the way through, and that there are others, closed at the upper end. A modern article on sponges will probably refer to the various holes of a sponge as canals, apertures, pores, cavities, funnels, or oscules, according to their shape or purpose. Pliny calls the holes by one name only, *fistulae*. Now Pliny knew, or took from his authorities, that sponges were animal, but it is sometimes impossible to make out whether he is speaking of the living sponge or of the domestic article. Most of XXXI. Ch. 47, deals with the latter, but the classification is apparently concerned with the former.

Pliny's second class, the female, is said to have *fistulae perpetuae*, but the third class to have *fistulae* that are very small and very numerous. The words of Pliny imply that his first and third classes have *fistulae* that are not *perpetuae*.

As a matter of fact, the oscules of all living sponges never close. Therefore, if *perpetuus* can mean "never-closing," and if Pliny has in mind sponges in their native state, he is attributing to a particular class a characteristic that really belongs to them all. The adjective *perpetuus*, however, is a strange one to use in this sense, as it means properly "long and unbroken."

It is probable that Pliny has written carelessly and vaguely, and in partial ignorance.

POPULAR MEDICINE IN ANCIENT ITALY

The origin of medicine is obscure. Some anthropologists, arguing from the customs of primitive peoples, tell us that it arose from magic. By that term are meant powers, which we should call supernatural, but to primitive man were quite normal, supposed to reside in certain objects, and capable of being put into action by those who know the proper procedure. Magic of this kind has played a large part in the evolution of medicine, but before the age of magic there may have been a period, perhaps a long one, when man, like a sick dog, treated himself instinctively if ill or in pain. Very soon in the age of magic appeared " medicine men," who did much to build up a system of ritual, incantations, amulets, and taboos, which reinforced or even replaced the vegetable or animal remedies. Out of this stage, there slowly evolved, as man's reasoning power grew, the stage of rational medicine, in which the medicine man was superseded by the professional physician or surgeon, although many of his duties were carried out by the head of the family. In this way arose the distinction, which even today has not disappeared, between professional, and folk or popular, medicine.

The best professionals of Greece, mostly by their own efforts but partly through the influence of other countries, especially Egypt, had by 400 B.C. entirely

569

discarded superstitious methods of healing. Two treatises [a] in the Hippocratic *Corpus* declare that all diseases are due to natural causes, and can be cured only by natural means. But traces of superstition are to be seen in the works of Celsus and Galen, and in popular medicine it flourished. The truth is that, however much the best physicians despised them, superstitious methods had their uses. A patient who is cheerful, and buoyed up by strong, even if false hopes, is more likely to do well than is a patient worried and depressed. If a man has complete faith in the efficacy of a completely inert compound, his chances of recovery are improved merely by the psychological effect of his belief. Herein lies at least one reason for the long vogue of medicines that we now know are physiologically useless. Magical ritual and incantations were often amusing, and always gave the impression that something of great importance was about to happen. The power of suggestion and auto-suggestion had full scope to act, especially among people who were far more credulous and superstitious than the present age of positive science.

Roman medicine for many generations was entirely popular, for the Romans never developed a scientific medicine of their own. Until 219 B.C., when the Greek physician Archagathus migrated to Rome from the Peloponnesus, they doctored themselves.[b] Cato's hatred of professional physicians, apparent in

[a] *Airs, Waters, Places* and *Sacred Disease.*
[b] Doctors from Magna Graecia certainly influenced, directly or indirectly, medical practice in the rest of Italy, but we know little about the details. At Croton was one of the first Greek medical schools.

the letter to his son, may have been unusually strong, but Pliny's dislike was almost as great, and marked disapproval is shown by Pliny Junior, Serenus, and pseudo-Apuleius. There were many low-class physicians in the Graeco-Roman world, for no tests were required before beginning a practice. These deserved all the blame bestowed upon them by their disappointed dupes; Pliny, however, picks out for his most venomous attack Asclepiades, who was really a good physician and highly praised by Celsus.

During and after the Roman conquest of Greece, there came to Italy great numbers of these poorly qualified men, who, desirous of making a living, pandered to the tastes and fancies of the self-doctoring Romans, supplying them with remedies of different sorts, but most of them useless except as faith cures. In this way there came to be known to the Romans a vast number of foreign drugs, most of which were perhaps never tried in Italy at all, but many of them appear to have become popular. How these new remedies were put on the market or " advertised " (as we might say) can be seen by reading the *Compositiones* of Scribonius Largus, a lower-grade doctor of perhaps a better type than the majority. He confesses to buying quack remedies from an African *muliercula* and a Roman *honesta matrona*, and one for pleurisy from a man who, to keep his prescription a secret, pretended to include ingredients which actually he never used.[a] He also bought from his friend Zopyrus of Gortyn *pro magno munere* an amulet to protect from hydrophobia—a piece of hyena skin wrapped in cloth.[b]

[a] See pp. 53, 10, 11, 41 of Helmreich's edition.
[b] See p. 70 (Helmreich).

POPULAR MEDICINE IN ANCIENT ITALY

But the man who introduced to the Romans most of the new or foreign remedies was Pliny himself, who in Books XX–XXXII gives perhaps several thousands. He did little, if any, independent research, but collected recipes, botanical and animal, from every available source, including some he professed to dislike. According to his own statement Pliny preferred herbal simples, but he prescribes without disapproval mixtures, animal remedies, remedies from professional doctors and even those of the Magi, whom he cordially hated. The grosser forms of superstition—draughts of blood and relics from the cross or gallows—aroused his scorn, but he places them on record, while amulets, ritual, and incantations, are described or mentioned, though often prefixed by " they say that," or " it is thought that." Pliny sometimes reports gossip, and forgets his professed aim to be utilitarian. In this jumble of so-called cures very little guidance is given to the harassed attendant in search of a remedy for a difficult case.

The *Natural History* is not a good practical textbook. So thought many who later wrote popular works on the same subject, several of which are extant. These picked out recipes that appealed to them from Pliny's book, adding some from other sources. By the time of Plinius Junior, who wrote what is probably the earliest of the extant epitomes, a great deal of the matter in the *Natural History* had become what may be called communal knowledge, so that direct borrowing from Pliny, although possible, should not necessarily be assumed. The " Pliny " just mentioned is the pen-name of one who wrote a *medicina Plinii* about A.D. 350. He was followed by Serenus

Sammonicus, the author of a didactic poem in 1107 hexameters, covering the whole ground in 64 sections, pseudo-Apuleius with his *Herbarius*, Sextus Placitus, who gives recipes only from animals and birds, and Marcellus Empiricus of Bordeaux. The dates of these four are uncertain, but are grouped around A.D. 400.

Animal remedies, as given by Pliny, are very often, perhaps usually, based on a simple magic, such as " like cures like." There is some magic in the plant remedies, but much more in those from animals. The reason may be that animals, more akin to man than plants, have a closer " sympathy " and a sharper " antipathy," two rather mysterious qualities which Pliny, influenced by some Greek thinker, believed to be the active principles in all cures. The magic of the medical Books is of a mild and inoffensive kind—ritual, incantations, amulets, neglect of rational doses for those with the magical numbers three, seven, nine, and so on.

A typical but imaginative Plinian cure might be to draw a ring round a plant with iron, gather it at night without letting it fall to the ground, say for what purpose and for whom it is gathered, and to administer three leaves or three cyathi of a decoction. In a dose of this kind there is " power " (*vis*), not only in the plant, but in the ritual, the words, and the number three.

Popular medicine in Italy can be better understood if contrasted with professional medicine, which among the Greeks had reached a very high standard by 400 B.C. At Alexandria a hundred years later a further advance was made, and Celsus wrote a textbook inspired by Alexandrian influence. By com-

paring the treatment of epilepsy or malaria in Celsus and Pliny we can throw some light on the question, especially if we remember that epilepsy frightened the ancients, and that malaria was obstinate or incurable. The professionals discarded all superstitious or magical remedies, and relied on regimen, rest, and warmth, using drugs (except purges and emetics) very sparingly. Popular medicine had recourse to any and every supposed remedy, however absurd and disgusting to our minds, and to amulets, incantations, and various other kinds of magic. What we call " shock " remedies were sometimes employed; one of the most striking, used in the treatment of another disease, was to duck the victims of hydrophobia unawares into cold water.

Some popular medicines used were really of therapeutic value, but most of them were chosen because of a fanciful resemblance or relationship to the disease, e.g. black hellebore for diseases caused by black bile. Very common were amulets, usually prophylactic, although curative became common in Italy in the first century A.D. A common type of amulet is to take the eye of a crab, the crab being allowed to go free, and to wear it as an amulet for diseases of the eye. The theory behind all this is that the crab's eye retains power to heal eyes so long as the crab lives; the eye amulet absorbs the eye trouble and transfers it to the mutilated animal, which usually dies, carrying with it the complaint.

Pliny did not like compound prescriptions, but Roman popular medicine had several, for in order to make sure of the proper ingredient a great number of them were often combined in a " blunderbuss," as in the famous antidote of Mithridates, which finally

had over seventy components. Conversely, when a remedy was found suitable for one complaint it was often assumed by false analogy that it would be good for many others. The outstanding example is betony, used for forty-seven ailments.

The main conclusion to be drawn is that popular cures, except in a few obvious cases, were faith cures. Faith is a powerful healer today; in ancient times, owing to the greater credulity of the age, it was probably a far more effective healer.

LIST OF DISEASES AND AFFECTIONS MENTIONED BY PLINY

To equate modern diseases with the names used by ancient physicians is a task full of uncertainty. In some cases indeed there is no difficulty; a disease may have such distinctive symptoms, and be so unlike any other, that its description in Celsus or Galen points clearly to one, and only one, diagnosis, examples being intermittent malarial fevers and the common cold. Pneumonia again in both Greek and Latin writings is usually easy to detect (although there is some chance of confusion with acute bronchitis), and so are also dropsy and pleurisy. Often, however, we can do no more than divide into groups: (1) diseases and (2) the ancient names of diseases, and then identify a group from one with a group from the other. Many quite different diseases are so alike symptomatically that identification can be established, even today, only by a microscopic examination conducted with a technique quite unknown to the ancients. Great care is needed with eye diseases and skin diseases, both of which were far more common in earlier days than they are with us, for dust was everywhere and disinfecting cleansing was practically unknown. The principle of grouping is nearly always the safest one to adopt; to attempt more is hazardous. For example, we have on the one hand *collectio, furunculus, panus, vomica* and *tumor* ; on the other we have " boil," " abscess," " gathering " and " carbuncle." The group of complaints covered by the Latin terms is nearly, if not quite, the same as that covered by the English, but any attempt to make more specific identification is attended with much uncertainty; perhaps *panus* is the only one we can isolate more completely.

577

LIST OF DISEASES

More important for our appreciation of antiquity than the identification of specific diseases is to ascertain which, if any, modern diseases were unknown in the Hellenistic age. Here the evidence, especially that relating to infectious fevers, is most disappointing. These fevers are endemic in the modern world, and figure largely in treatises on pathology. But the old medical writers—" Hippocrates," Celsus, Galen and the many compilers who succeeded Galen—do not describe, or give treatment for, small-pox, chicken-pox, measles, scarlatina, typhoid or even influenza. The most that can be said is that in isolated clinical histories or in chance aphoristic remarks one or other of them may be referred to; the evidence is strongest for diphtheria. Moreover, in the pseudo-Aristotelian *Problems* (VII 8) it is said that consumption, ophthalmia and the itch are infectious, but that fevers are not. It is difficult to believe that a people who knew that consumption is infectious would have called scarlatina non-infectious if it had been endemic among them.

The Romans borrowed many names of diseases from the Greeks. Usually, of course, the Latin word refers to the same disease as does the Greek, especially in the works of medical writers. But care must be exercised; λέπρα, for instance, seems to be much narrower than *lepra*.

Celsus is by far the most trustworthy authority to follow in identifying the diseases mentioned by Pliny, for both were Romans, both (probably) laymen and nearly contemporaries.

Aegilops.—A lacrimal fistula at the angle near the nose.

Albugo.—An unknown kind of white ulcer on the eye. In XXVI § 160 used of a head ulcer. The word occurs only in the Vulgate Bible and in Pliny.

Alopecia.—A disease in which the hair fell out. Meaning literally " fox mange," it is translated " mange " in the text. It is perhaps unsafe to limit it to the modern alopecia. Celsus (VI 4) has a brief account of it, saying that it occurred in the hair and beard. He distinguishes it from ὀφίασις, probably ringworm, for this had a winding shape, whereas *alopecia* " sub qualibet figura dilatatur."

Amphemerinos.—Quotidian malaria.

Angina.—An acute swelling in the neck, generally quinsy. A loose term like our " sore throat." Sometimes possibly diphtheria.

LIST OF DISEASES

Apostema.—Greek for abscess.

Argema.—A small white ulcer, partly on the cornea, partly on the sclerotic coat of the eye.

Articularius morbus.—This in XXII 34 is joined to *podagricus*, and so means probably not gout but arthritis.

Asthma.—Apparently only XXVI 34. See also XXV 82.

Atrophus.—" Wasting away," of all such conditions, of which phthisis is one.

Boa.—" A disease when the body is red with pimples," XXIV 53. See also XXVI 120. An exanthem not certainly identified. Shingles is localised. It cannot be, as Hardouin thought, measles, because that disease seems to have been first described by Rhazes.

Cachecta.—A patient who is in a very bad state of health; sometimes a " consumptive " patient is meant.

Cacoethes.—A Greek adjective applied to sores that are very difficult or impossible to cure; "malignant" is the nearest, but not quite exact, equivalent.

Calculus.—Stone or gravel in the bladder.

Caligo.—Dimness of the eyes, hard to distinguish from *nubecula* (film) and *caligatio* (mistiness).

Carbunculus.—In XXVI 5, 6 seems certainly to be anthrax, and Pliny's description resembles that of Celsus V 28, 1. The word was, however, used of minor affections; for example, *carbunculus oculi* is a stye, and it is often used of a bad abscess.

Carcinoma.—Superficial malignant disease, severe forms of which are called *cacoethe*. It seems impossible to distinguish, at least in Pliny, *carcinoma* from *ulcera cacoethe*, *phagedaena* and *gangraena*.

Cardiacus.—The adjective refers to either disease or patient. Sometimes a simple ailment, heartburn, is referred to, at other times a serious complaint, said by W. G. Spencer on Celsus III 19 to be a kind of syncope. In fact the reference may be to any ailment supposed to be connected with the heart.

Cephalaea.—Aretaeus (III 2) calls this a severe, chronic headache, and says that there are ἰδέαι μυρίαι. Persistent neuralgia, except when it means malarial headache, must ,be the complaint referred to.

Cerium.—Described by W. G. Spencer on Celsus V 28, 13 as a follicular abscess among hair. Its appearance—κηρίον means " honeycomb "—enables us to distinguish it from *panus* ; it was also often more severe.

Chiragra.—Gout or gouty pains in the hands. But see *podagra*.

Cholera.—Perhaps never Asiatic cholera, but *cholera nostras* and possibly certain types of dysentery and severe diarrhoea. The word is derived from χολή, "bile."

Clavus.—Wart, corn or callus.

Coeliacus morbus.—W. G. Spencer on Celsus IV 19, 1 (last note) says that the author appears to be de-

579

scribing pyloric spasm and intestinal atony. Cf. Aretaeus IV 7.

Collectio.—The most general term for a boil or abscess, a " gathering."

Colostratio.—Disease of babies caused by the first milk.

Colum.—Colitis, or inflammation of the colon.

Comitialis morbus.—Epilepsy and sometimes other fits.

Condyloma.—A small tumour in the anus due to inflammation. See Celsus VI 18, 8.

Convulsa.—Sprains.

Cotidiana. — Quotidian ague, malaria with fever occurring every day.

Destillatio.—A "running" cold in the head. Sometimes internal catarrh.

Duritia.—An induration, from whatever cause, in any part of the body.

Dysinteria.—Usually dysentery, but probably also severe diarrhoea, however caused.

Dyspnoea.—Difficulty of breathing, however caused.

Elephantiasis.—The usual name of leprosy. See XXVI 7 and 8, where it is said to have quickly died out in Italy.

Enterocele.—Hernia.

Epinyctis.—Either (1) a sore on the eye-lid or (2) an eruption caused by fleas or bugs.

Epiphora.—Running from the eyes as the result of some ailment.

Eruptio.—A bursting out of morbid matter, either through the skin or sometimes in other ways.

Extuberatio.—A fleshy excrescence, perhaps not morbid.

The word apparently occurs only in XXXI 104.

Febris.—Feverishness, or else one of the recognised types of malaria.

Fistula.—Practically synonymous with the modern term.

Flemina.—A severe congestion of blood around the ankles. It is neuter plural.

Fluctio and *fluxus.*—There seems to be little if any difference in the meaning of these words —any flow, but usually a morbid one. Pliny prefers *fluctio.*

Formicatio.—An irritating wart. See Celsus V 28, 14.

Furfur.—Scurf (anywhere).

Furunculus.—A boil, said by Celsus (V 28, 8) not to be dangerous, whereas Pliny (XXVI 125) says that it is sometimes *mortiferum malum.*

Gangraena.—Gangrene, hard to distinguish from *phagedaena* and *ulcera serpentia.*

Gemursa.—A disease the seat of which was between the toes. It is said by Pliny (XXVI 8) to have died out quickly in Italy. See Littré's note.

Glaucoma.—Opaqueness of the crystalline lens.

Gravedo.—The usual term for the common cold.

Gremia.—Rheum.

Hepaticus.—A sufferer from any liver complaint.

Herpes.—A spreading eruption on the skin.

Hydrocelicus.—A sufferer from hydrocele.

Hydropisis.—Dropsy.

Hypochysis.—Cataract.

Ictericus.—A sufferer from jaundice.

Ignis sacer.—Erysipelas. Per-

LIST OF DISEASES

haps also some form of eczema or lupus. Also = shingles.

Ileus.—Severe colic. Possibly appendicitis was included under this term.

Impetigo.—The Romans used this term of various kinds of eczema. Celsus (V 28, 17) mentions four, the last being incurable.

Impetus.—Inflammation or an inflamed swelling; Pliny has *impetus oculorum.* With the genitive of a word meaning a specific disease it denotes an attack of it.

Intertrigo.—Chafing, especially between the legs.

Ischias.—Sciatica.

Laterum dolor.—"Severe pain in the side," nearly always pleurisy.

Lentigo.—Freckles.

Leprae.—Seems to be used of any scaly disease of the skin; Pliny gives cures. There was a kind regarded as incurable, but this is not mentioned by Pliny, who has forty-six references, all to cures.

Lethargus (*lethargia*).—In Hippocrates probably the comatose form of pernicious malaria, but later perhaps also prolonged coma of any kind.

Lichen.—This is said by Pliny (XXVI 2–4) to be a new disease to Italy, usually beginning on the chin. Hence the name *mentagra* (chin disease). Littré diagnoses it as leprosy, but Pliny says (XXVI § 1) *sine dolore quidem illos, ac sine pernicie vitae.* This statement, as Pliny puts it, applies also to *carbunculus* and *elephantiasis*, but Pliny's own

account of these diseases is quite inconsistent with *sine pernicie.* So Pliny's remark is carelessly inaccurate, or applies only to *lichenes.*

Lippitudo.—Inflammation of the eye, generally ophthalmia.

Luxata.—Dislocations.

Malandria.—Pustules on the neck.

Melancholicus.—One suffering from melancholia, which included malarial cachexia and many melancholic conditions, even mere nervousness. In fact it included any disease supposed to be caused by "black bile" (μέλαινα χολή).

Mentagra.—In XXVI 2 called a lichen beginning on the chin. See *lichen.*

Nome (pl. *nomae*).—A spreading ulcer, much the same as *ulcus serpens.*

Nubecula.—A cloudy film on the eye, sometimes cataract.

Nyctalops.—One afflicted with night blindness.

Opisthotonus.—The form of tetanus in which the body curves backwards.

Orthopnoea.—Serious asthma, when the patient cannot breathe unless upright.

Panus.—Spencer in a note on Celsus V 18, 19 calls this a "superficial abscess in a hair follicle." It occurred chiefly on the scalp, on the groin and under the arm.

Paronychia (*-um*).—Whitlow.

Parotis.—A swelling of the glands by the ears. Some authorities think that it may have included mumps, which is described in Hippocrates, *Epidemics* 1.

581

LIST OF DISEASES

Perfrictio.—Sometimes a severe chill.

Peripleumonicus.—A sufferer from pneumonia.

Pernio.—Chilblain.

Pestilentia.—Plague; a term as vague as the English, but usually bubonic.

Phagedaena.—Gangrene, hard to distinguish from *gangraena.* In XXVI 100 an abnormal diseased appetite.

Phlegmon.—Inflammation beneath the skin.

Phreniticus.—Properly a sufferer from *phrenitis* or *phrenesis,* pernicious malaria accompanied by raving. It also refers to the symptom when not caused by malaria, for in post-Hippocratic medical works it often seems equivalent to "brain fever." Perhaps sometimes meningitis.

Phthiriasis.—Phthiriasis, skin disease caused by lice.

Phthisis.—Pulmonary consumption.

Pituita.—Excessive mucus, in any part of the body.

Pleuriticus.—A sufferer from pleurisy.

Plumbum in XXV 155, points to the leaden bluish colour of certain eye diseases. Serenus XIV 33: *si vero horrendum ducent glaucomata plumbum.*

Podagra.—Gout or gouty pains in the foot. Sometimes perhaps the result of lead poisoning. See Spencer's *Celsus* I 464. Pliny (XXVI 100) says that the disease was on the increase in his day. The word (often with *chiragra*) refers sometimes to pains caused by senile degeneration.

Porrigo.—Dandruff or scurf (on hairy parts).

Prurigo and *pruritus.*—Itch; the words can scarcely be discriminated, although perhaps *pruritus* tends to be used of the symptom, *prurigo* of the infection.

Psora.—Several skin diseases are included under this term among which are itch and perhaps leprosy.

Pterygium.—An inflammatory swelling at the inner angle of the lower eyelid; another name for it is *unguis.* It also means a whitlow.

Pusula.—Pustule or blister.

Quartana.—Quartan ague, or malaria occurring after intervals of two days. It was reckoned the mildest form of the disease.

Ramex.—Hernia.

Regius morbus.—Jaundice.

Rhagades.—Chaps.

Rheumatismus.—Catarrh, whether of the nose, throat or stomach.

Rosio.—Gnawing pain in the chest or bowels.

Rupta.—Torn muscles etc.

Scabies.—Not our scabies, which is caused by the itch mite, but described by Celsus (V 28, 16) as a hardening of the skin, which grows ruddy and bursts into pustules with itching ulceration. It includes many types of eczema. *Scabies* of the bladder, a disease of which the symptom was scaly concretions in the urine.

Scabritia.—Diseased roughness of fingers, nails, eyes, etc.

Scelotyrbe.—Lameness of the knee or ankle.

LIST OF DISEASES

Siriasis.—Probably some form of sunstroke.

Spasma.—Cramp.

Splenicus.—Suffering from enlarged or diseased spleen. Enlargement of the spleen is a common after-effect of repeated attacks of malaria.

Stegna.—See note on XXIII 120.

Stomacace.—Scurvy of the mouth.

Stomachicus.—It is doubtful whether this means " one with stomach trouble " or " one with disease of the oesophagus." It is a word not much used by medical writers, but Caelius Aurelianus has a section on disease of the oesophagus. Although the Romans distinguished (Celsus IV 1) stomach from oesophagus (*stomachus* can mean either), they appear to have described under the same name their morbid conditions. In English " stomach," at least in popular speech, is equally vague.

Stranguria.—Strangury.

Struma.—A scrofulous sore.

Suffusio.—Usually cataract.

Suspiriosus.—Asthmatic. Apparently a popular word, as it is rarely found in the medical writers.

Syntecticus.—One wasting away, from whatever cause.

Tertiana.—Tertian ague, malaria with an onset every other day.

Testa.—A brick-coloured spot on the face. See XXVI 163 and XXVIII 185.

Tetanus.—Tetanus. See Celsus IV 6, 1 with Spencer's notes on *opisthotonus* and *emprosthotonus*.

Tormina (neut. pl.).—A general word for colic. It also sometimes means strangury.

Tremulus.—One with morbid tremors, palsied. See XX 85 *paralyticis et tremulis.*

Tuber.—A hard tumour.

Tumor.—Any morbid swelling.

Tussis.—A cough—the complaint rather than the act.

Tympanicus.—One afflicted with tympanites, a kind of dropsy, which makes the belly swell.

Ulcus.—A favourite word with Pliny, usually used in the plural. *Ulcera manantia* are " running " sores, and *ulcera putrescentia (serpentia)* include gangrene and superficial malignant diseases.

Unguis.—Another name for *pterygium*, an inflammatory swelling at the inner angle of the lower eyelid.

Varix.—Varicose vein.

Varus.—A pimple on the face.

Verruca.—Wart, a less wide term than *clavus.*

Vertigo.—Vertigo, usually giddiness caused by illness.

Vitiligo.—This includes more than one kind of psoriasis. The Romans distinguished the dull white, the dark, and the bright white. Sometimes perhaps leprosy.

Vomica.—Abscess; any gathering of pus, but apparently larger than *furunculus.* It was sometimes internal, but *panus* was superficial.

Zoster.—This (" girdle disease ") was herpes round the waist, possibly shingles. Pliny calls it a form of erysipelas (*ignis sacer*), XXVI 121.

583

INDEX OF FISHES

Index of Fishes, including (marked *) sea-mammals,
Molluscs, Crustaceans, and other animals.

A

Acharne, XXXII 145; probably
Serranus gigas, Great Sea-
Perch.

Ac(c)ipenser, IX 60; XXXII
145, 153; *Acipenser sturio*,
Sturgeon.

*Achillium, IX 148 (cf. XXXI
125); a fine, soft Sponge.

*Actinophorae, XXXII 148;
some spiral univalve, perhaps
the mollusc *Aponais pes-pele-
cani*, Pelican's Foot.

Acus, IX 166; *Syngnathus acus*
and *rubescens*, Pipe-fish (not
Belone belone, Garfish).

Adonis, IX 70; Blenny, probably *Blennius Montagui*.

Alabeta, V 51; *Labeo niloticus*,
Lebis (Labis). The name
should be *alabes*. Pliny mis-
took ἀλάβητα for a nom. case.

Alopex, Alopecias, XXXII 145;
Alopias (Alopecias) vulpes,
Thresher Shark.

Amia, IX 49; *Sarda sarda* and
probably *Thynnus pelamys*,
Pelamid (a Tunny).

Anguilla, IX 4, 73 ff., 160, 189;
XXXI 36; XXXII 16, 138;
Anguilla anguilla, Eel.

Anthias, IX 180, 182; XXXII
13; a name applied to several
species of fish. It includes
(certainly in Pliny) *Anthias
anthias*, but also larger

fish, perhaps a large Tunny,
such as *Germo* (*Thynnus*)
alalunga: and three sorts of
anthias mentioned by Oppian,
possibly *Sciaena aquila*, *Corvina nigra* and *Umbrina
cirrosa* (or instead of *C. nigra*,
Serranus gigas, a Sea-perch
or *Polyprion americanus* Jew-
fish, Stone Bass). Pliny's
anthias may contain a confusion with *acanthias*, which is
Squalus acanthias, Picked (or
Piked) Dogfish, or *Centrina
Salviani*.

Aper (or *caper*), XI 267 *Parasilurus aristotelis*, a species of
catfish.

Aphye, see *Apua*.

*Aplysia, IX 150; a coarse
"unwashable" kind of Sponge,
not the mollusc *Aplysia
depilans* (Sea Hare).

Apua, aphye, IX 160; XXXI,
95; XXXII 145; the young
(small fry) of various species
of fish; also in particular
Engraulis encrasicholus, Anchovy.

Aquila, IX 78; perhaps *Mylio-
batis aquila*, Eagle Ray.

Araneus, IX 155, XXXII 145;
Trachinus draco, Weever, and
the like.

*?Arbor, IX 8; XXXII 144;
unknown, perhaps a huge

585

INDEX OF FISHES

INDEX OF FISHES

145; unknown, but perhaps *Mobula giorna*, Horned Ray, or the Grey shark, *Notidanus griseus*, or the Piper, *Trigla lyra*.

Corvus, XXXII 146; *Umbrina cirrhosa* or *Corvina nigra*.

*Coryphia (*Coluthia*), XXXII 147; small molluscs, such as Winkles and Top-shells.

*Cucumis, IX 3; one of the Echinoderma; a Sea-cucumber, Sea-gherkin, cf. XXXII 147.

*Curalium, XXXII 21–24; XXVIII 164; cf. XIII 135, 140; *Corallium rubrum*, Red Coral.

Cybion, XXXII 146; a Tunny of a certain age, or a cut or preparation from a Tunny.

Cynops, XXXII 147; unknown.

*Cynosdexia, XXXII 148; an Octopus.

Cyprinus, IX 58, unknown, unless *in mari* in 58 is an error; 162, *Cyprinus carpio*, Carp.

D

*Dactylus, IX 184, bivalve molluscs such as *Lithodomus lithophagus*, Date Shell, and *Pholas dactylus*, Piddock; and Tellen or Sunset-shells; XXXII 151= *Solen*.

*Delphinus, VIII 91; IX 19 ff., 50, 57; X 210, 235, 263: XVIII 361; *Delphinus delphis*, and other species of Dolphins, which are not fish. In VIII 91 the "dolphins" which tear open crocodiles are probably two species of fish of the Nile —*Synodontis schall*, Shall and *Schilbe mystus*, Shilbe.

588

*Donax, XXXII 103 = *Solen*.

Draco, IX 82; XXIV 180; XXVI 31; XXVII 50; XXXI 96; XXXII 44, 45, 47, 79, 148; *Trachinus draco*, Weever and allied species.

Drino, XXXII 145; unknown.

E

Echeneis, Echenais, IX 79; XXXII 2–6, 139, 148; *Echeneis remora*, and *E. naucrates*, Sucking Fish; in IX 79 it is a goby or blenny.

*Echinometra, IX 100; *Echinus acutus*, *E. melo*, and *Cidaris cidaris*, Sea-urchins.

*Echinus, IX 40, 99, 147, 164; XI 165; XVIII 361; XXVIII 67; XXXI 95; XXXII 58, 67, 72, 88, 96, 103, 106, 127, 130, 148; various Sea-urchins, esp. *Echinus esculentus* and *Strongylocentrotus lividus*.

*?Elephantus, IX 10, unknown; —hardly Walrus of the far North?; *XXXII 148, *Homarus gammarus*, Lobster, dark coloured.

Elops = *Acipenser*, IX 60, 169; XXXII 46; *Acipenser sturio*, Sturgeon.

Enhydris, any kind of eel; cf. *Ophidion*.

Epodes, XXXII 152; flatfish of uncertain identity.

Erythinus, IX 56, 166; XXXII 101, 139, 148, 152; certainly one of the perches, perhaps *Anthias anthias*.

Exocoetus, IX 70; *Blennius Montagui*, a type of Blenny.

F

Faber, see *Zaeus*.

INDEX OF FISHES

G

Galeos, XXXII 25; a Dogfish or a Shark.

Garos, XXXI 93; XXXII 148; *Smaris smaris*, picarel.

Gerricula, XXXII 148; *Smaris smaris*, picarel.

Girres, XXXII 148; *Smaris smaris*, picarel.

Gladius = *Xiphias*, IX 3, 54; XXXII 15, 145; *Xiphias gladius*, Sword-fish.

Glanis or *glanus*, IX 145, XXXII 128, 148; *Parasilurus aristotelis*, a species of catfish.

Glauciscus, XXXII 129, 148; unknown.

Glaucus, IX 58; XXXII 153; unknown; may be a Dogfish or a Shark.

*Glycymaris, XXXII 147; a mollusc, probably *Venus verrucosa*; certainly a Clam.

Gobio, IX 175; here perhaps *Baleophthalmus Boddaerti*; IX 176, perhaps the lung-fish; 177; here perhaps *Gobius exanthematicus*, cf. XXXII 146; various Gobies, especially *Gobius niger*; includes *Gobio gobio*, the fresh-water Gudgeon.

Gonger, see *Conger*.

H

*Halipleumon, XXXII 149 = *Pulmo*, a Jellyfish (*Medusa*).

Helacatenes, XXXII 149; (doubtful reading), perhaps sharks or dogfish.

*Helix, XXXII 147, a type of spiral univalve of uncertain identity.

Helops, XXXII 153; see *Elops*.

Hepar, XXXII 149; one of the larger marine gadoids, perhaps a species of Ling.

Hippocampus, XXXII 58, 67, 83, 93, 109, 113, 139, 149; cf. IX 3; *Hippocampus antiquorum*, Sea-horse.

*Hippos perhaps *hippeus*? (cf. Aristot. *H.A.* iv, 2, 3.) IX 97; *Ocypoda cursor*, Runner Crab; so also perhaps in XXXII 149.

Hippurus, IX 57; XXXII 149; *Coryphaena hippurus*, the " dolphin-fish."

Hirundo, IX 82; XXXII 149; *Exocoetus volitans*, Flying Blenny, or *Dactylopterus volitans*, Flying Gurnard.

*Holothurium, IX 154; an unknown zoophyte animal regarded as related to Sponges.

*Homo marinus, IX 10; XXXII 144; unknown; African Manatee?

Hyaena, XXXII 154; *Puntazzo puntazzo*, Puntazzo.

I

Ichthyocolla, XXXII 72; Great Sturgeon, *Acipenser huso*; in other passages isinglass, a glue made from the Sturgeon.

Ictinus, XXXII 149; probably *Dactylopterus volitans*, Flying Gurnard, or *Exocoetus volitans*, Flying Blenny.

[*Indian fish*, IX 71. These are especially *Anabas scandens*, Climbing Perch.]

Isox, IX 44; *Salmo salar*, Salmon.

Iulis, XXXII 94, 149; a Wrasse, probably *Coris julis*, Rainbow-Wrasse.

Iulus, XXXII 152; unknown.

589

INDEX OF FISHES

L

Lacertus marinus, XXXII 146, 149; *Pneumatophorus colias*, Spanish Mackerel, and *Trachurus trachurus*, Horse Mackerel = Scad.

Lamia, IX 78; a large Shark, such as *Carcharias carcharodon*, Great White Shark.

Lamirus, XXXII 149; perhaps *Pagellus erythrinus*, Becker.

[*Larius* and *Verbannus* (*Lakes*), fish in, IX 69; probably species of the Carp family, *Rutilus rutilus*, Roach; *Idus idus*, Ide; *Abramis brama*, Bream.]

Laser, XXXI 25, 44; unknown.

Lelepris, XXXII 149; some kind of Wrasse.

*Leo, XXXII 149; cf. IX 97; *Nephrops norvegicus*, Lion-crab.

*Lepas, XXXII 149; a Mediterranean Limpet, especially *Patella Lamarckii* or the like.

Lepus marinus, IX 155; XX 223; XXIII 108; XXIV 18, 20; XXV 125; XXVIII 74, 129, 158, 159; XXIX 104; XXXII 8, 9, 48, 54, 58, 59, 68, 70, 88, 104, 110, 135, 149; *Aplysia depilans*, Sea Hare (a " Sea Slug "). In IX 195 one of the spiny Porcupine-fish of the Indian Ocean is also referred to.

*Limax, IX 162; XXX 56, 79, 101, 139; generic term for slugs.

*Locusta, IX 95–6, 158, 164, 185; XI 152; XXXVI 89; *Palinurus elephas*, Crawfish.

*Lolligo, IX 83, 93, 158, 164; XI 215, 258; XVIII 361; XXXII 15, 149; *Loligo

vulgaris and other Squids, especially *Ommatostrephes sagittatus*, a large kind.

Lucerna, IX 82; = *Uranoscopus.*

Lupus, IX 57, 61, 169; X 193; XXXI 15; *Morone labrax,* Sea Basse; XXXI 95, *Engraulis encrasicolus,* Anchovy.

M

Maena, IX 81; XXVI 23, cf. 127; XXXI 83; XXXII 83, 88, 90, 100, 105, 107, 126, 128, 149; cf. 152; Mendole, *Maena maena,* M. *osbeckii,* and M. *jusculum.*

Maeotes, XXXII 149; cf. 146; in Pliny, apparently small horse-mackerel and young tunny or pelamid.

*Maia, IX 97; a large Crab, probably *Maia squinado* or else *Homola barbata*; possibly also *Lithodes Maia.*

[*Margarita, pearl, got from *Margaritifera margaritifera* = *Mytilus margaritiferus,* Pearl Oyster, IX 106 ff. Inferior pearls came from Mussels, Oysters, Pinnas and Freshwater Mussels.]

Marris (better *mario?*), IX 75; perhaps a type of sturgeon.

[*Melandrya,* IX 48; cuts or cutlets of μελάνδρυς, a kind of large Tunny.]

Melanurus, XXXII 17, 149, 152; Oblade, *Oblata melanura.*

Mena, see *Maena.*

Merula, XXXII 149; a species of Wrasse, perhaps *Coricus rostratus.*

Milvus = *Ictinus,* IX 82.

*Mitulus, Mytilus, IX 160; XXXII 95, 111, 149; *Mytilus edulis,* Mussel.

INDEX OF FISHES

INDEX OF FISHES

Piddock, *Pholas* or *Lithodomus*.

Ophidion, XXXII 109, 149; an Eel or a related fish; includes perhaps *Oxystomus serpens*.

Orbis, XXXII 14, 149, 150; probably a species of Globe-fish.

*Orca, IX 12–14; XXXII 144; probably *Orcinus orca*, Grampus, Killer Whale.

Orcynus, XXXII 149; a large specimen of a Tunny.

Orphus, IX 57; XXXII 152; either *Serranus gigas*, a Sea Perch or *Polyprion americanus*, Jew-fish.

Orthagoriscus, see *Porcus*.

*Ostrea or *Ostreae*, II 109; V 180; IX 40, 52, 154, 160, 161, 168; X 129, 189, 192, 195; XI 129, 139, 226; XXVIII 66; XXXI 96; XXXII 59, 60, 64, 93, 149; a general term for bivalve molluscs, but properly *Ostrea edulis*, Oyster. See especially II 109; IX 154, 168; X 129, 189, 192, 195; XI 139; XXVIII 66; XXXI 96; XXXII 59–65.

*Otia, XXXII 149; *Haliotis tuberculata*, Sea-Ear or Ormer.

*Ozaena, IX 89; an ill-smelling species of Octopus, probably *Eledone moschata* and possibly also *E. Aldrovandi*.

P

*Pagurus, IX 97; *Pagurus bernhardus*, and other Hermit Crabs.

[*Paphlagonia*, some fishes in, IX 178; probably *Cobitis fossilis*, a kind of Loach.]

[**Parasites* on fish, and other " Sea Fleas," and " Sea-lice," all Crustaceans, IX 154. See also *Scorpion-like parasites*; *Pediculi*; *Phthir*.]

Parus, XXXII 152; unknown.

Passer, IX 72; *Pleuronectes platessa*, Plaice, or else *Platichthys flesus*, Flounder.

Pastinaca, IX 155; XXII 146; XXVIII 162; XXXI 25, 44; XXXII 57, 79, 83, 133; *Trygon pastinaca*, Sting Ray.

*Pecten, IX 101, 103, 147, 160, 162; XI 139, 147; XXXII 103, 150; species of Scallop, especially *Chlamys = Pecten varius* and *C. Jacobaeus*.

*Pectunculus, IX 84; XXXII 70, 150; a small or young Scallop.

*Pediculi marini, XXXII 77, 89; apparently Sea-lice, small crustaceans.

Pelamys, IX 47; a year-old tunny; XXXII 105, 107, 146, 149, 150, 151; a species of Tunny, *Sarda sarda*, Pelamid; sometimes smaller species or very young Tunny.

*Peloris, XXXII 99, 147; probably *Psammobia vespertina*, Sunset-shell.

?**Pentadactyli*, XXXII 147; unknown.

Perca, XXXII 145; *Perca fluviatilis*, Perch, and *Paracentropristis scriba* and related species, Sea Perch; IX 57; XXXII 107, 116, 126, 130, *Paracentropristis scriba*.

*Percis? Pegris?, XXXII 150; unknown mollusc.

*Perna, Pin(n)a, IX 115, 142; XXXII 150, 154; a bivalve mollusc, *Pinna nobilis* or else *P. fragilis*, Pinna-shell.

Phagrus, phager, IX 57; XXXII 150, a species of Sea Bream,

592

INDEX OF FISHES

perhaps *Pagrus pagrus*; XXXII 113, probably *Hydrocyon forskalii*.

*Phocae = *Vituli marini*.

*Phthir, XXXII 150; not, it seems, as D'Arcy Thompson thought, *Echeneis remora* and *E. naucrates*, Sucking Fish; but some Sea-louse, a crustacean.

Phycis, IX 81; XXXII 150, a species of Wrasse, probably *Crenilabrus pavo*.

*Physeter, IX 8; XXXII 144, cf. IX 4; probably Sperm Whale, *Physeter catodon* = *macrocephalus*.

*Pin(n)a, see *Perna*.

*Pinoteres, IX 98; *Pagurus bernhardus* and other Hermit Crabs; also *Pinnotheres pinnotheres*, Pinna-Guard Crab; in IX 142 we have the Pinna-Guard Crab and also the carid *Pontonia pinnophylax* = *tyrrhena*; cf. XXXII 150.

Piscatrix, IX 143; *Lophius piscatorius*, Angler-fish.

Pistrix, XXXII 144; *Pristis antiquorum*, Saw-fish.

*Platanista, IX 46; *Platanista Gangetica*, Gangetic Dolphin, Susu.

*Polypus, IX 40, 71, 78, 83, 85–93, 158, 163, 185; X 194, 195; XI 133, 199, 225, 258; XXXII 12, 121, 150; species of Octopus, especially *Octopus vulgaris*.

Pompili (accompanying ships), IX 51, a shoal of Tunny; Pliny errs. Tunny-shoals do not follow ships. These were pilot-fish, wrongly identified as Tunny. XXXII 153, *Naucrates ductor*, Pilot-fish; IX 88 (where pompilus is a

mistake for pontilus = πουτίλος). **Argonauta argo*, Argonaut = Paper Nautilus.

Porculus marinus, IX = *Porcus*.

Porcus, XXXII 19, cf. 56, 150; *Centrina salviani*.

Pristis, IX 4, 8, 41; *Pristis antiquorum*, Sawfish; and other quite different fish, and even **Whales*.

Psetta, IX 57; *Pleuronectes* and *Platichthys* sp., Plaice and Flounder.

*Pulmo, IX 154; XXXII 102, 111, etc.; species of Jellyfish (*Medusa*).

*Purpurae, IX 124–141; see *Murex*.

R

Raia, IX 78, 144, 161; *Raja batis* and similar kinds of Skate or Ray.

Rana, IX 143; *Lophius piscatorius*, Angler-fish.

Rhine = *Squatus*, XXXII 150; *Squatina squatina*, Angelfish.

[*Rhinobatus*], IX 161; *Rhinobatos rhinobatos*, wrongly alleged to be a hybrid between Angel-fish and Skate.

Rhombus, IX 52, 72, 144, 169; XXXII 102, 145, 150; *Scophthalmus maximus*, Turbot.

Rota, IX 8; probably *Orthagoriscus mola*.

Rubellio, XXXII 138; probably *Pagellus erythrinus*, the Becker.

S

*Saepia, see *Sepia*.

Salax?, XXXII 151; unknown.

Salmo, IX 68; *Salmo salar*, Salmon.

593

INDEX OF FISHES

Salpa, IX 68, 162; XXXII 151; *Sarpa salpa*, Saupe.

Sarda, XXXII 46, *Sardina pilchardus*, Sardine or Pilchard; XXXII 151, a large *pelamys*, q.v.

Sargus, IX 65, 162, 182; XXXII 151; *Diplodus sargus*, Sargue, Sargo; and *D. vulgaris*.

Saurus, XXXII 89; *Trachurus trachurus*, Horse Mackerel.

Scarus, IX 62; XI 162; XXXII 11, 151; XXXVII 187; *Sparisoma cretense*, Parrot-Wrasse.

Sciadeus, XXXII 151; *Sciaena aquila*, Maigre and related species.

Sciaena, IX 57; XXXII 106, 151 = *Sciadeus*.

Scias, XXXII 151 = *Sciadeus*.

*Scilla = Squilla.

*Scolopendra, IX 145; XXXII 151; species of Nereid worm.

Scomber, IX 49; XXXII 151; *Scomber scombrus*, Mackerel.

Scorpaena = Scorpio.

Scorpio, XX 150; XXXII 44, 67, 70, 102, 127–128; *Scorpaena scrofa* and *S. porcus*, Sculpin.

*[Scorpion-like parasites on Tunny, *Brachiella thynni*; on Sword-fish, *Pennella filosa*, IX 54].

*Sepia, IX 83, 84, 93 (its eggs perhaps IX 3, uva); *Sepia officinalis* and other Cuttlefish.

Serra, IX 3; XXXII 145; *Pristis antiquorum*, Sawfish.

Silurus, V 51, *Lates niloticus*, Nile Perch; VI 205, unknown; IX 44, *Lates niloticus*; IX 45, *Silurus glanis*, Sheatfish; IX 58, 165, *Parasilurus aristotelis*; XVIII 293, unknown;

XXXII 90, 93, 94, 104, 111, 119, 125, 126, 131, probably all *Lates niloticus*; XXXII 145, unknown.

*Simones = Delphini.

Smaris (Zmaris), XXXII 108, 128; *Smaris smaris*, Picarel; and related species.

Smyrus, XXXII 151, see *Zmyrus*.

Solea, IX 52, 57, 72; XXXII 102, 151; *Pleuronectes solea*, Sole, and allied species.

*Solen, X 192; XI 139; XXXII 151; species of the bivalve mollusc Razor Shell, especially *Solen coarctatus*.

Sorus, XXXII 151; *Scombresox rondeletii*, Skipjack, Skipper.

Sphyraena, XXXII 154; *Sphyraena sphyraena*, Barracuda.

*Spondylus, XXXII 154; *Spondylus gaedaropus*, Thorny "Oyster."

*Spongea, IX 146, 150; XXXI 123–131; species of Sponge, especially *Spongia officinalis* and its variety *mollissima*.

Squalus, IX 78; smaller Dogfish and Sharks.

Squatina, IX 40, 78, 144, 161, 162; *Squatina squatina*, Angelfish.

Squatus, XXXII 150; = *Squatina*.

*Squilla, Scilla, IX 158; XI 152; XXXII 151, species of *Palaemon*, Prawn, and *Crangon*, Shrimp; IX 142, probably *Pontonia pinnophylax = tyrrhena*.

*Stellae marinae, IX 154, 183; XXXII 44, 151; various Starfish.

*Strombus, XXXII 117, 129, 151; some species of spiral-shelled mollusc.

594

INDEX OF FISHES

595

INDEX OF FISHES

Printed in Great Britain by
Fletcher & Son Ltd, Norwich

THE LOEB CLASSICAL LIBRARY

VOLUMES ALREADY PUBLISHED

Latin Authors

AMMIANUS MARCELLINUS. Translated by J. C. Rolfe. 3 Vols.

APULEIUS: THE GOLDEN ASS (METAMORPHOSES). W. Adlington (1566). Revised by S. Gaselee.

ST. AUGUSTINE: CITY OF GOD. 7 Vols. Vol. I. G. E. McCracken. Vol. II. and VII. W. M. Green. Vol. III. D. Wiesen. Vol. IV. P. Levine. Vol. V. E. M. Sanford and W. M. Green. Vol. VI. W. C. Greene.

ST. AUGUSTINE, CONFESSIONS OF. W. Watts (1631). 2 Vols.

ST. AUGUSTINE, SELECT LETTERS. J. H. Baxter.

AUSONIUS. H. G. Evelyn White. 2 Vols.

BEDE. J. E. King. 2 Vols.

BOETHIUS: TRACTS and DE CONSOLATIONE PHILOSOPHIAE. REV. H. F. Stewart and E. K. Rand. Revised by S. J. Tester.

CAESAR: ALEXANDRIAN, AFRICAN and SPANISH WARS. A. G. Way.

CAESAR: CIVIL WARS. A. G. Peskett.

CAESAR: GALLIC WAR. H. J. Edwards.

CATO: DE RE RUSTICA; VARRO: DE RE RUSTICA. H. B. Ash and W. D. Hooper.

CATULLUS. F. W. Cornish; TIBULLUS. J. B. Postgate; PERVIGILIUM VENERIS. J. W. Mackail.

CELSUS: DE MEDICINA. W. G. Spencer. 3 Vols.

CICERO: BRUTUS, and ORATOR. G. L. Hendrickson and H. M. Hubbell.

[CICERO]: AD HERENNIUM. H. Caplan.

CICERO: DE ORATORE, etc. 2 Vols. Vol. I. DE ORATORE, Books I. and II. E. W. Sutton and H. Rackham. Vol. II. DE ORATORE, Book III. De Fato; Paradoxa Stoicorum; De Partitione Oratoria. H. Rackham.

CICERO: DE FINIBUS. H. Rackham.

CICERO: DE INVENTIONE, etc. H. M. Hubbell.

CICERO: DE NATURA DEORUM and ACADEMICA. H. Rackham.

CICERO: DE OFFICIIS. Walter Miller.

CICERO: DE REPUBLICA and DE LEGIBUS: SOMNIUM SCIPIONIS. Clinton W. Keyes.

CICERO: DE SENECTUTE, DE AMICITIA, DE DIVINATIONE. W. A. Falconer.
CICERO: IN CATILINAM, PRO FLACCO, PRO MURENA, PRO SULLA. New version by C. Macdonald.
CICERO: LETTERS to ATTICUS. E. O. Winstedt. 3 Vols.
CICERO: LETTERS TO HIS FRIENDS. W. Glynn Williams, M. Cary, M. Henderson. 4 Vols.
CICERO: PHILIPPICS. W. C. A. Ker.
CICERO: PRO ARCHIA POST REDITUM, DE DOMO, DE HARUS- PICUM RESPONSIS, PRO PLANCIO. N. H. Watts.
CICERO: PRO CAECINA, PRO LEGE MANILIA, PRO CLUENTIO, PRO RABIRIO. H. Grose Hodge.
CICERO: PRO CAELIO, DE PROVINCIIS CONSULARIBUS, PRO BALBO. R. Gardner.
CICERO: PRO MILONE, IN PISONEM, PRO SCAURO, PRO FONTEIO, PRO RABIRIO POSTUMO, PRO MARCELLO, PRO LIGARIO, PRO REGE DEIOTARO. N. H. Watts.
CICERO: PRO QUINCTIO, PRO ROSCIO AMERINO, PRO ROSCIO COMOEDO, CONTRA RULLUM. J. H. Freese.
CICERO: PRO SESTIO, IN VATINIUM. R. Gardner.
CICERO: TUSCULAN DISPUTATIONS. J. E. King.
CICERO: VERRINE ORATIONS. L. H. G. Greenwood. 2 Vols.
CLAUDIAN. M. Platnauer. 2 Vols.
COLUMELLA: DE RE RUSTICA. DE ARBORIBUS. H. B. Ash, E. S. Forster and E. Heffner. 3 Vols.
CURTIUS, Q.: HISTORY OF ALEXANDER. J. C. Rolfe. 2 Vols.
FLORUS. E. S. Forster; and CORNELIUS NEPOS. J. C. Rolfe.
FRONTINUS: STRATAGEMS and AQUEDUCTS. C. E. Bennett and M. B. McElwain.
FRONTO: CORRESPONDENCE. C. R. Haines. 2 Vols.
GELLIUS, J. C. Rolfe. 3 Vols.
HORACE: ODES AND EPODES. C. E. Bennett.
HORACE: SATIRES, EPISTLES, ARS POETICA. H. R. Fairclough.
JEROME: SELECTED LETTERS. F. A. Wright.
JUVENAL and PERSIUS. G. G. Ramsay.
LIVY. B. O. Foster, F. G. Moore, Evan T. Sage, and A. C. Schlesinger and R. M. Geer (General Index). 14 Vols.
LUCAN. J. D. Duff.
LUCRETIUS. W. H. D. Rouse. Revised by M. F. Smith.
MARTIAL. W. C. A. Ker. 2 Vols.
MINOR LATIN POETS: from PUBLILIUS SYRUS TO RUTILIUS NAMATIANUS, including GRATTIUS, CALPURNIUS SICULUS, NEMESIANUS, AVIANUS, and others with "Aetna" and the "Phoenix." J. Wight Duff and Arnold M. Duff.
OVID: THE ART OF LOVE and OTHER POEMS. J. H. Mozley.
OVID: FASTI. Sir James G. Frazer.

OVID: HEROIDES and AMORES. Grant Showerman.
OVID: METAMORPHOSES. F. J. Miller. 2 Vols.
OVID: TRISTIA and EX PONTO. A. L. Wheeler.
PERSIUS. Cf. JUVENAL.
PETRONIUS. M. Heseltine; SENECA; APOCOLOCYNTOSIS. W. H. D. Rouse.
PHAEDRUS AND BABRIUS (Greek). B. E. Perry.
PLAUTUS. Paul Nixon. 5 Vols.
PLINY: LETTERS, PANEGYRICUS. Betty Radice. 2 Vols.
PLINY: NATURAL HISTORY. Vols. I.–V. and IX. H. Rackham. VI.–VIII. W. H. S. Jones. X. D. E. Eichholz. 10 Vols.
PROPERTIUS. H. E. Butler.
PRUDENTIUS. H. J. Thomson. 2 Vols.
QUINTILIAN. H. E. Butler. 4 Vols.
REMAINS OF OLD LATIN. E. H. Warmington. 4 Vols. Vol. I. (ENNIUS AND CAECILIUS.) Vol. II. (LIVIUS, NAEVIUS, PACUVIUS, ACCIUS.) Vol. III. (LUCILIUS and LAWS OF XII TABLES.) Vol. IV. (ARCHAIC INSCRIPTIONS.)
SALLUST. J. C. Rolfe.
SCRIPTORES HISTORIAE AUGUSTAE. D. Magie. 3 Vols.
SENECA, THE ELDER: CONTROVERSIAE, SUASORIAE. M. Winterbottom. 2 Vols.
SENECA: APOCOLOCYNTOSIS. Cf. PETRONIUS.
SENECA: EPISTULAE MORALES. R. M. Gummere. 3 Vols.
SENECA: MORAL ESSAYS. J. W. Basore. 3 Vols.
SENECA: TRAGEDIES. F. J. Miller. 2 Vols.
SENECA: NATURALES QUAESTIONES. T. H. Corcoran. 2 Vols.
SIDONIUS: POEMS and LETTERS. W. B. Anderson. 2 Vols.
SILIUS ITALICUS. J. D. Duff. 2 Vols.
STATIUS. J. H. Mozley. 2 Vols.
SUETONIUS. J. C. Rolfe. 2 Vols.
TACITUS: DIALOGUS. Sir Wm. Peterson. AGRICOLA and GERMANIA. Maurice Hutton. Revised by M. Winterbottom, R. M. Ogilvie, E. H. Warmington.
TACITUS: HISTORIES AND ANNALS. C. H. Moore and J. Jackson. 4 Vols.
TERENCE. John Sargeaunt. 2 Vols.
TERTULLIAN: APOLOGIA and DE SPECTACULIS. T. R. Glover. MINUCIUS FELIX. G. H. Rendall.
VALERIUS FLACCUS. J. H. Mozley.
VARRO: DE LINGUA LATINA. R. G. Kent. 2 Vols.
VELLEIUS PATERCULUS and RES GESTAE DIVI AUGUSTI. F. W. Shipley.
VIRGIL. H. R. Fairclough. 2 Vols.
VITRUVIUS: DE ARCHITECTURA. F. Granger. 2 Vols.

3

Greek Authors

ACHILLES TATIUS. S. Gaselee.

AELIAN: ON THE NATURE OF ANIMALS. A. F. Scholfield. 3 Vols.

AENEAS TACTICUS, ASCLEPIODOTUS and ONASANDER. The Illinois Greek Club.

AESCHINES. C. D. Adams.

AESCHYLUS. H. Weir Smyth. 2 Vols.

ALCIPHRON, AELIAN, PHILOSTRATUS: LETTERS. A. R. Benner and F. H. Fobes.

ANDOCIDES, ANTIPHON, Cf. MINOR ATTIC ORATORS.

APOLLODORUS. Sir James G. Frazer. 2 Vols.

APOLLONIUS RHODIUS. R. C. Seaton.

THE APOSTOLIC FATHERS. Kirsopp Lake. 2 Vols.

APPIAN: ROMAN HISTORY. Horace White. 4 Vols.

ARATUS. Cf. CALLIMACHUS.

ARISTIDES: ORATIONS. C. A. Behr. Vol. I.

ARISTOPHANES. Benjamin Bickley Rogers. 3 Vols. Verse trans.

ARISTOTLE: ART OF RHETORIC. J. H. Freese.

ARISTOTLE: ATHENIAN CONSTITUTION, EUDEMIAN ETHICS, VICES AND VIRTUES. H. Rackham.

ARISTOTLE: GENERATION OF ANIMALS. A. L. Peck.

ARISTOTLE: HISTORIA ANIMALIUM. A. L. Peck. Vols. I.–II.

ARISTOTLE: METAPHYSICS. H. Tredennick. 2 Vols.

ARISTOTLE: METEOROLOGICA. H. D. P. Lee.

ARISTOTLE: MINOR WORKS. W. S. Hett. On Colours, On Things Heard, On Physiognomies, On Plants, On Marvellous Things Heard, Mechanical Problems, On Indivisible Lines, On Situations and Names of Winds, On Melissus, Xenophanes, and Gorgias.

ARISTOTLE: NICOMACHEAN ETHICS. H. Rackham.

ARISTOTLE: OECONOMICA and MAGNA MORALIA. G. C. Armstrong; (with METAPHYSICS, Vol. II.).

ARISTOTLE: ON THE HEAVENS. W. K. C. Guthrie.

ARISTOTLE: ON THE SOUL. PARVA NATURALIA. ON BREATH. W. S. Hett.

ARISTOTLE: CATEGORIES, ON INTERPRETATION, PRIOR ANALYTICS. H. P. Cooke and H. Tredennick.

ARISTOTLE: POSTERIOR ANALYTICS, TOPICS. H. Tredennick and E. S. Forster.

ARISTOTLE: ON SOPHISTICAL REFUTATIONS.
On Coming to be and Passing Away, On the Cosmos. E. S. Forster and D. J. Furley.

ARISTOTLE: PARTS OF ANIMALS. A. L. Peck; MOTION AND PROGRESSION OF ANIMALS. E. S. Forster.

ARISTOTLE: PHYSICS. Rev. P. Wicksteed and F. M. Cornford. 2 Vols.

ARISTOTLE: POETICS and LONGINUS. W. Hamilton Fyfe; DEMETRIUS ON STYLE. W. Rhys Roberts.

ARISTOTLE: POLITICS. H. Rackham.

ARISTOTLE: PROBLEMS. W. S. Hett. 2 Vols.

ARISTOTLE: RHETORICA AD ALEXANDRUM (with PROBLEMS. Vol. II). H. Rackham.

ARRIAN: HISTORY OF ALEXANDER and INDICA. Rev. E. Iliffe Robson. 2 Vols.

ATHENAEUS: DEIPNOSOPHISTAE. C. B. Gulick. 7 Vols.

BABRIUS AND PHAEDRUS (Latin). B. E. Perry.

ST. BASIL: LETTERS. R. J. Deferrari. 4 Vols.

CALLIMACHUS: FRAGMENTS. C. A. Trypanis. MUSAEUS: HERO AND LEANDER. T. Gelzer and C. Whitman.

CALLIMACHUS, Hymns and Epigrams, and LYCOPHRON. A. W. Mair; ARATUS. G. R. Mair.

CLEMENT OF ALEXANDRIA. Rev. G. W. Butterworth.

COLLUTHUS. Cf. OPPIAN.

DAPHNIS AND CHLOE. Thornley's Translation revised by J. M. Edmonds: and PARTHENIUS. S. Gaselee.

DEMOSTHENES I.: OLYNTHIACS, PHILIPPICS and MINOR ORATIONS. I.–XVII. AND XX. J. H. Vince.

DEMOSTHENES II.: DE CORONA and DE FALSA LEGATIONE. C. A. Vince and J. H. Vince.

DEMOSTHENES III.: MEIDIAS, ANDROTION, ARISTOCRATES, TIMOCRATES and ARISTOGEITON, I. AND II. J. H. Vince.

DEMOSTHENES IV.–VI.: PRIVATE ORATIONS and IN NEAERAM. A. T. Murray.

DEMOSTHENES VII.: FUNERAL SPEECH, EROTIC ESSAY, EXORDIA and LETTERS. N. W. and N. J. DeWitt.

DIO CASSIUS: ROMAN HISTORY. E. Cary. 9 Vols.

DIO CHRYSOSTOM. J. W. Cohoon and H. Lamar Crosby. 5 Vols.

DIODORUS SICULUS. 12 Vols. Vols. I.–VI. C. H. Oldfather. Vol. VII. C. L. Sherman. Vol. VIII. C. B. Welles. Vols. IX. and X. R. M. Geer. Vol. XI. F. Walton. Vol. XII. F. Walton. General Index. R. M. Geer.

DIOGENES LAERTIUS. R. D. Hicks. 2 Vols. New Introduction by H. S. Long.

DIONYSIUS OF HALICARNASSUS: ROMAN ANTIQUITIES Spelman's translation revised by E. Cary. 7 Vols.

DIONYSIUS OF HALICARNASSUS: CRITICAL ESSAYS. S. Usher. 2 Vols.

EPICTETUS. W. A. Oldfather. 2 Vols.

EURIPIDES. A. S. Way. 4 Vols. Verse trans.

EUSEBIUS: ECCLESIASTICAL HISTORY. Kirsopp Lake and J. E. L. Oulton. 2 Vols.

GALEN: ON THE NATURAL FACULTIES. A. J. Brock.

THE GREEK ANTHOLOGY. W. R. Paton. 5 Vols.

GREEK ELEGY AND IAMBUS with the ANACREONTEA. J. M. Edmonds. 2 Vols.

THE GREEK BUCOLIC POETS (THEOCRITUS, BION, MOSCHUS). J. M. Edmonds.

GREEK MATHEMATICAL WORKS. Ivor Thomas. 2 Vols.

HERODES. Cf. THEOPHRASTUS: CHARACTERS.

HERODIAN. C. R. Whittaker. 2 Vols.

HERODOTUS. A. D. Godley. 4 Vols.

HESIOD AND THE HOMERIC HYMNS. H. G. Evelyn White.

HIPPOCRATES and the FRAGMENTS OF HERACLEITUS. W. H. S. Jones and E. T. Withington. 4 Vols.

HOMER: ILIAD. A. T. Murray. 2 Vols.

HOMER: ODYSSEY. A. T. Murray. 2 Vols.

ISAEUS. E. W. Forster.

ISOCRATES. George Norlin and LaRue Van Hook. 3 Vols.

[ST. JOHN DAMASCENE]: BARLAAM AND IOASAPH. Rev. G. R. Woodward, Harold Mattingly and D. M. Lang.

JOSEPHUS. 9 Vols. Vols. I.–IV. H. Thackeray. Vol. V. H. Thackeray and R. Marcus. Vols. VI.–VII. R. Marcus. Vol. VIII. R. Marcus and Allen Wikgren. Vol. IX. L. H. Feldman.

JULIAN. Wilmer Cave Wright. 3 Vols.

LIBANIUS. A. F. Norman. Vol. I.

LUCIAN. 8 Vols. Vols. I.–V. A. M. Harmon. Vol. VI. K. Kilburn. Vols. VII.–VIII. M. D. Macleod.

LYCOPHRON. Cf. CALLIMACHUS.

LYRA GRAECA. J. M. Edmonds. 3 Vols.

LYSIAS. W. R. M. Lamb.

MANETHO. W. G. Waddell: PTOLEMY: TETRABIBLOS. F. E. Robbins.

MARCUS AURELIUS. C. R. Haines.

MENANDER. F. G. Allison.

MINOR ATTIC ORATORS (ANTIPHON, ANDOCIDES, LYCURGUS, DEMADES, DINARCHUS, HYPERIDES). K. J. Maidment and J. O. Burtt. 2 Vols.

MUSAEUS: HERO AND LEANDER. Cf. CALLIMACHUS.

NONNOS: DIONYSIACA. W. H. D. Rouse. 3 Vols.

OPPIAN, COLLUTHUS, TRYPHIODORUS. A. W. Mair.

PAPYRI. NON-LITERARY SELECTIONS. A. S. Hunt and C. C. Edgar. 2 Vols. LITERARY SELECTIONS (Poetry). D. L. Page.

PARTHENIUS. Cf. DAPHNIS and CHLOE.

PAUSANIAS: DESCRIPTION OF GREECE. W. H. S. Jones. 4 Vols. and Companion Vol. arranged by R. E. Wycherley.

PHILO. 10 Vols. Vols. I.–V. F. H. Colson and Rev. G. H. Whitaker. Vols. VI.–IX. F. H. Colson. Vol. X. F. H. Colson and the Rev. J. W. Earp.

PHILO: two supplementary Vols. (*Translation only.*) Ralph Marcus.

PHILOSTRATUS: THE LIFE OF APOLLONIUS OF TYANA. F. C. Conybeare. 2 Vols.

PHILOSTRATUS: IMAGINES; CALLISTRATUS: DESCRIPTIONS. A. Fairbanks.

PHILOSTRATUS and EUNAPIUS: LIVES OF THE SOPHISTS. Wilmer Cave Wright.

PINDAR. Sir J. E. Sandys.

PLATO: CHARMIDES, ALCIBIADES, HIPPARCHUS, THE LOVERS, THEAGES, MINOS and EPINOMIS. W. R. M. Lamb.

PLATO: CRATYLUS, PARMENIDES, GREATER HIPPIAS, LESSER HIPPIAS. H. N. Fowler.

PLATO: EUTHYPHRO, APOLOGY, CRITO, PHAEDO, PHAEDRUS. H. N. Fowler.

PLATO: LACHES, PROTAGORAS, MENO, EUTHYDEMUS. W. R. M. Lamb.

PLATO: LAWS. Rev. R. G. Bury. 2 Vols.

PLATO: LYSIS, SYMPOSIUM, GORGIAS. W. R. M. Lamb.

PLATO: REPUBLIC. Paul Shorey. 2 Vols.

PLATO: STATESMAN, PHILEBUS. H. N. Fowler; Ion. W. R. M. Lamb.

PLATO: THEAETETUS and SOPHIST. H. N. Fowler.

PLATO: TIMAEUS, CRITIAS, CLITOPHO, MENEXENUS, EPISTULAE. Rev. R. G. Bury.

PLOTINUS: A. H. Armstrong. Vols. I.–III.

PLUTARCH: MORALIA. 17 Vols. Vols. I.–V. F. C. Babbitt. Vol. VI. W. C. Helmbold. Vols. VII. and XIV. P. H. De Lacy and B. Einarson. Vol. VIII. P. A. Clement and H. B. Hoffleit. Vol. IX. E. L. Minar, Jr., F. H. Sandbach, W. C. Helmbold. Vol. X. H. N. Fowler. Vol. XI. L. Pearson and F. H. Sandbach. Vol. XII. H. Cherniss and W. C. Helmbold. Vol. XV. F. H. Sandbach.

PLUTARCH: THE PARALLEL LIVES. B. Perrin. 11 Vols.

POLYBIUS. W. R. Paton. 6 Vols.

PROCOPIUS: HISTORY OF THE WARS. H. B. Dewing. 7 Vols.

PTOLEMY: TETRABIBLOS. Cf. MANETHO.

QUINTUS SMYRNAEUS. A. S. Way. Verse trans.

SEXTUS EMPIRICUS. Rev. R. G. Bury. 4 Vols.

SOPHOCLES. F. Storr. 2 Vols. Verse trans.

STRABO: GEOGRAPHY. Horace L. Jones. 8 Vols.

THEOPHRASTUS: CHARACTERS. J. M. Edmonds. HERODES, etc. A. D. Knox

THEOPHRASTUS: ENQUIRY INTO PLANTS. Sir Arthur Hort, Bart. 2 Vols.
THUCYDIDES. C. F. Smith. 4 Vols.
TRYPHIODORUS. Cf. OPPIAN.
XENOPHON: CYROPAEDIA. Walter Miller. 2 Vols.
XENOPHON: HELLENICA. C. L. Brownson. 2 Vols.
XENOPHON: ANABASIS. C. L. Brownson.
XENOPHON: MEMORABILIA AND OECONOMICUS. E. C. Marchant. SYMPOSIUM AND APOLOGY. O. J. Todd.
XENOPHON: SCRIPTA MINORA. E. C. Marchant and G. W. Bowersock.

IN PREPARATION

Greek Authors

ARRIAN I. New version by P. Brunt.
PLUTARCH: MORALIA XIII 1–2. H. Cherniss.
THEOPHRASTUS: DE CAUSIS PLANTARUM. G. K. K. Link and B. Einarson.

Latin Authors

MANILIUS. G. P. Goold.

DESCRIPTIVE PROSPECTUS ON APPLICATION

CAMBRIDGE, MASS.   HARVARD UNIVERSITY PRESS
LONDON   WILLIAM HEINEMANN LTD